INTRODUCTION TO
NONLINEAR OPTIMIZATION

MOS-SIAM Series on Optimization

This series is published jointly by the Mathematical Optimization Society and the Society for Industrial and Applied Mathematics. It includes research monographs, books on applications, textbooks at all levels, and tutorials. Besides being of high scientific quality, books in the series must advance the understanding and practice of optimization. They must also be written clearly and at an appropriate level for the intended audience.

Editor-in-Chief
Katya Scheinberg
Lehigh University

Editorial Board
Santanu S. Dey, *Georgia Institute of Technology*
Maryam Fazel, *University of Washington*
Andrea Lodi, *University of Bologna*
Arkadi Nemirovski, *Georgia Institute of Technology*
Stefan Ulbrich, *Technische Universität Darmstadt*
Luis Nunes Vicente, *University of Coimbra*
David Williamson, *Cornell University*
Stephen J. Wright, *University of Wisconsin*

Series Volumes

Beck, Amir, *Introduction to Nonlinear Optimization: Theory, Algorithms, and Applications with MATLAB*

Attouch, Hedy, Buttazzo, Giuseppe, and Michaille, Gérard, *Variational Analysis in Sobolev and BV Spaces: Applications to PDEs and Optimization, Second Edition*

Shapiro, Alexander, Dentcheva, Darinka, and Ruszczynski, Andrzej, *Lectures on Stochastic Programming: Modeling and Theory, Second Edition*

Locatelli, Marco and Schoen, Fabio, *Global Optimization: Theory, Algorithms, and Applications*

De Loera, Jesús A., Hemmecke, Raymond, and Köppe, Matthias, *Algebraic and Geometric Ideas in the Theory of Discrete Optimization*

Blekherman, Grigoriy, Parrilo, Pablo A., and Thomas, Rekha R., editors, *Semidefinite Optimization and Convex Algebraic Geometry*

Delfour, M. C., *Introduction to Optimization and Semidifferential Calculus*

Ulbrich, Michael, *Semismooth Newton Methods for Variational Inequalities and Constrained Optimization Problems in Function Spaces*

Biegler, Lorenz T., *Nonlinear Programming: Concepts, Algorithms, and Applications to Chemical Processes*

Shapiro, Alexander, Dentcheva, Darinka, and Ruszczynski, Andrzej, *Lectures on Stochastic Programming: Modeling and Theory*

Conn, Andrew R., Scheinberg, Katya, and Vicente, Luis N., *Introduction to Derivative-Free Optimization*

Ferris, Michael C., Mangasarian, Olvi L., and Wright, Stephen J., *Linear Programming with MATLAB*

Attouch, Hedy, Buttazzo, Giuseppe, and Michaille, Gérard, *Variational Analysis in Sobolev and BV Spaces: Applications to PDEs and Optimization*

Wallace, Stein W. and Ziemba, William T., editors, *Applications of Stochastic Programming*

Grötschel, Martin, editor, *The Sharpest Cut: The Impact of Manfred Padberg and His Work*

Renegar, James, *A Mathematical View of Interior-Point Methods in Convex Optimization*

Ben-Tal, Aharon and Nemirovski, Arkadi, *Lectures on Modern Convex Optimization: Analysis, Algorithms, and Engineering Applications*

Introduction to Nonlinear Optimization
Theory, Algorithms, and Applications with MATLAB

Amir Beck
Technion–Israel Institute of Technology
Kfar Saba, Israel

Society for Industrial and Applied Mathematics
Philadelphia

Mathematical Optimization Society
Philadelphia

Universities Press

INTRODUCTION TO NONLINEAR OPTIMIZATION: THEORY, ALGORITHMS, AND APPLICATIONS WITH MATLAB

UNIVERSITIES PRESS (INDIA) PRIVATE LIMITED

Registered Office
3-6-747/1/A & 3-6-754/1, Himayatnagar, Hyderabad 500 029, Telangana, India
info@universitiespress.com; www.universitiespress.com

Distributed by
Orient Blackswan Private Limited

Registered Office
3-6-752 Himayatnagar, Hyderabad 500 029, Telangana, India

Other Offices
Bengaluru, Bhopal, Chennai, Guwahati, Hyderabad, Jaipur, Kolkata, Lucknow, Mumbai, New Delhi, Noida, Patna, Vijayawada

Original American edition published by SIAM: Society for Industrial and Applied Mathematics, Philadelphia, Pennsylvania
Copyright © 2014. All rights reserved.
This edition is authorised for sale only in India, Pakistan, Nepal, Bhutan, Bangladesh and Sri Lanka.

Trademarked names may be used in this book without the inclusion of a trademark symbol. These names are used in an editorial context only; no infringement of trademark is intended.

MATLAB is a registered trademark of The MathWorks, Inc. For MATLAB product information, please contact The MathWorks, Inc., 3 Apple Hill Drive, Natick, MA, 01760-2098 USA, 508-647-7000, Fax: 508-647-7001 *info@mathworks.com, www.mathworks.com.*

First published in India by Universities Press (India) Private Limited 2018

ISBN 978 93 86235 35 0

Printed at
Glorious Printers, Delhi

Published by
Universities Press (India) Private Limited
3-6-747/1/A & 3-6-754/1, Himayatnagar
Hyderabad 500 029, Telangana, India

For
My wife Nili
My daughters Noy and Vered
My parents Nili and Itzhak

Contents

Preface		**xi**
1	**Mathematical Preliminaries**	**1**
	1.1 The Space $\mathbb{R}^n$	1
	1.2 The Space $\mathbb{R}^{m \times n}$	2
	1.3 Inner Products and Norms	2
	1.4 Eigenvalues and Eigenvectors	5
	1.5 Basic Topological Concepts	6
	Exercises	10
2	**Optimality Conditions for Unconstrained Optimization**	**13**
	2.1 Global and Local Optima	13
	2.2 Classification of Matrices	17
	2.3 Second Order Optimality Conditions	23
	2.4 Global Optimality Conditions	30
	2.5 Quadratic Functions	32
	Exercises	34
3	**Least Squares**	**37**
	3.1 "Solution" of Overdetermined Systems	37
	3.2 Data Fitting	39
	3.3 Regularized Least Squares	41
	3.4 Denoising	42
	3.5 Nonlinear Least Squares	45
	3.6 Circle Fitting	45
	Exercises	47
4	**The Gradient Method**	**49**
	4.1 Descent Directions Methods	49
	4.2 The Gradient Method	52
	4.3 The Condition Number	58
	4.4 Diagonal Scaling	63
	4.5 The Gauss–Newton Method	67
	4.6 The Fermat–Weber Problem	68
	4.7 Convergence Analysis of the Gradient Method	73
	Exercises	79
5	**Newton's Method**	**83**
	5.1 Pure Newton's Method	83

		5.2	Damped Newton's Method	88
		5.3	The Cholesky Factorization	89
			Exercises	94
6	**Convex Sets**			**97**
		6.1	Definition and Examples	97
		6.2	Algebraic Operations with Convex Sets	100
		6.3	The Convex Hull	101
		6.4	Convex Cones	104
		6.5	Topological Properties of Convex Sets	108
		6.6	Extreme Points	111
			Exercises	113
7	**Convex Functions**			**117**
		7.1	Definition and Examples	117
		7.2	First Order Characterizations of Convex Functions	119
		7.3	Second Order Characterization of Convex Functions	123
		7.4	Operations Preserving Convexity	125
		7.5	Level Sets of Convex Functions	130
		7.6	Continuity and Differentiability of Convex Functions	132
		7.7	Extended Real-Valued Functions	135
		7.8	Maxima of Convex Functions	137
		7.9	Convexity and Inequalities	139
			Exercises	141
8	**Convex Optimization**			**147**
		8.1	Definition	147
		8.2	Examples	149
		8.3	The Orthogonal Projection Operator	156
		8.4	CVX	158
			Exercises	166
9	**Optimization over a Convex Set**			**169**
		9.1	Stationarity	169
		9.2	Stationarity in Convex Problems	173
		9.3	The Orthogonal Projection Revisited	173
		9.4	The Gradient Projection Method	175
		9.5	Sparsity Constrained Problems	183
			Exercises	189
10	**Optimality Conditions for Linearly Constrained Problems**			**191**
		10.1	Separation and Alternative Theorems	191
		10.2	The KKT conditions	195
		10.3	Orthogonal Regression	203
			Exercises	205
11	**The KKT Conditions**			**207**
		11.1	Inequality Constrained Problems	207
		11.2	Inequality and Equality Constrained Problems	210
		11.3	The Convex Case	213
		11.4	Constrained Least Squares	218

	11.5	Second Order Optimality Conditions	222
	11.6	Optimality Conditions for the Trust Region Subproblem	227
	11.7	Total Least Squares	230
	Exercises		233

12 Duality   237
    12.1   Motivation and Definition 237
    12.2   Strong Duality in the Convex Case 241
    12.3   Examples 247
    Exercises 270

Bibliographic Notes   275

Bibliography   277

Index   281

Preface

This book emerged from the idea that an optimization training should include three basic components: a strong theoretical and algorithmic foundation, familiarity with various applications, and the ability to apply the theory and algorithms on actual "real-life" problems. The book is intended to be the basis of such an extensive training. The mathematical development of the main concepts in nonlinear optimization is done rigorously, where a special effort was made to keep the proofs as simple as possible. The results are presented gradually and accompanied with many illustrative examples. Since the aim is not to give an encyclopedic overview, the focus is on the most useful and important concepts. The theory is complemented by numerous discussions on applications from various scientific fields such as signal processing, economics and localization. Some basic algorithms are also presented and studied to provide some flavor of this important aspect of optimization. Many topics are demonstrated by MATLAB programs, and ideally, the interested reader will find satisfaction in the ability of actually solving problems on his or her own. The book contains several topics that, compared to other classical textbooks, are treated differently. The following are some examples of the less common issues.

- The treatment of stationarity is comprehensive and discusses this important notion in the presence of sparsity constraints.

- The concept of "hidden convexity" is discussed and illustrated in the context of the trust region subproblem.

- The MATLAB toolbox CVX is explored and used.

- The gradient mapping and its properties are studied and used in the analysis of the gradient projection method.

- Second order necessary optimality conditions are treated using a descent direction approach.

- Applications such as circle fitting, Chebyshev center, the Fermat–Weber problem, denoising, clustering, total least squares, and orthogonal regression are studied both theoretically and algorithmically, illustrating concepts such as duality. MATLAB programs are used to show how the theory can be implemented.

The book is intended for students and researchers with a basic background in advanced calculus and linear algebra, but no prior knowledge of optimization theory is assumed. The book contains more than 170 exercises, which can be used to deepen the understanding of the material. The MATLAB functions described throughout the book can be found at

```
www.siam.org/books/mo19.
```

The outline of the book is as follows. **Chapter 1** recalls some of the important concepts in linear algebra and calculus that are essential for the understanding of the mathematical developments in the book. **Chapter 2** focuses on local and global optimality conditions for smooth unconstrained problems. Quadratic functions are also introduced along with their basic properties. Linear and nonlinear least squares problems are introduced and studied in **Chapter 3**. Several applications such as data fitting, denoising, and circle fitting are discussed. The gradient method is introduced and studied in **Chapter 4**. The chapter also contains a discussion on descent direction methods and various stepsize strategies. Extensions such as the scaled gradient method and damped Gauss–Newton are considered. The connection between the gradient method and Weiszfeld's method for solving the Fermat–Weber problem is established. Newton's method is discussed in **Chapter 5**. Convex sets and functions along with their basic properties are the subjects of **Chapters 6 and 7**. Convex optimization problems are introduced in **Chapter 8**, which also includes a variety of applications as well as CVX demonstrations. **Chapter 9** focuses on several important topics related to optimization problems over convex sets: stationarity, gradient mappings, and the gradient projection method. The chapter ends with results on sparsity constrained problems, illuminating the different type of results obtained when the underlying set is not convex. The derivation of the KKT optimality conditions from the separation and alternative theorems is the subject of **Chapter 10**, where only linearly constrained problems are considered. The extension of the KKT conditions to problems with nonlinear constraints is discussed in **Chapter 11**, which also considers the second order necessary conditions. Applications of the conditions to the trust region and total least squares problems are studied. The book ends with a discussion on duality in **Chapter 12**. Strong duality under convexity assumptions is established. This chapter also includes a large amount of examples, applications, and MATLAB illustrations.

I would like to thank Dror Pan and Luba Tetruashvili for reading the book and for their helpful remarks. It has been a pleasure working with the SIAM staff, namely with Bruce Bailey, Elizabeth Greenspan, Sara Murphy, Gina Rinelli, and Kelly Thomas.

Chapter 1

Mathematical Preliminaries

In this short chapter we will review some important notions and results from calculus, linear algebra, and topology that will be frequently used throughout the book. This chapter is not intended to be, by any means, a comprehensive treatment of these subjects, and the interested reader can find more material in advanced calculus and linear algebra books.

1.1 ▪ The Space $\mathbb{R}^n$

The vector space $\mathbb{R}^n$ is the set of n-dimensional column vectors with real components endowed with the component-wise addition operator

$$\begin{pmatrix} x_1 \\ x_2 \\ \vdots \\ x_n \end{pmatrix} + \begin{pmatrix} y_1 \\ y_2 \\ \vdots \\ y_n \end{pmatrix} = \begin{pmatrix} x_1 + y_1 \\ x_2 + y_2 \\ \vdots \\ x_n + y_n \end{pmatrix}$$

and the scalar-vector product

$$\lambda \begin{pmatrix} x_1 \\ x_2 \\ \vdots \\ x_n \end{pmatrix} = \begin{pmatrix} \lambda x_1 \\ \lambda x_2 \\ \vdots \\ \lambda x_n \end{pmatrix},$$

where in the above $x_1, x_2, \ldots, x_n, \lambda$ are real numbers. Throughout the book we will be mainly interested in problems over $\mathbb{R}^n$, although other vector spaces will be considered in a few cases. We will denote the standard basis of $\mathbb{R}^n$ by $\mathbf{e}_1, \mathbf{e}_2, \ldots, \mathbf{e}_n$, where $\mathbf{e}_i$ is the n-length column vector whose ith component is one while all the others are zeros. The column vectors of all ones and all zeros will be denoted by $\mathbf{e}$ and $\mathbf{0}$, respectively, where the length of the vectors will be clear from the context.

Important Subsets of $\mathbb{R}^n$

The *nonnegative orthant* is the subset of $\mathbb{R}^n$ consisting of all vectors in $\mathbb{R}^n$ with nonnegative components and is denoted by $\mathbb{R}^n_+$:

$$\mathbb{R}^n_+ = \{(x_1, x_2, \ldots, x_n)^T : x_1, x_2, \ldots, x_n \geq 0\}.$$

Similarly, the *positive orthant* consists of all the vectors in $\mathbb{R}^n$ with positive components and is denoted by $\mathbb{R}^n_{++}$:

$$\mathbb{R}^n_{++} = \{(x_1, x_2, \ldots, x_n)^T : x_1, x_2, \ldots, x_n > 0\}.$$

For given $\mathbf{x}, \mathbf{y} \in \mathbb{R}^n$, the *closed line segment* between $\mathbf{x}$ and $\mathbf{y}$ is a subset of $\mathbb{R}^n$ denoted by $[\mathbf{x}, \mathbf{y}]$ and defined as

$$[\mathbf{x}, \mathbf{y}] = \{\mathbf{x} + \alpha(\mathbf{y} - \mathbf{x}) : \alpha \in [0, 1]\}.$$

The *open line segment* $(\mathbf{x}, \mathbf{y})$ is similarly defined as

$$(\mathbf{x}, \mathbf{y}) = \{\mathbf{x} + \alpha(\mathbf{y} - \mathbf{x}) : \alpha \in (0, 1)\}$$

when $\mathbf{x} \neq \mathbf{y}$ and is the empty set $\emptyset$ when $\mathbf{x} = \mathbf{y}$. The *unit-simplex*, denoted by Δ_n, is the subset of $\mathbb{R}^n$ comprising all nonnegative vectors whose sum is 1:

$$\Delta_n = \{\mathbf{x} \in \mathbb{R}^n : \mathbf{x} \geq \mathbf{0}, \mathbf{e}^T \mathbf{x} = 1\}.$$

1.2 • The Space $\mathbb{R}^{m \times n}$

The set of all real-valued matrices of order $m \times n$ is denoted by $\mathbb{R}^{m \times n}$. Some special matrices that will be frequently used are the $n \times n$ identity matrix denoted by $\mathbf{I}_n$ and the $m \times n$ zeros matrix denoted by $\mathbf{0}_{m \times n}$. We will frequently omit the subscripts of these matrices when the dimensions will be clear from the context.

1.3 • Inner Products and Norms

Inner Products

We begin with the formal definition of an *inner product*.

Definition 1.1 (inner product). *An* **inner product** *on $\mathbb{R}^n$ is a map $\langle \cdot, \cdot \rangle : \mathbb{R}^n \times \mathbb{R}^n \to \mathbb{R}$ with the following properties:*

1. **(symmetry)** $\langle \mathbf{x}, \mathbf{y} \rangle = \langle \mathbf{y}, \mathbf{x} \rangle$ *for any* $\mathbf{x}, \mathbf{y} \in \mathbb{R}^n$.
2. **(additivity)** $\langle \mathbf{x}, \mathbf{y} + \mathbf{z} \rangle = \langle \mathbf{x}, \mathbf{y} \rangle + \langle \mathbf{x}, \mathbf{z} \rangle$ *for any* $\mathbf{x}, \mathbf{y}, \mathbf{z} \in \mathbb{R}^n$.
3. **(homogeneity)** $\langle \lambda \mathbf{x}, \mathbf{y} \rangle = \lambda \langle \mathbf{x}, \mathbf{y} \rangle$ *for any* $\lambda \in \mathbb{R}$ *and* $\mathbf{x}, \mathbf{y} \in \mathbb{R}^n$.
4. **(positive definiteness)** $\langle \mathbf{x}, \mathbf{x} \rangle \geq 0$ *for any* $\mathbf{x} \in \mathbb{R}^n$ *and* $\langle \mathbf{x}, \mathbf{x} \rangle = 0$ *if and only if* $\mathbf{x} = \mathbf{0}$.

Example 1.2. Perhaps the most widely used inner product is the so-called *dot product* defined by

$$\langle \mathbf{x}, \mathbf{y} \rangle = \mathbf{x}^T \mathbf{y} = \sum_{i=1}^n x_i y_i \text{ for any } \mathbf{x}, \mathbf{y} \in \mathbb{R}^n.$$

Since this is in a sense the "standard" inner product, we will by default assume—unless explicitly stated otherwise—that the underlying inner product is the dot product. ∎

Example 1.3. The dot product is not the only possible inner product on $\mathbb{R}^n$. For example, let $\mathbf{w} \in \mathbb{R}^n_{++}$. Then it is easy to show that the following weighted dot product is also an inner product:

$$\langle \mathbf{x}, \mathbf{y} \rangle_\mathbf{w} = \sum_{i=1}^n w_i x_i y_i. \quad \blacksquare$$

1.3. Inner Products and Norms

Vector Norms

Definition 1.4 (norm). *A norm $\|\cdot\|$ on $\mathbb{R}^n$ is a function $\|\cdot\| : \mathbb{R}^n \to \mathbb{R}$ satisfying the following:*

1. *(nonnegativity)* $\|\mathbf{x}\| \geq 0$ *for any* $\mathbf{x} \in \mathbb{R}^n$ *and* $\|\mathbf{x}\| = 0$ *if and only if* $\mathbf{x} = 0$.
2. *(positive homogeneity)* $\|\lambda \mathbf{x}\| = |\lambda| \|\mathbf{x}\|$ *for any* $\mathbf{x} \in \mathbb{R}^n$ *and* $\lambda \in \mathbb{R}$.
3. *(triangle inequality)* $\|\mathbf{x} + \mathbf{y}\| \leq \|\mathbf{x}\| + \|\mathbf{y}\|$ *for any* $\mathbf{x}, \mathbf{y} \in \mathbb{R}^n$.

One natural way to generate a norm on $\mathbb{R}^n$ is to take any inner product $\langle \cdot, \cdot \rangle$ on $\mathbb{R}^n$ and define the associated norm

$$\|\mathbf{x}\| \equiv \sqrt{\langle \mathbf{x}, \mathbf{x} \rangle} \quad \text{for all } \mathbf{x} \in \mathbb{R}^n,$$

which can be easily seen to be a norm. If the inner product is the dot product, then the associated norm is the so-called *Euclidean norm* or l_2-*norm*:

$$\|\mathbf{x}\|_2 = \sqrt{\sum_{i=1}^{n} x_i^2} \quad \text{for all } \mathbf{x} \in \mathbb{R}^n.$$

By default, the underlying norm on $\mathbb{R}^n$ is $\|\cdot\|_2$, and the subscript 2 will be frequently omitted. The Euclidean norm belongs to the class of l_p norms (for $p \geq 1$) defined by

$$\|\mathbf{x}\|_p \equiv \sqrt[p]{\sum_{i=1}^{n} |x_i|^p}.$$

The restriction $p \geq 1$ is necessary since for $0 < p < 1$, the function $\|\cdot\|_p$ is actually not a norm (see Exercise 1.1). Another important norm is the l_∞ norm given by

$$\|\mathbf{x}\|_\infty \equiv \max_{i=1,2,\ldots,n} |x_i|.$$

Unsurprisingly, it can be shown (see Exercise 1.2) that

$$\|\mathbf{x}\|_\infty = \lim_{p \to \infty} \|\mathbf{x}\|_p.$$

An important inequality connecting the dot product of two vectors and their norms is the Cauchy–Schwarz inequality, which will be used frequently throughout the book.

Lemma 1.5 (Cauchy-Schwarz inequality). *For any* $\mathbf{x}, \mathbf{y} \in \mathbb{R}^n$,

$$|\mathbf{x}^T \mathbf{y}| \leq \|\mathbf{x}\|_2 \cdot \|\mathbf{y}\|_2.$$

Equality is satisfied if and only if $\mathbf{x}$ *and* $\mathbf{y}$ *are linearly dependent.*

Matrix Norms

Similarly to vector norms, we can define the concept of a *matrix norm*.

Definition 1.6. *A norm $\|\cdot\|$ on $\mathbb{R}^{m \times n}$ is a function $\|\cdot\| : \mathbb{R}^{m \times n} \to \mathbb{R}$ satisfying the following:*

1. **(nonnegativity)** $\|\mathbf{A}\| \geq 0$ *for any* $\mathbf{A} \in \mathbb{R}^{m \times n}$ *and* $\|\mathbf{A}\| = 0$ *if and only if* $\mathbf{A} = \mathbf{0}$.
2. **(positive homogeneity)** $\|\lambda \mathbf{A}\| = |\lambda| \cdot \|\mathbf{A}\|$ *for any* $\mathbf{A} \in \mathbb{R}^{m \times n}$ *and* $\lambda \in \mathbb{R}$.
3. **(triangle inequality)** $\|\mathbf{A} + \mathbf{B}\| \leq \|\mathbf{A}\| + \|\mathbf{B}\|$ *for any* $\mathbf{A}, \mathbf{B} \in \mathbb{R}^{m \times n}$.

Many examples of matrix norms are generated by using the concept of induced norms, which we now describe. Given a matrix $\mathbf{A} \in \mathbb{R}^{m \times n}$ and two norms $\|\cdot\|_a$ and $\|\cdot\|_b$ on $\mathbb{R}^n$ and $\mathbb{R}^m$, respectively, the *induced matrix norm* $\|\mathbf{A}\|_{a,b}$ is defined by

$$\|\mathbf{A}\|_{a,b} = \max_{\mathbf{x}} \{\|\mathbf{A}\mathbf{x}\|_b : \|\mathbf{x}\|_a \leq 1\}.$$

It can be shown that the above definition implies that for any $\mathbf{x} \in \mathbb{R}^n$ the inequality

$$\|\mathbf{A}\mathbf{x}\|_b \leq \|\mathbf{A}\|_{a,b} \|\mathbf{x}\|_a$$

holds. An induced matrix norm is indeed a norm in the sense that it satisfies the three properties required from a matrix norm (see Definition 1.6): nonnegativity, positive homogeneity, and the triangle inequality. We refer to the matrix norm $\|\cdot\|_{a,b}$ as the (a,b)-norm. When $a = b$ (for example, when the two vector norms are l_a norms), we will simply refer to it as an a-norm and omit one of the subscripts in its notation; that is, the notation is $\|\cdot\|_a$ instead of $\|\cdot\|_{a,a}$.

Example 1.7 (spectral norm). If $\|\cdot\|_a = \|\cdot\|_b = \|\cdot\|_2$, then the induced norm of a matrix $\mathbf{A} \in \mathbb{R}^{m \times n}$ is the maximum singular value of $\mathbf{A}$:

$$\|\mathbf{A}\|_2 = \|\mathbf{A}\|_{2,2} = \sqrt{\lambda_{\max}(\mathbf{A}^T \mathbf{A})} \equiv \sigma_{\max}(\mathbf{A}).$$

Since the Euclidean norm is the "standard" vector norm, the induced norm, namely the spectral norm, will be the standard matrix norm, and thus the subscripts of this norm will usually be omitted. ■

Example 1.8 (1-norm). When $\|\cdot\|_a = \|\cdot\|_b = \|\cdot\|_1$, the induced matrix norm of a matrix $\mathbf{A} \in \mathbb{R}^{m \times n}$ is given by

$$\|\mathbf{A}\|_1 = \max_{j=1,2,\ldots,n} \sum_{i=1}^{m} |A_{i,j}|.$$

This norm is also called the *maximum absolute column sum* norm. ■

Example 1.9 (∞-norm). When $\|\cdot\|_a = \|\cdot\|_b = \|\cdot\|_\infty$, then the induced matrix norm of a matrix $\mathbf{A} \in \mathbb{R}^{m \times n}$ is given by

$$\|\mathbf{A}\|_\infty = \max_{i=1,2,\ldots,m} \sum_{j=1}^{n} |A_{i,j}|.$$

This norm is also called the *maximum absolute row sum norm*. ■

An example of a matrix norm that is not an *induced* norm is the *Frobenius norm* defined by

$$\|\mathbf{A}\|_F = \sqrt{\sum_{i=1}^{m} \sum_{j=1}^{n} A_{ij}^2}, \quad \mathbf{A} \in \mathbb{R}^{m \times n}.$$

1.4 • Eigenvalues and Eigenvectors

Let $\mathbf{A} \in \mathbb{R}^{n \times n}$. Then a nonzero vector $\mathbf{v} \in \mathbb{C}^n$ is called an *eigenvector* of $\mathbf{A}$ if there exists a $\lambda \in \mathbb{C}$ (the complex field) for which

$$\mathbf{A}\mathbf{v} = \lambda \mathbf{v}.$$

The scalar λ is the *eigenvalue* corresponding to the eigenvector $\mathbf{v}$. In general, real-valued matrices can have complex eigenvalues, but it is well known that all the eigenvalues of symmetric matrices are real. The eigenvalues of a symmetric $n \times n$ matrix $\mathbf{A}$ are denoted by

$$\lambda_1(\mathbf{A}) \geq \lambda_2(\mathbf{A}) \geq \cdots \geq \lambda_n(\mathbf{A}).$$

The maximum eigenvalue is also denoted by $\lambda_{\max}(\mathbf{A})(= \lambda_1(\mathbf{A}))$ and the minimum eigenvalue is also denoted by $\lambda_{\min}(\mathbf{A})(= \lambda_n(\mathbf{A}))$. One of the most useful results related to eigenvalues is the spectral decomposition theorem, which states that any symmetric matrix $\mathbf{A}$ has an orthonormal basis of eigenvectors.

Theorem 1.10 (spectral decomposition theorem). *Let $\mathbf{A} \in \mathbb{R}^{n \times n}$ be an $n \times n$ symmetric matrix. Then there exists an orthogonal matrix $\mathbf{U} \in \mathbb{R}^{n \times n}$ ($\mathbf{U}^T\mathbf{U} = \mathbf{U}\mathbf{U}^T = \mathbf{I}$) and a diagonal matrix $\mathbf{D} = \mathrm{diag}(d_1, d_2, \ldots, d_n)$ for which*

$$\mathbf{U}^T \mathbf{A} \mathbf{U} = \mathbf{D}. \tag{1.1}$$

The columns of the matrix $\mathbf{U}$ in the factorization (1.1) constitute an orthonormal basis comprised of eigenvectors of $\mathbf{A}$ and the diagonal elements of $\mathbf{D}$ are the corresponding eigenvalues. A direct result of the spectral decomposition theorem is that the trace and the determinant of a matrix can be expressed via its eigenvalues:

$$\mathrm{Tr}(\mathbf{A}) = \sum_{i=1}^{n} \lambda_i(\mathbf{A}), \tag{1.2}$$

$$\det(\mathbf{A}) = \prod_{i=1}^{n} \lambda_i(\mathbf{A}). \tag{1.3}$$

Another important consequence of the spectral decomposition theorem is the bounding of the so-called *Rayleigh quotient*. For a symmetric matrix $\mathbf{A} \in \mathbb{R}^{n \times n}$, the Rayleigh quotient is defined by

$$R_{\mathbf{A}}(\mathbf{x}) = \frac{\mathbf{x}^T \mathbf{A} \mathbf{x}}{\|\mathbf{x}\|^2} \text{ for any } \mathbf{x} \neq \mathbf{0}.$$

We can now use the spectral decomposition theorem to prove the following lemma providing lower and upper bounds on the Rayleigh quotient.

Lemma 1.11. *Let $\mathbf{A} \in \mathbb{R}^{n \times n}$ be symmetric. Then*

$$\lambda_{\min}(\mathbf{A}) \leq R_{\mathbf{A}}(\mathbf{x}) \leq \lambda_{\max}(\mathbf{A}) \text{ for any } \mathbf{x} \neq \mathbf{0}.$$

Proof. By the spectral decomposition theorem there exists an orthogonal matrix $\mathbf{U} \in \mathbb{R}^{n \times n}$ and a diagonal matrix $\mathbf{D} = \mathrm{diag}(d_1, d_2, \ldots, d_n)$ such that $\mathbf{U}^T \mathbf{A} \mathbf{U} = \mathbf{D}$. We can assume without the loss of generality that the diagonal elements of $\mathbf{D}$, which are the eigenvalues of $\mathbf{A}$, are ordered nonincreasingly: $d_1 \geq d_2 \geq \cdots \geq d_n$, where $d_1 = \lambda_{\max}(\mathbf{A})$ and

$d_n = \lambda_{\min}(\mathbf{A})$. Making the change of variables $\mathbf{x} = \mathbf{U}\mathbf{y}$ and noting that $\mathbf{U}$ is a nonsingular matrix, we obtain that

$$\max_{\mathbf{x} \neq 0} \frac{\mathbf{x}^T \mathbf{A} \mathbf{x}}{\|\mathbf{x}\|^2} = \max_{\mathbf{y} \neq 0} \frac{\mathbf{y}^T \mathbf{U}^T \mathbf{A} \mathbf{U} \mathbf{y}}{\|\mathbf{U}\mathbf{y}\|^2} = \max_{\mathbf{y} \neq 0} \frac{\mathbf{y}^T \mathbf{D} \mathbf{y}}{\|\mathbf{y}\|^2} = \max_{\mathbf{y} \neq 0} \frac{\sum_{i=1}^n d_i y_i^2}{\sum_{i=1}^n y_i^2}.$$

Since $d_i \leq d_1$ for all $i = 1, 2, \ldots, n$, it follows that $\sum_{i=1}^n d_i y_i^2 \leq d_1 (\sum_{i=1}^n y_i^2)$, and hence

$$R_{\mathbf{A}}(\mathbf{x}) = \max_{\mathbf{x} \neq 0} \frac{\mathbf{x}^T \mathbf{A} \mathbf{x}}{\|\mathbf{x}\|^2} \leq \max_{\mathbf{y} \neq 0} \frac{d_1(\sum_{i=1}^n y_i^2)}{\sum_{i=1}^n y_i^2} = d_1 = \lambda_{\max}(\mathbf{A}).$$

The inequality $R_{\mathbf{A}}(\mathbf{x}) \geq \lambda_{\min}(\mathbf{A})$ follows by a similar argument. $\square$

The lower and upper bounds on the Rayleigh quotient given in the last lemma are attained at eigenvectors corresponding to the minimal and maximal eigenvalues respectively. Indeed, if $\mathbf{v}$ and $\mathbf{w}$ are eigenvectors corresponding to the minimal and maximal eigenvalues respectively, then

$$R_{\mathbf{A}}(\mathbf{v}) = \frac{\mathbf{v}^T \mathbf{A} \mathbf{v}}{\|\mathbf{v}\|^2} = \frac{\lambda_{\min}(\mathbf{A}) \|\mathbf{v}\|^2}{\|\mathbf{v}\|^2} = \lambda_{\min}(\mathbf{A}),$$

$$R_{\mathbf{A}}(\mathbf{w}) = \frac{\mathbf{w}^T \mathbf{A} \mathbf{w}}{\|\mathbf{w}\|^2} = \frac{\lambda_{\max}(\mathbf{A}) \|\mathbf{w}\|^2}{\|\mathbf{w}\|^2} = \lambda_{\max}(\mathbf{A}).$$

The above facts are summarized in the following lemma.

Lemma 1.12. *Let $\mathbf{A} \in \mathbb{R}^{n \times n}$ be symmetric. Then*

$$\min_{\mathbf{x} \neq 0} R_{\mathbf{A}}(\mathbf{x}) = \lambda_{\min}(\mathbf{A}), \tag{1.4}$$

and the eigenvectors of $\mathbf{A}$ corresponding to the minimal eigenvalue are minimizers of problem (1.4). *In addition,*

$$\max_{\mathbf{x} \neq 0} R_{\mathbf{A}}(\mathbf{x}) = \lambda_{\max}(\mathbf{A}), \tag{1.5}$$

and the eigenvectors of $\mathbf{A}$ corresponding to the maximal eigenvalue are maximizers of problem (1.5).

1.5 • Basic Topological Concepts

We begin with the definition of a ball.

Definition 1.13 (open ball, closed ball). *The* **open ball** *with center $\mathbf{c} \in \mathbb{R}^n$ and radius r is denoted by $B(\mathbf{c}, r)$ and defined by*

$$B(\mathbf{c}, r) = \{\mathbf{x} \in \mathbb{R}^n : \|\mathbf{x} - \mathbf{c}\| < r\}.$$

The **closed ball** *with center $\mathbf{c}$ and radius r is denoted by $B[\mathbf{c}, r]$ and defined by*

$$B[\mathbf{c}, r] = \{\mathbf{x} \in \mathbb{R}^n : \|\mathbf{x} - \mathbf{c}\| \leq r\}.$$

1.5. Basic Topological Concepts

Note that the norm used in the definition of the ball is not necessarily the Euclidean norm. As usual, if the norm is not specified, we will assume by default that it is the Euclidean norm. The ball $B(\mathbf{c}, r)$ for some arbitrary $r > 0$ is also referred to as a *neighborhood* of $\mathbf{c}$. The first topological notion we define is that of an *interior point* of a set. This is a point which has a neighborhood contained in the set.

Definition 1.14 (interior points). *Given a set $U \subseteq \mathbb{R}^n$, a point $\mathbf{c} \in U$ is an* **interior point** *of U if there exists $r > 0$ for which $B(\mathbf{c}, r) \subseteq U$.*

The set of all interior points of a given set U is called the *interior* of the set and is denoted by $\mathrm{int}(U)$:

$$\mathrm{int}(U) = \{\mathbf{x} \in U : B(\mathbf{x}, r) \subseteq U \text{ for some } r > 0\}.$$

Example 1.15. Following are some examples of interiors of sets which were previously discussed:

$$\mathrm{int}(\mathbb{R}^n_+) = \mathbb{R}^n_{++},$$
$$\mathrm{int}(B[\mathbf{c}, r]) = B(\mathbf{c}, r) \quad (\mathbf{c} \in \mathbb{R}^n, r \in \mathbb{R}_{++}).$$

Definition 1.16 (open sets). *An* **open set** *is a set that contains only interior points. In other words, $U \subseteq \mathbb{R}^n$ is an open set if*

for every $\mathbf{x} \in U$ there exists $r > 0$ such that $B(\mathbf{x}, r) \subseteq U$.

Examples of open sets are $\mathbb{R}^n$, open balls (hence the name), and the positive orthant $\mathbb{R}^n_{++}$. A known result is that a union of any number of open sets is an open set and the intersection of a finite number of open sets is open.

Definition 1.17 (closed sets). *A set $U \subseteq \mathbb{R}^n$ is said to be* **closed** *if it contains all the limits of convergent sequences of points in U; that is, U is closed if for every sequence of points $\{\mathbf{x}_i\}_{i \geq 1} \subseteq U$ satisfying $\mathbf{x}_i \to \mathbf{x}^*$ as $i \to \infty$, it holds that $\mathbf{x}^* \in U$.*

A known property is that a set U is closed if and only if its complement U^c is open (see Exercise 1.15). Examples of closed sets are the closed ball $B[\mathbf{c}, r]$, closed lines segments, the nonnegative orthant $\mathbb{R}^n_+$, and the unit simplex Δ_n. The space $\mathbb{R}^n$ is both closed and open. An important and useful result states that level sets, as well as contour sets, of continuous functions are closed. This is stated in the following proposition.

Proposition 1.18 (closedness of level and contour sets of continuous functions). *Let f be a continuous function defined over a closed set $S \subseteq \mathbb{R}^n$. Then for any $\alpha \in \mathbb{R}$ the sets*

$$Lev(f, \alpha) = \{\mathbf{x} \in S : f(\mathbf{x}) \leq \alpha\},$$
$$Con(f, \alpha) = \{\mathbf{x} \in S : f(\mathbf{x}) = \alpha\}$$

are closed.

Definition 1.19 (boundary points). *Given a set $U \subseteq \mathbb{R}^n$, a* **boundary point** *of U is a point $\mathbf{x} \in \mathbb{R}^n$ satisfying the following: any neighborhood of $\mathbf{x}$ contains at least one point in U and at least one point in its complement U^c.*

The set of all boundary points of a set U is denoted by $\mathrm{bd}(U)$ and is called *the boundary of U.*

Example 1.20. Some examples of boundary sets are

$$\mathrm{bd}(B(\mathbf{c},r)) = \mathrm{bd}(B[\mathbf{c},r]) = \{\mathbf{x} \in \mathbb{R}^n : \|\mathbf{x}-\mathbf{c}\| = r\} \quad (\mathbf{c} \in \mathbb{R}^n, r \in \mathbb{R}_{++}),$$
$$\mathrm{bd}(\mathbb{R}^n_{++}) = \mathrm{bd}(\mathbb{R}^n_+) = \{\mathbf{x} \in \mathbb{R}^n_+ : \exists i : x_i = 0\},$$
$$\mathrm{bd}(\mathbb{R}^n) = \emptyset. \quad \blacksquare$$

The *closure* of a set $U \subseteq \mathbb{R}^n$ is denoted by $\mathrm{cl}(U)$ and is defined to be the smallest closed set containing U:

$$\mathrm{cl}(U) = \bigcap \{T : U \subseteq T, T \text{ is closed}\}.$$

The closure set is indeed a closed set as an intersection of closed sets. Another equivalent definition of $\mathrm{cl}(U)$ is given by

$$\mathrm{cl}(U) = U \cup \mathrm{bd}(U).$$

Example 1.21.

$$\mathrm{cl}(\mathbb{R}^n_{++}) = \mathbb{R}^n_+,$$
$$\mathrm{cl}(B(\mathbf{c},r)) = B[\mathbf{c},r] \quad (\mathbf{c} \in \mathbb{R}^n, r \in \mathbb{R}_{++}),$$
$$\mathrm{cl}((\mathbf{x},\mathbf{y})) = [\mathbf{x},\mathbf{y}] \quad (\mathbf{x},\mathbf{y} \in \mathbb{R}^n, \mathbf{x} \neq \mathbf{y}). \quad \blacksquare$$

Definition 1.22 (boundedness and compactness).

1. A set $U \subseteq \mathbb{R}^n$ is called **bounded** if there exists $M > 0$ for which $U \subseteq B(\mathbf{0},M)$.

2. A set $U \subseteq \mathbb{R}^n$ is called **compact** if it is closed and bounded.

Examples of compact sets are closed balls and line segments. The positive orthant is not compact since it is unbounded, and open balls are not compact since they are not closed.

1.5.1 ▪ Differentiability

Let f be a function defined on a set $S \subseteq \mathbb{R}^n$. Let $\mathbf{x} \in \mathrm{int}(S)$ and let $\mathbf{0} \neq \mathbf{d} \in \mathbb{R}^n$. If the limit

$$\lim_{t \to 0^+} \frac{f(\mathbf{x}+t\mathbf{d}) - f(\mathbf{x})}{t}$$

exists, then it is called *the directional derivative* of f at $\mathbf{x}$ along the direction $\mathbf{d}$ and is denoted by $f'(\mathbf{x};\mathbf{d})$. For any $i = 1,2,\ldots,n$, if the limit

$$\frac{\partial f}{\partial x_i}(\mathbf{x}) = \lim_{t \to 0} \frac{f(\mathbf{x}+t\mathbf{e}_i) - f(\mathbf{x})}{t}$$

exists, then it is called *the ith partial derivative* of f at $\mathbf{x}$.

If all the partial derivatives of a function f exist at a point $\mathbf{x} \in \mathbb{R}^n$, then the *gradient* of f at $\mathbf{x}$ is defined to be the column vector consisting of all the partial derivatives:

$$\nabla f(\mathbf{x}) = \begin{pmatrix} \frac{\partial f}{\partial x_1}(\mathbf{x}) \\ \frac{\partial f}{\partial x_2}(\mathbf{x}) \\ \vdots \\ \frac{\partial f}{\partial x_n}(\mathbf{x}) \end{pmatrix}.$$

A function f defined on an open set $U \subseteq \mathbb{R}^n$ is called *continuously differentiable over U* if all the partial derivatives exist and are continuous on U. The definition of continuous differentiability can also be extended to nonopen sets by using the convention that a function f is said to be continuously differentiable over a set C if there exists an open set U containing C on which the function is also defined and continuously differentiable. In the setting of continuous differentiability, we have the following important formula for the directional derivative:

$$f'(\mathbf{x}; \mathbf{d}) = \nabla f(\mathbf{x})^T \mathbf{d}$$

for all $\mathbf{x} \in U$ and $\mathbf{d} \in \mathbb{R}^n$. It can also be shown in this setting of continuous differentiability that the following approximation result holds.

Proposition 1.23. *Let $f : U \to \mathbb{R}$ be defined on an open set $U \subseteq \mathbb{R}^n$. Suppose that f is continuously differentiable over U. Then*

$$\lim_{\mathbf{d} \to 0} \frac{f(\mathbf{x}+\mathbf{d}) - f(\mathbf{x}) - \nabla f(\mathbf{x})^T \mathbf{d}}{\|\mathbf{d}\|} = 0 \text{ for all } \mathbf{x} \in U.$$

Another way to write the above result is as follows:

$$f(\mathbf{y}) = f(\mathbf{x}) + \nabla f(\mathbf{x})^T (\mathbf{y} - \mathbf{x}) + o(\|\mathbf{y} - \mathbf{x}\|),$$

where $o(\cdot) : \mathbb{R}_+ \to \mathbb{R}$ is a one-dimensional function satisfying $\frac{o(t)}{t} \to 0$ as $t \to 0^+$.

The partial derivatives $\frac{\partial f}{\partial x_i}$ are themselves real-valued functions that can be partially differentiated. The (i,j)th partial derivatives of f at $\mathbf{x} \in U$ (if it exists) is defined by

$$\frac{\partial^2 f}{\partial x_i \partial x_j}(\mathbf{x}) = \frac{\partial \left(\frac{\partial f}{\partial x_j}\right)}{\partial x_i}(\mathbf{x}).$$

A function f defined on an open set $U \subseteq \mathbb{R}^n$ is called *twice continuously differentiable over U* if all the second order partial derivatives exist and are continuous over U. Under the assumption of twice continuous differentiability, the second order partial derivatives are symmetric, meaning that for any $i \neq j$ and any $\mathbf{x} \in U$

$$\frac{\partial^2 f}{\partial x_i \partial x_j}(\mathbf{x}) = \frac{\partial^2 f}{\partial x_j \partial x_i}(\mathbf{x}).$$

The *Hessian* of f at a point $\mathbf{x} \in U$ is the $n \times n$ matrix

$$\nabla^2 f(\mathbf{x}) = \begin{pmatrix} \frac{\partial^2 f}{\partial x_1^2}(\mathbf{x}) & \frac{\partial^2 f}{\partial x_1 \partial x_2}(\mathbf{x}) & \cdots & \frac{\partial^2 f}{\partial x_1 \partial x_n}(\mathbf{x}) \\ \frac{\partial^2 f}{\partial x_2 \partial x_1}(\mathbf{x}) & \frac{\partial^2 f}{\partial x_2^2}(\mathbf{x}) & & \vdots \\ \vdots & \vdots & & \vdots \\ \frac{\partial^2 f}{\partial x_n \partial x_1}(\mathbf{x}) & \frac{\partial^2 f}{\partial x_n \partial x_2}(\mathbf{x}) & \cdots & \frac{\partial^2 f}{\partial x_n^2}(\mathbf{x}) \end{pmatrix},$$

where all the second order partial derivatives are evaluated at $\mathbf{x}$. Since f is twice continuously differentiable over U, the Hessian matrix is symmetric. There are two main approximation results (linear and quadratic) which are direct consequences of Taylor's approximation theorem that will be used frequently in the book and are thus recalled here.

Theorem 1.24 (linear approximation theorem). *Let $f : U \to \mathbb{R}$ be a twice continuously differentiable function over an open set $U \subseteq \mathbb{R}^n$, and let $\mathbf{x} \in U, r > 0$ satisfy $B(\mathbf{x}, r) \subseteq U$. Then for any $\mathbf{y} \in B(\mathbf{x}, r)$ there exists $\xi \in [\mathbf{x}, \mathbf{y}]$ such that*

$$f(\mathbf{y}) = f(\mathbf{x}) + \nabla f(\mathbf{x})^T (\mathbf{y} - \mathbf{x}) + \frac{1}{2}(\mathbf{y} - \mathbf{x})^T \nabla^2 f(\xi)(\mathbf{y} - \mathbf{x}).$$

Theorem 1.25 (quadratic approximation theorem). *Let $f : U \to \mathbb{R}$ be a twice continuously differentiable function over an open set $U \subseteq \mathbb{R}^n$, and let $\mathbf{x} \in U, r > 0$ satisfy $B(\mathbf{x}, r) \subseteq U$. Then for any $\mathbf{y} \in B(\mathbf{x}, r)$*

$$f(\mathbf{y}) = f(\mathbf{x}) + \nabla f(\mathbf{x})^T (\mathbf{y} - \mathbf{x}) + \frac{1}{2}(\mathbf{y} - \mathbf{x})^T \nabla^2 f(\mathbf{x})(\mathbf{y} - \mathbf{x}) + o(\|\mathbf{y} - \mathbf{x}\|^2).$$

Exercises

1.1. Show that $\|\cdot\|_{1/2}$ is not a norm.

1.2. Prove that for any $\mathbf{x} \in \mathbb{R}^n$ one has

$$\|\mathbf{x}\|_\infty = \lim_{p \to \infty} \|\mathbf{x}\|_p.$$

1.3. Show that for any $\mathbf{x}, \mathbf{y}, \mathbf{z} \in \mathbb{R}^n$

$$\|\mathbf{x} - \mathbf{z}\| \leq \|\mathbf{x} - \mathbf{y}\| + \|\mathbf{y} - \mathbf{z}\|.$$

1.4. Prove the Cauchy–Schwarz inequality (Lemma 1.5). Show that equality holds if and only if the vectors $\mathbf{x}$ and $\mathbf{y}$ are linearly dependent.

1.5. Suppose that $\mathbb{R}^m$ and $\mathbb{R}^n$ are equipped with norms $\|\cdot\|_b$ and $\|\cdot\|_a$, respectively. Show that the induced matrix norm $\|\cdot\|_{a,b}$ satisfies the triangle inequality. That is, for any $\mathbf{A}, \mathbf{B} \in \mathbb{R}^{m \times n}$ the inequality

$$\|\mathbf{A} + \mathbf{B}\|_{a,b} \leq \|\mathbf{A}\|_{a,b} + \|\mathbf{B}\|_{a,b}$$

holds.

Exercises

1.6. Let $\|\cdot\|$ be a norm on $\mathbb{R}^n$. Show that the norm function $f(\mathbf{x}) = \|\mathbf{x}\|$ is a continuous function over $\mathbb{R}^n$.

1.7. **(attainment of the maximum in the induced norm definition).** Suppose that $\mathbb{R}^m$ and $\mathbb{R}^n$ are equipped with norms $\|\cdot\|_b$ and $\|\cdot\|_a$, respectively, and let $\mathbf{A} \in \mathbb{R}^{m \times n}$. Show that there exists $\mathbf{x} \in \mathbb{R}^n$ such that $\|\mathbf{x}\|_a \leq 1$ and $\|\mathbf{A}\mathbf{x}\|_b = \|\mathbf{A}\|_{a,b}$.

1.8. Suppose that $\mathbb{R}^m$ and $\mathbb{R}^n$ are equipped with norms $\|\cdot\|_b$ and $\|\cdot\|_a$, respectively. Show that the induced matrix norm $\|\cdot\|_{a,b}$ can be computed by the formula

$$\|\mathbf{A}\|_{a,b} = \max_{\mathbf{x}} \{\|\mathbf{A}\mathbf{x}\|_b : \|\mathbf{x}\|_a = 1\}.$$

1.9. Suppose that $\mathbb{R}^m$ and $\mathbb{R}^n$ are equipped with norms $\|\cdot\|_b$ and $\|\cdot\|_a$, respectively. Show that the induced matrix norm $\|\cdot\|_{a,b}$ can be computed by the formula

$$\|\mathbf{A}\|_{a,b} = \max_{\mathbf{x} \neq 0} \frac{\|\mathbf{A}\mathbf{x}\|_b}{\|\mathbf{x}\|_a}.$$

1.10. Let $\mathbf{A} \in \mathbb{R}^{m \times n}, \mathbf{B} \in \mathbb{R}^{n \times k}$ and assume that $\mathbb{R}^m, \mathbb{R}^n$, and $\mathbb{R}^k$ are equipped with the norms $\|\cdot\|_c, \|\cdot\|_b$, and $\|\cdot\|_a$, respectively. Prove that

$$\|\mathbf{A}\mathbf{B}\|_{a,c} \leq \|\mathbf{A}\|_{b,c} \|\mathbf{B}\|_{a,b}.$$

1.11. Prove the formula of the ∞-matrix norm given in Example 1.9.

1.12. Let $\mathbf{A} \in \mathbb{R}^{m \times n}$. Prove that

(i) $\frac{1}{\sqrt{n}} \|\mathbf{A}\|_\infty \leq \|\mathbf{A}\|_2 \leq \sqrt{m} \|\mathbf{A}\|_\infty$,

(ii) $\frac{1}{\sqrt{m}} \|\mathbf{A}\|_1 \leq \|\mathbf{A}\|_2 \leq \sqrt{n} \|\mathbf{A}\|_1$.

1.13. Let $\mathbf{A} \in \mathbb{R}^{m \times n}$. Show that

(i) $\|\mathbf{A}\| = \|\mathbf{A}^T\|$ (here $\|\cdot\|$ is the spectral norm),

(ii) $\|\mathbf{A}\|_F^2 = \sum_{i=1}^n \lambda_i(\mathbf{A}^T \mathbf{A})$.

1.14. Let $\mathbf{A} \in \mathbb{R}^{n \times n}$ be a symmetric matrix. Show that

$$\max_{\mathbf{x}} \{\mathbf{x}^T \mathbf{A} \mathbf{x} : \|\mathbf{x}\|^2 = 1\} = \lambda_{\max}(\mathbf{A}).$$

1.15. Prove that a set $U \subseteq \mathbb{R}^n$ is closed if and only if its complement U^c is open.

1.16. (i) Let $\{A_i\}_{i \in I}$ be a collection of open sets where I is a given index set. Show that $\bigcup_{i \in I} A_i$ is an open set. Show that if I is finite, then $\bigcap_{i \in I} A_i$ is open.

(ii) Let $\{A_i\}_{i \in I}$ be a collection of closed sets where I is a given index set. Show that $\bigcap_{i \in I} A_i$ is a closed set. Show that if I is finite, then $\bigcup_{i \in I} A_i$ is closed.

1.17. Give an example of open sets $A_i, i \in I$ for which $\bigcap_{i \in I} A_i$ is not open.

1.18. Let $A, B \subseteq \mathbb{R}^n$. Prove that $\mathrm{cl}(A \cap B) \subseteq \mathrm{cl}(A) \cap \mathrm{cl}(B)$. Give an example in which the inclusion is proper.

1.19. Let $A, B \subseteq \mathbb{R}^n$. Prove that $\mathrm{int}(A \cap B) = \mathrm{int}(A) \cap \mathrm{int}(B)$ and that $\mathrm{int}(A) \cup \mathrm{int}(B) \subseteq \mathrm{int}(A \cup B)$. Show an example in which the latter inclusion is proper.

Chapter 2
Optimality Conditions for Unconstrained Optimization

2.1 • Global and Local Optima

Although our main interest in this section is to discuss minimum and maximum points of a function over the entire space, we will nonetheless present the more general definition of global minimum and maximum points of a function over a given set.

Definition 2.1 (global minimum and maximum). *Let $f : S \rightarrow \mathbb{R}$ be defined on a set $S \subseteq \mathbb{R}^n$. Then*

1. $\mathbf{x}^* \in S$ *is called a* **global minimum point** *of f over S if $f(\mathbf{x}) \geq f(\mathbf{x}^*)$ for any $\mathbf{x} \in S$,*

2. $\mathbf{x}^* \in S$ *is called a* **strict global minimum point** *of f over S if $f(\mathbf{x}) > f(\mathbf{x}^*)$ for any $\mathbf{x}^* \neq \mathbf{x} \in S$,*

3. $\mathbf{x}^* \in S$ *is called a* **global maximum point** *of f over S if $f(\mathbf{x}) \leq f(\mathbf{x}^*)$ for any $\mathbf{x} \in S$,*

4. $\mathbf{x}^* \in S$ *is called a* **strict global maximum point** *of f over S if $f(\mathbf{x}) < f(\mathbf{x}^*)$ for any $\mathbf{x}^* \neq \mathbf{x} \in S$.*

The set S on which the optimization of f is performed is also called *the feasible set*, and any point $\mathbf{x} \in S$ is called *a feasible solution*. We will frequently omit the adjective "global" and just use the terminology "minimum point" and "maximum point." It is also customary to refer to a global minimum point as a *minimizer* or a *global minimizer* and to a global maximum point as a *maximizer* or a *global maximizer*.

A vector $\mathbf{x}^* \in S$ is called a *global optimum* of f over S if it is either a global minimum or a global maximum. The *maximal value of f over S* is defined as the supremum of f over S:

$$\max\{f(\mathbf{x}) : \mathbf{x} \in S\} = \sup\{f(\mathbf{x}) : \mathbf{x} \in S\}.$$

If $\mathbf{x}^* \in S$ is a global maximum of f over S, then the maximum value of f over S is $f(\mathbf{x}^*)$. Similarly the *minimal value of f over S* is the infimum of f over S,

$$\min\{f(\mathbf{x}) : \mathbf{x} \in S\} = \inf\{f(\mathbf{x}) : \mathbf{x} \in S\},$$

and is equal to $f(\mathbf{x}^*)$ when $\mathbf{x}^*$ is a global minimum of f over S. Note that in this book we will not use the sup/inf notation but rather use only the min/max notation, where the usage of this notation does not imply that the maximum or minimum is actually attained. As opposed to global maximum and minimum points, minimal and maximal values are always unique. There could be several global minimum points, but there could be only one minimal value. The set of all global minimizers of f over S is denoted by

$$\text{argmin}\{f(\mathbf{x}) : \mathbf{x} \in S\},$$

and the set of all global maximizers of f over S is denoted by

$$\text{argmax}\{f(\mathbf{x}) : \mathbf{x} \in S\}.$$

Note that the notation "$f : S \to \mathbb{R}$" means in particular that S is the *domain* of f, that is, the subset of $\mathbb{R}^n$ on which f is defined. In Definition 2.1 the minimization and maximization is over the domain of the function. However, later on we will also deal with functions $f : S \to \mathbb{R}$ and discuss problems of finding global optima points with respect to a subset of the domain.

Example 2.2. Consider the two-dimensional linear function $f(x,y) = x + y$ defined over the unit ball $S = B[\mathbf{0}, 1] = \{(x, y)^T : x^2 + y^2 \leq 1\}$. Then by the Cauchy–Schwarz inequality we have for any $(x, y)^T \in S$

$$x + y = \begin{pmatrix} x & y \end{pmatrix} \begin{pmatrix} 1 \\ 1 \end{pmatrix} \leq \sqrt{x^2 + y^2} \sqrt{1^2 + 1^2} \leq \sqrt{2}.$$

Therefore, the maximal value of f over S is upper bounded by $\sqrt{2}$. On the other hand, the upper bound $\sqrt{2}$ is attained at $(x, y) = (\frac{1}{\sqrt{2}}, \frac{1}{\sqrt{2}})$. It is not difficult to see that this is the only point that attains this value, and thus $(\frac{1}{\sqrt{2}}, \frac{1}{\sqrt{2}})$ is the strict global maximum point of f over S, and the maximal value is $\sqrt{2}$. A similar argument shows that $(-\frac{1}{\sqrt{2}}, -\frac{1}{\sqrt{2}})$ is the strict global minimum point of f over S, and the minimal value is $-\sqrt{2}$. ∎

Example 2.3. Consider the two-dimensional function

$$f(x, y) = \frac{x + y}{x^2 + y^2 + 1}$$

defined over the entire space $\mathbb{R}^2$. The contour and surface plots of the function are given in Figure 2.1. This function has two optima points: a global maximizer $(x, y) = (\frac{1}{\sqrt{2}}, \frac{1}{\sqrt{2}})$ and a global minimizer $(x, y) = (-\frac{1}{\sqrt{2}}, -\frac{1}{\sqrt{2}})$. The proof of these facts will be given in Example 2.36. The maximal value of the function is

$$\frac{\frac{1}{\sqrt{2}} + \frac{1}{\sqrt{2}}}{(\frac{1}{\sqrt{2}})^2 + (\frac{1}{\sqrt{2}})^2 + 1} = \frac{1}{\sqrt{2}},$$

and the minimal value is $-\frac{1}{\sqrt{2}}$. ∎

Our main task will usually be to find and study global minimum or maximum points; however, most of the theoretical results only characterize *local* minima and maxima which are optimal points with respect to a neighborhood of the point of interest. The exact definitions follow.

2.1. Global and Local Optima

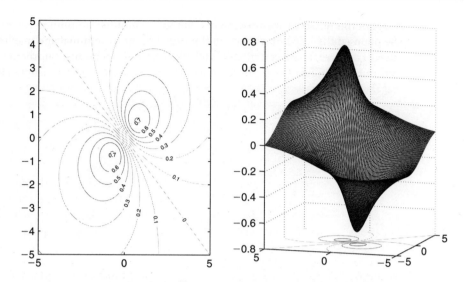

Figure 2.1. *Contour and surface plots of $f(x,y) = \frac{x+y}{x^2+y^2+1}$.*

Definition 2.4 (local minima and maxima). *Let $f : S \to \mathbb{R}$ be defined on a set $S \subseteq \mathbb{R}^n$. Then*

1. $\mathbf{x}^* \in S$ *is called a* **local minimum point** *of f over S if there exists $r > 0$ for which $f(\mathbf{x}^*) \leq f(\mathbf{x})$ for any $\mathbf{x} \in S \cap B(\mathbf{x}^*, r)$,*

2. $\mathbf{x}^* \in S$ *is called a* **strict local minimum point** *of f over S if there exists $r > 0$ for which $f(\mathbf{x}^*) < f(\mathbf{x})$ for any $\mathbf{x}^* \neq \mathbf{x} \in S \cap B(\mathbf{x}^*, r)$,*

3. $\mathbf{x}^* \in S$ *is called a* **local maximum point** *of f over S if there exists $r > 0$ for which $f(\mathbf{x}^*) \geq f(\mathbf{x})$ for any $\mathbf{x} \in S \cap B(\mathbf{x}^*, r)$,*

4. $\mathbf{x}^* \in S$ *is called a* **strict local maximum point** *of f over S if there exists $r > 0$ for which $f(\mathbf{x}^*) > f(\mathbf{x})$ for any $\mathbf{x}^* \neq \mathbf{x} \in S \cap B(\mathbf{x}^*, r)$.*

Of course, a global minimum (maximum) point is also a local minimum (maximum) point. As with global minimum and maximum points, we will also use the terminology *local minimizer* and *local maximizer* for local minimum and maximum points, respectively.

Example 2.5. Consider the one-dimensional function

$$f(x) = \begin{cases} (x-1)^2 + 2, & -1 \leq x \leq 1, \\ 2, & 1 \leq x \leq 2, \\ -(x-2)^2 + 2, & 2 \leq x \leq 2.5, \\ (x-3)^2 + 1.5, & 2.5 \leq x \leq 4, \\ -(x-5)^2 + 3.5, & 4 \leq x \leq 6, \\ -2x + 14.5, & 6 \leq x \leq 6.5, \\ 2x - 11.5, & 6.5 \leq x \leq 8, \end{cases}$$

described in Figure 2.2 and defined over the interval $[-1, 8]$. The point $x = -1$ is a strict global maximum point. The point $x = 1$ is a nonstrict local minimum point. All the points in the interval $(1, 2)$ are nonstrict local minimum points as well as nonstrict local maximum points. The point $x = 2$ is a local maximum point. The point $x = 3$ is a strict local minimum, and a non-strict global minimum point. The point $x = 5$ is a strict local maximum and $x = 6.5$ is a strict local minimum, which is a nonstrict global minimum point. Finally, $x = 8$ is a strict local maximum point. Note that, as already mentioned, $x = 3$ and $x = 6.5$ are both global minimum points of the function, and despite the fact that they are strict local minima, they are nonstrict global minimum points. ∎

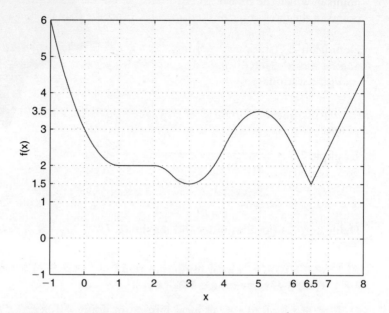

Figure 2.2. *Local and global optimum points of a one-dimensional function.*

First Order Optimality Condition

A well-known result is that for a one-dimensional function f defined and differentiable over an interval (a,b), if a point $x^* \in (a,b)$ is a local maximum or minimum, then $f'(x^*) = 0$. This is also known as Fermat's theorem. The multidimensional extension of this result states that the gradient is zero at local optimum points. We refer to such an optimality condition as a *first order optimality condition*, as it is expressed in terms of the first order derivatives. In what follows, we will also discuss *second order optimality conditions* that use in addition information on the second order (partial) derivatives.

Theorem 2.6 (first order optimality condition for local optima points). *Let $f : U \to \mathbb{R}$ be a function defined on a set $U \subseteq \mathbb{R}^n$. Suppose that $\mathbf{x}^* \in \mathrm{int}(U)$ is a local optimum point and that all the partial derivatives of f exist at $\mathbf{x}^*$. Then $\nabla f(\mathbf{x}^*) = 0$.*

Proof. Let $i \in \{1, 2, \ldots, n\}$ and consider the one-dimensional function $g(t) = f(\mathbf{x}^* + t\mathbf{e}_i)$. Note that g is differentiable at $t = 0$ and that $g'(0) = \frac{\partial f}{\partial x_i}(\mathbf{x}^*)$. Since $\mathbf{x}^*$ is a local optimum point of f, it follows that $t = 0$ is a local optimum of g, which immediately implies that $g'(0) = 0$. The latter equality is exactly the same as $\frac{\partial f}{\partial x_i}(\mathbf{x}^*) = 0$. Since this is true for any $i \in \{1, 2, \ldots, n\}$, the result $\nabla f(\mathbf{x}^*) = 0$ follows. □

Note that the proof of the first order optimality conditions for multivariate functions strongly relies on the first order optimality conditions for one-dimensional functions.

Theorem 2.6 presents a *necessary* optimality condition: the gradient vanishes at all local optimum points, which are interior points of the domain of the function; however, the reverse claim is not true—there could be points which are not local optimum points, whose gradient is zero. For example, the derivative of the one-dimensional function $f(x) = x^3$ is zero at $x = 0$, but this point is neither a local minimum nor a local maximum. Since points in which the gradient vanishes are the only candidates for local optima among the points in the interior of the domain of the function, they deserve an explicit definition.

Definition 2.7 (stationary points). *Let $f : U \to \mathbb{R}$ be a function defined on a set $U \subseteq \mathbb{R}^n$. Suppose that $\mathbf{x}^* \in \text{int}(U)$ and that f is differentiable over some neighborhood of $\mathbf{x}^*$. Then $\mathbf{x}^*$ is called a* **stationary point** *of f if $\nabla f(\mathbf{x}^*) = 0$.*

Theorem 2.6 essentially states that local optimum points are necessarily stationary points.

Example 2.8. Consider the one-dimensional quartic function $f(x) = 3x^4 - 20x^3 + 42x^2 - 36x$. To find its local and global optima points over $\mathbb{R}$, we first find all its stationary points. Since $f'(x) = 12x^3 - 60x^2 + 84x - 36 = 12(x-1)^2(x-3)$, it follows that $f'(x) = 0$ for $x = 1, 3$. The derivative $f'(x)$ does not change its sign when passing through $x = 1$—it is negative before and after—and thus $x = 1$ is not a local or global optimum point. On the other hand, the derivative does change its sign from negative to positive while passing through $x = 3$, and thus it is a local minimum point. Since the function must have a global minimum by the property that $f(x) \to \infty$ as $|x| \to \infty$, it follows that $x = 3$ is the global minimum point. ∎

2.2 • Classification of Matrices

In order to be able to characterize the second order optimality conditions, which are expressed via the Hessian matrix, the notion of "positive definiteness" must be defined.

Definition 2.9 (positive definiteness).

1. *A symmetric matrix $\mathbf{A} \in \mathbb{R}^{n \times n}$ is called* **positive semidefinite**, *denoted by $\mathbf{A} \succeq 0$, if $\mathbf{x}^T \mathbf{A} \mathbf{x} \geq 0$ for every $\mathbf{x} \in \mathbb{R}^n$.*

2. *A symmetric matrix $\mathbf{A} \in \mathbb{R}^{n \times n}$ is called* **positive definite**, *denoted by $\mathbf{A} \succ 0$, if $\mathbf{x}^T \mathbf{A} \mathbf{x} > 0$ for every $0 \neq \mathbf{x} \in \mathbb{R}^n$.*

In this book a positive definite or semidefinite matrix is always assumed to be symmetric.

Positive definiteness of a matrix does not mean that its components are positive, as the following examples illustrate.

Example 2.10. Let
$$\mathbf{A} = \begin{pmatrix} 2 & -1 \\ -1 & 1 \end{pmatrix}.$$
Then for any $\mathbf{x} = (x_1, x_2)^T \in \mathbb{R}^2$
$$\mathbf{x}^T \mathbf{A} \mathbf{x} = (x_1, x_2) \begin{pmatrix} 2 & -1 \\ -1 & 1 \end{pmatrix} \begin{pmatrix} x_1 \\ x_2 \end{pmatrix} = 2x_1^2 - 2x_1 x_2 + x_2^2 = x_1^2 + (x_1 - x_2)^2 \geq 0.$$

Thus, $\mathbf{A}$ is positive semidefinite. In fact, since $x_1^2+(x_1-x_2)^2 = 0$ if and only if $x_1 = x_2 = 0$, it follows that $\mathbf{A}$ is positive definite. This example illustrates the fact that a positive definite matrix might have negative components. ∎

Example 2.11. Let
$$\mathbf{A} = \begin{pmatrix} 1 & 2 \\ 2 & 1 \end{pmatrix}.$$

This matrix, whose components are all positive, is not positive definite since for $\mathbf{x} = (1,-1)^T$
$$\mathbf{x}^T \mathbf{A} \mathbf{x} = (1,-1) \begin{pmatrix} 1 & 2 \\ 2 & 1 \end{pmatrix} \begin{pmatrix} 1 \\ -1 \end{pmatrix} = -2. \quad \blacksquare$$

Although, as illustrated in the above examples, not all the components of a positive definite matrix need to be positive, the following result shows that the signs of the *diagonal* components of a positive definite matrix are positive.

Lemma 2.12. *Let $\mathbf{A} \in \mathbb{R}^{n \times n}$ be a positive definite matrix. Then the diagonal elements of $\mathbf{A}$ are positive.*

Proof. Since $\mathbf{A}$ is positive definite, it follows that $\mathbf{e}_i^T \mathbf{A} \mathbf{e}_i > 0$ for any $i \in \{1,2,\ldots,n\}$, which by the fact that $\mathbf{e}_i^T \mathbf{A} \mathbf{e}_i = A_{ii}$ implies the result. □

A similar argument shows that the diagonal elements of a positive semidefinite matrix are nonnegative.

Lemma 2.13. *Let $\mathbf{A}$ be a positive semidefinite matrix. Then the diagonal elements of $\mathbf{A}$ are nonnegative.*

Closely related to the notion of positive (semi)definiteness are the notions of negative (semi)definiteness and indefiniteness.

Definition 2.14 (negative (semi)definiteness, indefiniteness).

1. *A symmetric matrix $\mathbf{A} \in \mathbb{R}^{n \times n}$ is called **negative semidefinite**, denoted by $\mathbf{A} \preceq 0$, if $\mathbf{x}^T \mathbf{A} \mathbf{x} \leq 0$ for every $\mathbf{x} \in \mathbb{R}^n$.*

2. *A symmetric matrix $\mathbf{A} \in \mathbb{R}^{n \times n}$ is called **negative definite**, denoted by $\mathbf{A} \prec 0$, if $\mathbf{x}^T \mathbf{A} \mathbf{x} < 0$ for every $0 \neq \mathbf{x} \in \mathbb{R}^n$.*

3. *A symmetric matrix $\mathbf{A} \in \mathbb{R}^{n \times n}$ is called **indefinite** if there exist $\mathbf{x}$ and $\mathbf{y} \in \mathbb{R}^n$ such that $\mathbf{x}^T \mathbf{A} \mathbf{x} > 0$ and $\mathbf{y}^T \mathbf{A} \mathbf{y} < 0$.*

It follows immediately from the above definition that a matrix $\mathbf{A}$ is negative (semi)definite if and only if $-\mathbf{A}$ is positive (semi)definite. Therefore, we can prove and state all the results for positive (semi)definite matrices and the corresponding results for negative (semi)definite matrices will follow immediately. For example, the following result on negative definite and negative semidefinite matrices is a direct consequence of Lemmata 2.12 and 2.13.

2.2. Classification of Matrices

Lemma 2.15.

(a) *Let $\mathbf{A}$ be a negative definite matrix. Then the diagonal elements of $\mathbf{A}$ are negative.*

(b) *Let $\mathbf{A}$ be a negative semidefinite matrix. Then the diagonal elements of $\mathbf{A}$ are nonpositive.*

When the diagonal of a matrix contains both positive and negative elements, then the matrix is indefinite. The reverse claim is not correct.

Lemma 2.16. *Let $\mathbf{A}$ be a symmetric $n \times n$ matrix. If there exist positive and negative elements in the diagonal of $\mathbf{A}$, then $\mathbf{A}$ is indefinite.*

Proof. Let $i,j \in \{1,2,\ldots,n\}$ be indices such that $A_{i,i} > 0$ and $A_{j,j} < 0$. Then

$$\mathbf{e}_i^T \mathbf{A} \mathbf{e}_i = A_{i,i} > 0, \quad \mathbf{e}_j^T \mathbf{A} \mathbf{e}_j = A_{j,j} < 0,$$

and hence $\mathbf{A}$ is indefinite. □

In addition, a matrix is indefinite if and only if it is neither positive semidefinite nor negative semidefinite. It is not an easy task to check the definiteness of a matrix by using the definition given above. Therefore, our main task will be to find a useful characterization of positive (semi)definite matrices. It turns out that a complete characterization can be given in terms of the eigenvalues of the matrix.

Theorem 2.17 (eigenvalue characterization theorem). *Let $\mathbf{A}$ be a symmetric $n \times n$ matrix. Then*

(a) *$\mathbf{A}$ is positive definite if and only if all its eigenvalues are positive,*

(b) *$\mathbf{A}$ is positive semidefinite if and only if all its eigenvalues are nonnegative,*

(c) *$\mathbf{A}$ is negative definite if and only if all its eigenvalues are negative,*

(d) *$\mathbf{A}$ is negative semidefinite if and only if all its eigenvalues are nonpositive,*

(e) *$\mathbf{A}$ is indefinite if and only if it has at least one positive eigenvalue and at least one negative eigenvalue.*

Proof. We will prove part (a). The other parts follow immediately or by similar arguments. Since $\mathbf{A}$ is symmetric, it follows by the spectral decomposition theorem (Theorem 1.10) that there exist an orthogonal matrix $\mathbf{U}$ and a diagonal matrix $\mathbf{D} = \mathrm{diag}(d_1, d_2, \ldots, d_n)$ whose diagonal elements are the eigenvalues of $\mathbf{A}$, for which $\mathbf{U}^T \mathbf{A} \mathbf{U} = \mathbf{D}$. Making the linear transformation of variables $\mathbf{x} = \mathbf{U}\mathbf{y}$, we obtain that

$$\mathbf{x}^T \mathbf{A} \mathbf{x} = \mathbf{y}^T \mathbf{U}^T \mathbf{A} \mathbf{U} \mathbf{y} = \mathbf{y}^T \mathbf{D} \mathbf{y} = \sum_{i=1}^n d_i y_i^2.$$

We can therefore conclude by the nonsingularity of $\mathbf{U}$ that $\mathbf{x}^T \mathbf{A} \mathbf{x} > 0$ for any $\mathbf{x} \neq \mathbf{0}$ if and only if

$$\sum_{i=1}^n d_i y_i^2 > 0 \text{ for any } \mathbf{y} \neq \mathbf{0}. \tag{2.1}$$

Therefore, we need to prove that (2.1) holds if and only if $d_i > 0$ for all i. Indeed, if (2.1) holds then for any $i \in \{1, 2, \ldots, n\}$, plugging $\mathbf{y} = \mathbf{e}_i$ in the inequality implies that $d_i > 0$. On the other hand, if $d_i > 0$ for any i, then surely for any nonzero vector $\mathbf{y}$ one has $\sum_{i=1}^n d_i y_i^2 > 0$, meaning that (2.1) holds. $\square$

Since the trace and determinant of a symmetric matrix are the sum and product of its eigenvalues respectively, a simple consequence of the eigenvalue characterization theorem is the following.

Corollary 2.18. *Let $\mathbf{A}$ be a positive semidefinite (definite) matrix. Then $\mathrm{Tr}(\mathbf{A})$ and $\det(\mathbf{A})$ are nonnegative (positive).*

Since the eigenvalues of a diagonal matrix are its diagonal elements, it follows that the sign of a diagonal matrix is determined by the signs of the elements in its diagonal.

Lemma 2.19 (sign of diagonal matrices). *Let $\mathbf{D} = \mathrm{diag}(d_1, d_2, \ldots, d_n)$. Then*

(a) $\mathbf{D}$ *is positive definite if and only if $d_i > 0$ for all i,*

(b) $\mathbf{D}$ *is positive semidefinite if and only if $d_i \geq 0$ for all i,*

(c) $\mathbf{D}$ *is negative definite if and only if $d_i < 0$ for all i,*

(d) $\mathbf{D}$ *is negative semidefinite if and only if $d_i \leq 0$ for all i,*

(e) $\mathbf{D}$ *is indefinite if and only if there exist i, j such that $d_i > 0$ and $d_j < 0$.*

The eigenvalues of a matrix give full information on the sign of the matrix. However, we would like to find other simpler methods for detecting positive (semi)definiteness. We begin with an extremely simple rule for 2×2 matrices stating that for 2×2 matrices the condition in Corollary 2.18 is necessary and sufficient.

Proposition 2.20. *Let $\mathbf{A}$ be a symmetric 2×2 matrix. Then $\mathbf{A}$ is positive semidefinite (definite) if and only if $\mathrm{Tr}(\mathbf{A}), \det(\mathbf{A}) \geq 0$ ($\mathrm{Tr}(\mathbf{A}), \det(\mathbf{A}) > 0$).*

Proof. We will prove the result for positive semidefiniteness. The result for positive definiteness follows from similar arguments. By the eigenvalue characterization theorem (Theorem 2.17), it follows that $\mathbf{A}$ is positive semidefinite if and only if $\lambda_1(\mathbf{A}) \geq 0$ and $\lambda_2(\mathbf{A}) \geq 0$. The result now follows from the simple fact that for any two real number $a, b \in \mathbb{R}$ one has $a, b \geq 0$ if and only if $a + b \geq 0$ and $a \cdot b \geq 0$. Therefore, $\mathbf{A}$ is positive semidefinite if and only if $\lambda_1(\mathbf{A}) + \lambda_2(\mathbf{A}) = \mathrm{Tr}(\mathbf{A}) \geq 0$ and $\lambda_1(\mathbf{A})\lambda_2(\mathbf{A}) = \det(\mathbf{A}) \geq 0$. $\square$

Example 2.21. Consider the matrices

$$\mathbf{A} = \begin{pmatrix} 4 & 1 \\ 1 & 3 \end{pmatrix}, \qquad \mathbf{B} = \begin{pmatrix} 1 & 1 & 1 \\ 1 & 1 & 1 \\ 1 & 1 & 0.1 \end{pmatrix}.$$

The matrix $\mathbf{A}$ is positive definite since

$$\mathrm{Tr}(\mathbf{A}) = 4 + 3 = 7 > 0, \quad \det(\mathbf{A}) = 4 \cdot 3 - 1 \cdot 1 = 11 > 0.$$

2.2. Classification of Matrices

As for the matrix $\mathbf{B}$, we have

$$\mathrm{Tr}(\mathbf{B}) = 1 + 1 + 0.1 = 2.1 > 0, \qquad \det(\mathbf{B}) = 0.$$

However, despite the fact that the trace and the determinant of $\mathbf{B}$ are nonnegative, we cannot conclude that the matrix is positive semidefinite since Proposition 2.20 is valid only for 2×2 matrices. In this specific example we can show (even without computing the eigenvalues) that $\mathbf{B}$ is indefinite. Indeed,

$$\mathbf{e}_1^T \mathbf{B} \mathbf{e}_1 > 0,$$
$$(\mathbf{e}_2 - \mathbf{e}_3)^T \mathbf{B} (\mathbf{e}_2 - \mathbf{e}_3) = -0.9 < 0. \qquad \blacksquare$$

For any positive semidefinite matrix $\mathbf{A}$, we can define the square root matrix $\mathbf{A}^{\frac{1}{2}}$ in the following way. Let $\mathbf{A} = \mathbf{U}\mathbf{D}\mathbf{U}^T$ be the spectral decomposition of $\mathbf{A}$; that is, $\mathbf{U}$ is an orthogonal matrix, and $\mathbf{D} = \mathrm{diag}(d_1, d_2, \ldots, d_n)$ is a diagonal matrix whose diagonal elements are the eigenvalues of $\mathbf{A}$. Since $\mathbf{A}$ is positive semidefinite, we have that $d_1, d_2, \ldots, d_n \geq 0$, and we can define

$$\mathbf{A}^{\frac{1}{2}} = \mathbf{U}\mathbf{E}\mathbf{U}^T,$$

where $\mathbf{E} = \mathrm{diag}(\sqrt{d_1}, \sqrt{d_2}, \ldots, \sqrt{d_n})$. Obviously,

$$\mathbf{A}^{\frac{1}{2}} \mathbf{A}^{\frac{1}{2}} = \mathbf{U}\mathbf{E}\mathbf{U}^T \mathbf{U}\mathbf{E}\mathbf{U}^T = \mathbf{U}\mathbf{E}\mathbf{E}\mathbf{U}^T = \mathbf{U}\mathbf{D}\mathbf{U}^T = \mathbf{A}.$$

The matrix $\mathbf{A}^{\frac{1}{2}}$ is also called *the positive semidefinite square root*. A well-known test for positive definiteness is the *principal minors criterion*. Given an $n \times n$ matrix, the determinant of the upper left $k \times k$ submatrix is called *the kth principal minor* and is denoted by $D_k(\mathbf{A})$. For example, the principal minors of the 3×3 matrix

$$\mathbf{A} = \begin{pmatrix} a_{11} & a_{12} & a_{13} \\ a_{21} & a_{22} & a_{23} \\ a_{31} & a_{32} & a_{33} \end{pmatrix}$$

are

$$D_1(\mathbf{A}) = a_{11}, \quad D_2(\mathbf{A}) = \det\begin{pmatrix} a_{11} & a_{12} \\ a_{21} & a_{22} \end{pmatrix}, \quad D_3(\mathbf{A}) = \det\begin{pmatrix} a_{11} & a_{12} & a_{13} \\ a_{21} & a_{22} & a_{23} \\ a_{31} & a_{32} & a_{33} \end{pmatrix}.$$

The principal minors criterion states that a symmetric matrix is positive definite if and only if all its principal minors are positive.

Theorem 2.22 (principal minors criterion). *Let $\mathbf{A}$ be an $n \times n$ symmetric matrix. Then $\mathbf{A}$ is positive definite if and only if $D_1(\mathbf{A}) > 0, D_2(\mathbf{A}) > 0, \ldots, D_n(\mathbf{A}) > 0$.*

Note that the principal minors criterion is a tool for detecting positive definiteness of a matrix. It cannot be used in order detect positive *semi*definiteness.

Example 2.23. Let

$$\mathbf{A} = \begin{pmatrix} 4 & 2 & 3 \\ 2 & 3 & 2 \\ 3 & 2 & 4 \end{pmatrix}, \quad \mathbf{B} = \begin{pmatrix} 2 & 2 & 2 \\ 2 & 2 & 2 \\ 2 & 2 & -1 \end{pmatrix}, \quad \mathbf{C} = \begin{pmatrix} -4 & 1 & 1 \\ 1 & -4 & 1 \\ 1 & 1 & -4 \end{pmatrix}.$$

The matrix **A** is positive definite since

$$D_1(\mathbf{A}) = 4 > 0, \quad D_2(\mathbf{A}) = \det\begin{pmatrix} 4 & 2 \\ 2 & 3 \end{pmatrix} = 8 > 0, \quad D_3(\mathbf{A}) = \det\begin{pmatrix} 4 & 2 & 3 \\ 2 & 3 & 2 \\ 3 & 2 & 4 \end{pmatrix} = 13 > 0.$$

The principal minors of **B** are nonnegative: $D_1(\mathbf{B}) = 2, D_2(\mathbf{B}) = D_3(\mathbf{B}) = 0$; however, since they are not positive, the principal minors criterion does not provide any information on the sign of the matrix other than the fact that it is not positive definite. Since the matrix has both positive and negative diagonal elements, it is in fact indefinite (see Lemma 2.16). As for the matrix **C**, we will show that it is negative definite. For that, we will use the principal minors criterion to prove that $-\mathbf{C}$ is positive definite:

$$D_1(-\mathbf{C}) = 4 > 0, \quad D_2(-\mathbf{C}) = \det\begin{pmatrix} 4 & -1 \\ -1 & 4 \end{pmatrix} = 15 > 0,$$

$$D_3(-\mathbf{C}) = \det\begin{pmatrix} 4 & -1 & -1 \\ -1 & 4 & -1 \\ -1 & -1 & 4 \end{pmatrix} = 50 > 0. \quad \blacksquare$$

An important class of matrices that are known to be positive semidefinite is the class of *diagonally dominant* matrices.

Definition 2.24 (diagonally dominant matrices). *Let* **A** *be a symmetric* $n \times n$ *matrix. Then*

1. **A** *is called* **diagonally dominant** *if*

$$|A_{ii}| \geq \sum_{j \neq i} |A_{ij}|$$

   for all $i = 1, 2, \ldots, n$,

2. **A** *is called* **strictly diagonally dominant** *if*

$$|A_{ii}| > \sum_{j \neq i} |A_{ij}|$$

   for all $i = 1, 2, \ldots, n$.

We will now show that diagonally dominant matrices with nonnegative diagonal elements are positive semidefinite and that strictly diagonally dominant matrices with positive diagonal elements are positive definite.

Theorem 2.25 (positive (semi)definiteness of diagonally dominant matrices).

(a) *Let* **A** *be a symmetric* $n \times n$ *diagonally dominant matrix whose diagonal elements are nonnegative. Then* **A** *is positive semidefinite.*

(b) *Let* **A** *be a symmetric* $n \times n$ *strictly diagonally dominant matrix whose diagonal elements are positive. Then* **A** *is positive definite.*

Proof. (a) Suppose in contradiction that there exists a negative eigenvalue λ of **A**, and let **u** be a corresponding eigenvector. Let $i \in \{1, 2, \ldots, n\}$ be an index for which $|u_i|$ is maximal among $|u_1|, |u_2|, \ldots, |u_n|$. Then by the equality $\mathbf{A}\mathbf{u} = \lambda \mathbf{u}$ we have

$$|A_{ii} - \lambda| \cdot |u_i| = \left| \sum_{j \neq i} A_{ij} u_j \right| \leq \left(\sum_{j \neq i} |A_{ij}| \right) |u_i| \leq |A_{ii}| |u_i|,$$

implying that $|A_{ii} - \lambda| \leq |A_{ii}|$, which is a contradiction to the negativity of λ and the nonnegativity of A_{ii}.

(b) Since by part (a) we know that $\mathbf{A}$ is positive semidefinite, all we need to show is that $\mathbf{A}$ has no zero eigenvalues. Suppose in contradiction that there is a zero eigenvalue, meaning that there is a vector $\mathbf{u} \neq \mathbf{0}$ such that $\mathbf{A}\mathbf{u} = \mathbf{0}$. Then, similarly to the proof of part (a), let $i \in \{1, 2, \ldots, n\}$ be an index for which $|u_i|$ is maximal among $|u_1|, |u_2|, \ldots, |u_n|$, and we obtain

$$|A_{ii}| \cdot |u_i| = \left| \sum_{j \neq i} A_{ij} u_j \right| \leq \left(\sum_{j \neq i} |A_{ij}| \right) |u_i| < |A_{ii}| |u_i|,$$

which is obviously impossible, establishing the fact that $\mathbf{A}$ is positive definite. $\square$

2.3 • Second Order Optimality Conditions

We begin by stating the necessary second order optimality condition.

Theorem 2.26 (necessary second order optimality condition). *Let $f : U \to \mathbb{R}$ be a function defined on an open set $U \subseteq \mathbb{R}^n$. Suppose that f is twice continuously differentiable over U and that $\mathbf{x}^*$ is a stationary point. Then the following hold:*

(a) *If $\mathbf{x}^*$ is a local minimum point of f over U, then $\nabla^2 f(\mathbf{x}^*) \succeq \mathbf{0}$.*

(b) *If $\mathbf{x}^*$ is a local maximum point of f over U, then $\nabla^2 f(\mathbf{x}^*) \preceq \mathbf{0}$.*

Proof. (a) Since $\mathbf{x}^*$ is a local minimum point, there exists a ball $B(\mathbf{x}^*, r) \subseteq U$ for which $f(\mathbf{x}) \geq f(\mathbf{x}^*)$ for all $\mathbf{x} \in B(\mathbf{x}^*, r)$. Let $\mathbf{d} \in \mathbb{R}^n$ be a nonzero vector. For any $0 < \alpha < \frac{r}{\|\mathbf{d}\|}$, we have $\mathbf{x}_\alpha^* \equiv \mathbf{x}^* + \alpha \mathbf{d} \in B(\mathbf{x}^*, r)$, and hence for any such α

$$f(\mathbf{x}_\alpha^*) \geq f(\mathbf{x}^*). \tag{2.2}$$

On the other hand, by the linear approximation theorem (Theorem 1.24), it follows that there exists a vector $\mathbf{z}_\alpha \in [\mathbf{x}^*, \mathbf{x}_\alpha^*]$ such that

$$f(\mathbf{x}_\alpha^*) - f(\mathbf{x}^*) = \nabla f(\mathbf{x}^*)^T (\mathbf{x}_\alpha^* - \mathbf{x}^*) + \frac{1}{2} (\mathbf{x}_\alpha^* - \mathbf{x}^*)^T \nabla^2 f(\mathbf{z}_\alpha)(\mathbf{x}_\alpha^* - \mathbf{x}^*).$$

Since $\mathbf{x}^*$ is a stationary point of f, and by the definition of $\mathbf{x}_\alpha^*$, the latter equation reduces to

$$f(\mathbf{x}_\alpha^*) - f(\mathbf{x}^*) = \frac{\alpha^2}{2} \mathbf{d}^T \nabla^2 f(\mathbf{z}_\alpha) \mathbf{d}. \tag{2.3}$$

Combining (2.2) and (2.3), it follows that for any $\alpha \in (0, \frac{r}{\|\mathbf{d}\|})$ the inequality $\mathbf{d}^T \nabla^2 f(\mathbf{z}_\alpha) \mathbf{d} \geq 0$ holds. Finally, using the fact that $\mathbf{z}_\alpha \to \mathbf{x}^*$ as $\alpha \to 0^+$, and the continuity of the Hessian, we obtain that $\mathbf{d}^T \nabla^2 f(\mathbf{x}^*) \mathbf{d} \geq 0$. Since the latter inequality holds for any $\mathbf{d} \in \mathbb{R}^n$, the desired result is established.

(b) The proof of (b) follows immediately by employing the result of part (a) on the function $-f$. $\square$

The latter result is a necessary condition for local optimality. The next theorem states a sufficient condition for strict local optimality.

Theorem 2.27 (sufficient second order optimality condition). *Let $f : U \to \mathbb{R}$ be a function defined on an open set $U \subseteq \mathbb{R}^n$. Suppose that f is twice continuously differentiable over U and that $\mathbf{x}^*$ is a stationary point. Then the following hold:*

(a) *If $\nabla^2 f(\mathbf{x}^*) \succ 0$, then $\mathbf{x}^*$ is a strict local minimum point of f over U.*

(b) *If $\nabla^2 f(\mathbf{x}^*) \prec 0$, then $\mathbf{x}^*$ is a strict local maximum point of f over U.*

Proof. We will prove part (a). Part (b) follows by considering the function $-f$. Suppose then that $\mathbf{x}^*$ is a stationary point satisfying $\nabla^2 f(\mathbf{x}^*) \succ 0$. Since the Hessian is continuous, it follows that there exists a ball $B(\mathbf{x}^*, r) \subseteq U$ for which $\nabla^2 f(\mathbf{x}) \succ 0$ for any $\mathbf{x} \in B(\mathbf{x}^*, r)$. By the linear approximation theorem (Theorem 1.24), it follows that for any $\mathbf{x} \in B(\mathbf{x}^*, r)$, there exists a vector $\mathbf{z}_\mathbf{x} \in [\mathbf{x}^*, \mathbf{x}]$ (and hence $\mathbf{z}_\mathbf{x} \in B(\mathbf{x}^*, r)$) for which

$$f(\mathbf{x}) - f(\mathbf{x}^*) = \frac{1}{2}(\mathbf{x} - \mathbf{x}^*)^T \nabla^2 f(\mathbf{z}_\mathbf{x})(\mathbf{x} - \mathbf{x}^*). \qquad (2.4)$$

Since $\nabla^2 f(\mathbf{z}_\mathbf{x}) \succ 0$, it follows by (2.4) that for any $\mathbf{x} \in B(\mathbf{x}^*, r)$ such that $\mathbf{x} \neq \mathbf{x}^*$, the inequality $f(\mathbf{x}) > f(\mathbf{x}^*)$ holds, implying that $\mathbf{x}^*$ is a strict local minimum point of f over U. $\square$

Note that the sufficient condition implies the stronger property of *strict* local optimality. However, positive definiteness of the Hessian matrix is not a necessary condition for strict local optimality. For example, the one-dimensional function $f(x) = x^4$ over $\mathbb{R}$ has a strict local minimum at $x = 0$, but $f''(0)$ is not positive. Another important concept is that of a *saddle point*.

Definition 2.28 (saddle point). *Let $f : U \to \mathbb{R}$ be a function defined on an open set $U \subseteq \mathbb{R}^n$. Suppose that f is continuously differentiable over U. A stationary point $\mathbf{x}^*$ is called a* **saddle point** *of f over U if it is neither a local minimum point nor a local maximum point of f over U.*

A sufficient condition for a stationary point to be a saddle point in terms of the properties of the Hessian is given in the next result.

Theorem 2.29 (sufficient condition for a saddle point). *Let $f : U \to \mathbb{R}$ be a function defined on an open set $U \subseteq \mathbb{R}^n$. Suppose that f is twice continuously differentiable over U and that $\mathbf{x}^*$ is a stationary point. If $\nabla^2 f(\mathbf{x}^*)$ is an indefinite matrix, then $\mathbf{x}^*$ is a saddle point of f over U.*

Proof. Since $\nabla^2 f(\mathbf{x}^*)$ is indefinite, it has at least one positive eigenvalue $\lambda > 0$, corresponding to a normalized eigenvector which we will denote by $\mathbf{v}$. Since U is an open set, it follows that there exists a positive real $r > 0$ such that $\mathbf{x}^* + \alpha \mathbf{v} \in U$ for any $\alpha \in (0, r)$. By the quadratic approximation theorem (Theorem 1.25) and recalling that $\nabla f(\mathbf{x}^*) = 0$, we have that there exists a function $g : \mathbb{R}_{++} \to \mathbb{R}$ satisfying

$$\frac{g(t)}{t} \to 0 \text{ as } t \to 0, \tag{2.5}$$

such that for any $\alpha \in (0, r)$

$$f(\mathbf{x}^* + \alpha \mathbf{v}) = f(\mathbf{x}^*) + \frac{\alpha^2}{2} \mathbf{v}^T \nabla^2 f(\mathbf{x}^*) \mathbf{v} + g(\alpha^2 \|\mathbf{v}\|^2)$$
$$= f(\mathbf{x}^*) + \frac{\lambda \alpha^2}{2} \|\mathbf{v}\|^2 + g(\|\mathbf{v}\|^2 \alpha^2).$$

Since $\|\mathbf{v}\| = 1$, the latter can be rewritten as

$$f(\mathbf{x}^* + \alpha \mathbf{v}) = f(\mathbf{x}^*) + \frac{\lambda \alpha^2}{2} + g(\alpha^2).$$

By (2.5) it follows that there exists an $\varepsilon_1 \in (0, r)$ such that $g(\alpha^2) > -\frac{\lambda}{2} \alpha^2$ for all $\alpha \in (0, \varepsilon_1)$, and hence $f(\mathbf{x}^* + \alpha \mathbf{v}) > f(\mathbf{x}^*)$ for all $\alpha \in (0, \varepsilon_1)$. This shows that $\mathbf{x}^*$ *cannot* be a local maximum point of f over U. A similar argument—exploiting an eigenvector of $\nabla^2 f(\mathbf{x}^*)$ corresponding to a negative eigenvalue—shows that $\mathbf{x}^*$ cannot be a local minimum point of f over U, establishing the desired result that $\mathbf{x}^*$ is a saddle point. □

Another important issue is the one of deciding on whether a function actually has a global minimizer or maximizer. This is the issue of *attainment* or *existence*. A very well known result is due to Weierstrass, stating that a continuous function attains its minimum and maximum over a compact set.

Theorem 2.30 (Weierstrass theorem). *Let f be a continuous function defined over a nonempty and compact set $C \subseteq \mathbb{R}^n$. Then there exists a global minimum point of f over C and a global maximum point of f over C.*

When the underlying set is not compact, the Weierstrass theorem does not guarantee the attainment of the solution, but certain properties of the function f can imply attainment of the solution even in the noncompact setting. One example of such a property is *coerciveness*.

Definition 2.31 (coerciveness). *Let $f : \mathbb{R}^n \to \mathbb{R}$ be a continuous function defined over $\mathbb{R}^n$. The function f is called* **coercive** *if*

$$\lim_{\|\mathbf{x}\| \to \infty} f(\mathbf{x}) = \infty.$$

The important property of coercive functions that will be frequently used in this book is that a coercive function always attains a global minimum point on any closed set.

Theorem 2.32 (attainment under coerciveness). *Let $f : \mathbb{R}^n \to \mathbb{R}$ be a continuous and coercive function and let $S \subseteq \mathbb{R}^n$ be a nonempty closed set. Then f has a global minimum point over S.*

Proof. Let $\mathbf{x}_0 \in S$ be an arbitrary point in S. Since the function is coercive, it follows that there exists an $M > 0$ such that

$$f(\mathbf{x}) > f(\mathbf{x}_0) \text{ for any } \mathbf{x} \text{ such that } \|\mathbf{x}\| > M. \tag{2.6}$$

Since any global minimizer $\mathbf{x}^*$ of f over S satisfies $f(\mathbf{x}^*) \leq f(\mathbf{x}_0)$, it follows from (2.6) that the set of global minimizers of f over S is the same as the set of global minimizers of f over $S \cap B[\mathbf{0}, M]$. The set $S \cap B[\mathbf{0}, M]$ is compact and nonempty, and thus by the Weierstrass theorem, there exists a global minimizer of f over $S \cap B[\mathbf{0}, M]$ and hence also over S. $\square$

Example 2.33. Consider the function $f(x_1, x_2) = x_1^2 + x_2^2$ over the set

$$C = \{(x_1, x_2) : x_1 + x_2 \leq -1\}.$$

The set C is not bounded, and thus the Weierstrass theorem does not guarantee the existence of a global minimizer of f over C, but since f is coercive and C is closed, Theorem 2.32 does guarantee the existence of such a global minimizer. It is also not a difficult task to find the global minimum point in this example. There are two options: In one option the global minimum point is in the interior of C, and in that case by Theorem 2.6 $\nabla f(\mathbf{x}) = 0$, meaning that $\mathbf{x} = \mathbf{0}$, which is impossible since the zeros vector is not in C. The other option is that the global minimum point is attained at the boundary of C given by $\mathrm{bd}(C) = \{(x_1, x_2) : x_1 + x_2 = -1\}$. We can then substitute $x_1 = -x_2 - 1$ into the objective function and recast the problem as the one-dimensional optimization problem of minimizing $g(x_2) = (-1 - x_2)^2 + x_2^2$ over $\mathbb{R}$. Since $g'(x_2) = 2(1 + x_2) + 2x_2$, it follows that g' has a single root, which is $x_2 = -0.5$, and hence $x_1 = -0.5$. Since $(x_1, x_2) = (-0.5, -0.5)$ is the only candidate for a global minimum point, and since there *must* be at least one global minimizer, it follows that $(x_1, x_2) = (-0.5, -0.5)$ is the global minimum point of f over C. ∎

Example 2.34. Consider the function

$$f(x_1, x_2) = 2x_1^3 + 3x_2^2 + 3x_1^2 x_2 - 24 x_2$$

over $\mathbb{R}^2$. Let us find all the stationary points of f over $\mathbb{R}^2$ and classify them. First,

$$\nabla f(\mathbf{x}) = \begin{pmatrix} 6x_1^2 + 6x_1 x_2 \\ 6x_2 + 3x_1^2 - 24 \end{pmatrix}.$$

Therefore, the stationary points are those satisfying

$$6x_1^2 + 6x_1 x_2 = 0,$$
$$6x_2 + 3x_1^2 - 24 = 0.$$

The first equation is the same as $6x_1(x_1 + x_2) = 0$, meaning that either $x_1 = 0$ or $x_1 + x_2 = 0$. If $x_1 = 0$, then by the second equation $x_2 = 4$. If $x_1 + x_2 = 0$, then substituting $x_1 = -x_2$ in the second equation yields the equation $3x_1^2 - 6x_1 - 24 = 0$ whose solutions are $x_1 = 4, -2$. Overall, the stationary points of the function f are $(0, 4), (4, -4), (-2, 2)$. The Hessian of f is given by

$$\nabla^2 f(x_1, x_2) = 6 \begin{pmatrix} 2x_1 + x_2 & x_1 \\ x_1 & 1 \end{pmatrix}.$$

For the stationary point $(0, 4)$ we have

$$\nabla^2 f(0, 4) = 6 \begin{pmatrix} 4 & 0 \\ 0 & 1 \end{pmatrix},$$

2.3. Second Order Optimality Conditions

which is a positive definite matrix as a diagonal matrix with positive components in the diagonal (see Lemma 2.19). Therefore, $(0,4)$ is a strict local minimum point. It is not a global minimum point since such a point does not exist as the function f is not bounded below:
$$f(x_1, 0) = 2x_1^3 \to -\infty \text{ as } x_1 \to -\infty.$$

The Hessian of f at $(4, -4)$ is
$$\nabla^2 f(4, -4) = 6 \begin{pmatrix} 4 & 4 \\ 4 & 1 \end{pmatrix}.$$

Since $\det(\nabla^2 f(4, -4)) = -6^2 \cdot 12 < 0$, it follows that the Hessian at $(4, -4)$ is indefinite. Indeed, the determinant is the product of its two eigenvalues – one must be positive and the other negative. Therefore, by Theorem 2.29, $(4, -4)$ is a saddle point. The Hessian at $(-2, 2)$ is
$$\nabla^2 f(-2, 2) = 6 \begin{pmatrix} -2 & -2 \\ -2 & 1 \end{pmatrix},$$

which is indefinite by the fact that it has both positive and negative elements on its diagonal. Therefore, $(-2, 2)$ is a saddle point. To summarize, $(0, 4)$ is a strict local minimum point of f, and $(4, -4), (-2, 2)$ are saddle points. ∎

Example 2.35. Let
$$f(x_1, x_2) = (x_1^2 + x_2^2 - 1)^2 + (x_2^2 - 1)^2.$$
The gradient of f is given by
$$\nabla f(\mathbf{x}) = 4 \begin{pmatrix} (x_1^2 + x_2^2 - 1) x_1 \\ (x_1^2 + x_2^2 - 1) x_2 + (x_2^2 - 1) x_2 \end{pmatrix}.$$

The stationary points are those satisfying
$$(x_1^2 + x_2^2 - 1) x_1 = 0, \tag{2.7}$$
$$(x_1^2 + x_2^2 - 1) x_2 + (x_2^2 - 1) x_2 = 0. \tag{2.8}$$

By equation (2.7), there are two cases: either $x_1 = 0$, and then by equation (2.8) x_2 is equal to one of the values $0, 1, -1$; the second option is that $x_1^2 + x_2^2 = 1$, and then by equation (2.8), we have that $x_2 = 0, \pm 1$ and hence x_1 is $\pm 1, 0$ respectively. Overall, there are 5 stationary points: $(0,0), (1,0), (-1,0), (0,1), (0,-1)$. The Hessian of the function is
$$\nabla^2 f(\mathbf{x}) = 4 \begin{pmatrix} 3x_1^2 + x_2^2 - 1 & 2x_1 x_2 \\ 2x_1 x_2 & x_1^2 + 6x_2^2 - 2 \end{pmatrix}.$$

Since
$$\nabla^2 f(0,0) = 4 \begin{pmatrix} -1 & 0 \\ 0 & -2 \end{pmatrix} \prec 0,$$
it follows that $(0,0)$ is a strict local maximum point. By the fact that $f(x_1, 0) = (x_1^2 - 1)^2 + 1 \to \infty$ as $x_1 \to \infty$, the function is not bounded above and thus $(0,0)$ is not a global maximum point. Also,
$$\nabla^2 f(1, 0) = \nabla^2 f(-1, 0) = 4 \begin{pmatrix} 2 & 0 \\ 0 & -1 \end{pmatrix},$$

which is an indefinite matrix and hence $(1,0)$ and $(-1,0)$ are saddle points. Finally,

$$\nabla^2 f(0,1) = \nabla^2 f(0,-1) = 4\begin{pmatrix} 0 & 0 \\ 0 & 4 \end{pmatrix} \succeq 0.$$

The fact that the Hessian matrices of f at $(0,1)$ and $(0,-1)$ are positive semidefinite is not enough in order to conclude that these are local minimum points; they might be saddle points. However, in this case it is not difficult to see that $(0,1)$ and $(0,-1)$ are in fact *global minimum points* since $f(0,1) = f(0,-1) = 0$, and the function is lower bounded by zero. Note that since there are two global minimum points, they are *nonstrict* global minima, but they actually are *strict* local minimum points since each has a neighborhood in which it is the unique minimizer. The contour and surface plots of the function are plotted in Figure 2.3. ∎

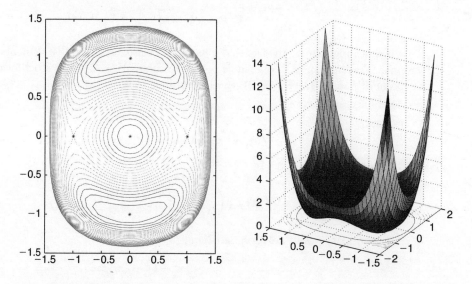

Figure 2.3. *Contour and surface plots of $f(x_1, x_2) = (x_1^2 + x_2^2 - 1)^2 + (x_2^2 - 1)^2$. The five stationary points $(0,0), (0,1), (0,-1), (1,0), (-1,0)$ are denoted by asterisks. The points $(0,-1), (0,1)$ are strict local minimum points as well as global minimum points, $(0,0)$ is a local maximum point, and $(-1,0), (1,0)$ are saddle points.*

Example 2.36. Returning to Example 2.3, we will now investigate the stationary points of the function

$$f(x,y) = \frac{x+y}{x^2+y^2+1}.$$

The gradient of the function is

$$\nabla f(x,y) = \frac{1}{(x^2+y^2+1)^2} \begin{pmatrix} (x^2+y^2+1) - 2(x+y)x \\ (x^2+y^2+1) - 2(x+y)y \end{pmatrix}.$$

Therefore, the stationary points of the function are those satisfying

$$-x^2 - 2xy + y^2 = -1,$$
$$x^2 - 2xy - y^2 = -1.$$

2.3. Second Order Optimality Conditions

Adding the two equations yields the equation $xy = \frac{1}{2}$. Subtracting the two equations yields $x^2 = y^2$, which, along with the fact that x and y have the same sign, implies that $x = y$. Therefore, we obtain that

$$x^2 = \frac{1}{2}$$

whose solutions are $x = \pm \frac{1}{\sqrt{2}}$. Thus, the function has two stationary points which are $(\frac{1}{\sqrt{2}}, \frac{1}{\sqrt{2}})$ and $(-\frac{1}{\sqrt{2}}, -\frac{1}{\sqrt{2}})$. We will now prove that the statement given in Example 2.3 is correct: $(\frac{1}{\sqrt{2}}, \frac{1}{\sqrt{2}})$ is the global maximum point of f over $\mathbb{R}^2$ and $(-\frac{1}{\sqrt{2}}, -\frac{1}{\sqrt{2}})$ is the global minimum point of f over $\mathbb{R}^2$. Indeed, note that $f(\frac{1}{\sqrt{2}}, \frac{1}{\sqrt{2}}) = \frac{1}{\sqrt{2}}$. In addition, for any $(x,y)^T \in \mathbb{R}^2$,

$$f(x,y) = \frac{x+y}{x^2+y^2+1} \leq \sqrt{2}\frac{\sqrt{x^2+y^2}}{x^2+y^2+1} \leq \sqrt{2}\max_{t \geq 0}\frac{t}{t^2+1},$$

where the first inequality follows from the Cauchy-Schwarz inequality. Since for any $t \geq 0$, the inequality $t^2 + 1 \geq 2t$ holds, we have that $f(x,y) \leq \frac{1}{\sqrt{2}}$ for any $(x,y)^T \in \mathbb{R}^2$. Therefore, $(\frac{1}{\sqrt{2}}, \frac{1}{\sqrt{2}})$ attains the maximal value of the function and is therefore the global maximum point of f over $\mathbb{R}^2$. A similar argument shows that $(-\frac{1}{\sqrt{2}}, -\frac{1}{\sqrt{2}})$ is the global minimum point of f over $\mathbb{R}^2$. ∎

Example 2.37. Let

$$f(x_1, x_2) = -2x_1^2 + x_1 x_2^2 + 4x_1^4.$$

The gradient of the function is

$$\nabla f(\mathbf{x}) = \begin{pmatrix} -4x_1 + x_2^2 + 16x_1^3 \\ 2x_1 x_2 \end{pmatrix},$$

and the stationary points are those satisfying the system of equations

$$-4x_1 + x_2^2 + 16x_1^3 = 0,$$
$$2x_1 x_2 = 0.$$

By the second equation, we obtain that either $x_1 = 0$ or $x_2 = 0$ (or both). If $x_1 = 0$, then by the first equation, we have $x_2 = 0$. If $x_2 = 0$, then by the first equation $-4x_1 + 16x_1^3 = 0$ and thus $x_1(-1 + 4x_1^2) = 0$, so that x_1 is one of the three values $0, 0.5, -0.5$. The function therefore has three stationary points: $(0,0), (0.5,0), (-0.5,0)$. The Hessian of f is

$$\nabla^2 f(x_1, x_2) = \begin{pmatrix} -4 + 48x_1^2 & 2x_2 \\ 2x_2 & 2x_1 \end{pmatrix}.$$

Since

$$\nabla^2 f(0.5, 0) = \begin{pmatrix} 8 & 0 \\ 0 & 1 \end{pmatrix} \succ 0,$$

it follows that $(0.5, 0)$ is a strict local minimum point of f. It is not a *global* minimum point since the function is not bounded below: $f(-1, x_2) = 2 - x_2^2 \to -\infty$ as $x_2 \to \infty$. As for the stationary point $(-0.5, 0)$,

$$\nabla^2 f(-0.5, 0) = \begin{pmatrix} 8 & 0 \\ 0 & -1 \end{pmatrix},$$

and hence since the Hessian is indefinite, $(-0.5, 0)$ is a saddle point of f. Finally, the Hessian of f at the stationary point $(0,0)$ is

$$\nabla^2 f(0,0) = \begin{pmatrix} -4 & 0 \\ 0 & 0 \end{pmatrix}.$$

The fact that the Hessian is negative semidefinite implies that $(0,0)$ is either a local maximum point or a saddle point of f. Note that

$$f(\alpha^4, \alpha) = -2\alpha^8 + \alpha^6 + 4\alpha^{16} = \alpha^6(-2\alpha^2 + 1 + 4\alpha^{10}).$$

It is easy to see that for a small enough $\alpha > 0$, the above expression is positive. Similarly,

$$f(-\alpha^4, \alpha) = -2\alpha^8 - \alpha^6 + 4\alpha^{16} = \alpha^6(-2\alpha^2 - 1 + 4\alpha^{10}),$$

and the above expression is negative for small enough $\alpha > 0$. This means that $(0,0)$ is a saddle point since at any one of its neighborhoods, we can find points with smaller values than $f(0,0) = 0$ and points with larger values than $f(0,0) = 0$. The surface and contour plots of the function are given in Figure 2.4. ∎

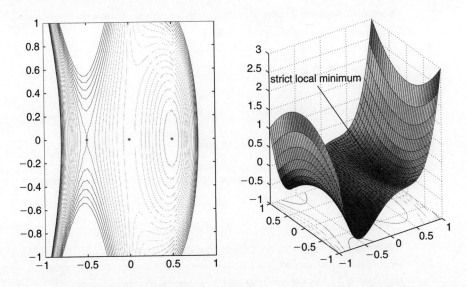

Figure 2.4. *Contour and surface plots of $f(x_1, x_2) = -2x_1^2 + x_1 x_2^2 + 4x_1^4$. The three stationary point $(0,0), (0.5, 0), (-0.5, 0)$ are denoted by asterisks. The point $(0.5, 0)$ is a strict local minimum, while $(0,0)$ and $(-0.5, 0)$ are saddle points.*

2.4 ▪ Global Optimality Conditions

The conditions described in the last section can only guarantee—at best—local optimality of stationary points since they exploit only local information: the values of the gradient and the Hessian at a given point. Conditions that ensure global optimality of points must use global information. For example, when the Hessian of the function is always positive semidefinite, all the stationary points are also global minimum points. Later on, we will refer to this property as convexity.

2.4. Global Optimality Conditions

Theorem 2.38. *Let f be a twice continuously differentiable function defined over $\mathbb{R}^n$. Suppose that $\nabla^2 f(\mathbf{x}) \succeq 0$ for any $\mathbf{x} \in \mathbb{R}^n$. Let $\mathbf{x}^* \in \mathbb{R}^n$ be a stationary point of f. Then $\mathbf{x}^*$ is a global minimum point of f.*

Proof. By the linear approximation theorem (Theorem 1.24), it follows that for any $\mathbf{x} \in \mathbb{R}^n$, there exists a vector $\mathbf{z_x} \in [\mathbf{x}^*, \mathbf{x}]$ for which

$$f(\mathbf{x}) - f(\mathbf{x}^*) = \frac{1}{2}(\mathbf{x} - \mathbf{x}^*)^T \nabla^2 f(\mathbf{z_x})(\mathbf{x} - \mathbf{x}^*).$$

Since $\nabla^2 f(\mathbf{z_x}) \succeq 0$, we have that $f(\mathbf{x}) \geq f(\mathbf{x}^*)$, establishing the fact that $\mathbf{x}^*$ is a global minimum point of f. $\square$

Example 2.39. Let

$$f(\mathbf{x}) = x_1^2 + x_2^2 + x_3^2 + x_1 x_2 + x_1 x_3 + x_2 x_3 + (x_1^2 + x_2^2 + x_3^2)^2.$$

Then

$$\nabla f(\mathbf{x}) = \begin{pmatrix} 2x_1 + x_2 + x_3 + 4x_1(x_1^2 + x_2^2 + x_3^2) \\ 2x_2 + x_1 + x_3 + 4x_2(x_1^2 + x_2^2 + x_3^2) \\ 2x_3 + x_1 + x_2 + 4x_3(x_1^2 + x_2^2 + x_3^2) \end{pmatrix}.$$

Obviously, $\mathbf{x} = \mathbf{0}$ is a stationary point. We will show that it is a global minimum point. The Hessian of f is

$$\nabla^2 f(\mathbf{x}) = \begin{pmatrix} 2 + 4(x_1^2 + x_2^2 + x_3^2) + 8x_1^2 & 1 + 8x_1 x_2 & 1 + 8x_1 x_3 \\ 1 + 8x_1 x_2 & 2 + 4(x_1^2 + x_2^2 + x_3^2) + 8x_2^2 & 1 + 8x_2 x_3 \\ 1 + 8x_1 x_3 & 1 + 8x_2 x_3 & 2 + 4(x_1^2 + x_2^2 + x_3^2) + 8x_3^2 \end{pmatrix}.$$

The Hessian is positive semidefinite since it can be written as the sum

$$\nabla^2 f(\mathbf{x}) = \mathbf{A} + \mathbf{B}(\mathbf{x}) + \mathbf{C}(\mathbf{x}),$$

where

$$\mathbf{A} = \begin{pmatrix} 2 & 1 & 1 \\ 1 & 2 & 1 \\ 1 & 1 & 2 \end{pmatrix}, \quad \mathbf{B}(\mathbf{x}) = 4(x_1^2 + x_2^2 + x_3^2)\mathbf{I}_3, \quad \mathbf{C}(\mathbf{x}) = \begin{pmatrix} 8x_1^2 & 8x_1 x_2 & 8x_1 x_3 \\ 8x_1 x_3 & 8x_2^2 & 8x_2 x_3 \\ 8x_1 x_3 & 8x_2 x_3 & 8x_3^2 \end{pmatrix}.$$

The above three matrices are positive semidefinite for any $\mathbf{x} \in \mathbb{R}^3$; indeed, $\mathbf{A} \succeq 0$ since it is diagonally dominant with positive diagonal elements. The matrix $\mathbf{B}(\mathbf{x})$, as a nonnegative multiplier of the identity matrix, is positive semidefinite, and finally,

$$\mathbf{C}(\mathbf{x}) = 8\mathbf{x}\mathbf{x}^T,$$

and hence $\mathbf{C}(\mathbf{x})$ is positive semidefinite as a positive multiply of the matrix $\mathbf{x}\mathbf{x}^T$, which is positive semidefinite (see Exercise 2.6). To summarize, $\nabla^2 f(\mathbf{x})$ is positive semidefinite as the sum of three positive semidefinite matrices (simple extension of Exercise 2.4), and hence, by Theorem 2.38, $\mathbf{x} = \mathbf{0}$ is a global minimum point of f over $\mathbb{R}^3$. ∎

2.5 • Quadratic Functions

Quadratic functions are an important class of functions that are useful in the modeling of many optimization problems. We will now define and derive some of the basic results related to this important class of functions.

Definition 2.40. *A* **quadratic function** *over* $\mathbb{R}^n$ *is a function of the form*

$$f(\mathbf{x}) = \mathbf{x}^T \mathbf{A}\mathbf{x} + 2\mathbf{b}^T\mathbf{x} + c, \qquad (2.9)$$

where $\mathbf{A} \in \mathbb{R}^{n \times n}$ *is symmetric,* $\mathbf{b} \in \mathbb{R}^n$, *and* $c \in \mathbb{R}$.

We will frequently refer to the matrix $\mathbf{A}$ in (2.9) as the matrix *associated* with the quadratic function f. The gradient and Hessian of a quadratic function have simple analytic formulas:

$$\nabla f(\mathbf{x}) = 2\mathbf{A}\mathbf{x} + 2\mathbf{b}, \qquad (2.10)$$
$$\nabla^2 f(\mathbf{x}) = 2\mathbf{A}. \qquad (2.11)$$

By the above formulas we can deduce several important properties of quadratic functions, which are associated with their stationary points.

Lemma 2.41. *Let* $f(\mathbf{x}) = \mathbf{x}^T \mathbf{A}\mathbf{x} + 2\mathbf{b}^T\mathbf{x} + c$, *where* $\mathbf{A} \in \mathbb{R}^{n \times n}$ *is symmetric,* $\mathbf{b} \in \mathbb{R}^n$, *and* $c \in \mathbb{R}$. *Then*

(a) $\mathbf{x}$ *is a stationary point of* f *if and only if* $\mathbf{A}\mathbf{x} = -\mathbf{b}$,

(b) *if* $\mathbf{A} \succeq \mathbf{0}$, *then* $\mathbf{x}$ *is a global minimum point of* f *if and only if* $\mathbf{A}\mathbf{x} = -\mathbf{b}$,

(c) *if* $\mathbf{A} \succ \mathbf{0}$, *then* $\mathbf{x} = -\mathbf{A}^{-1}\mathbf{b}$ *is a strict global minimum point of* f.

Proof. (a) The proof of (a) follows immediately from the formula of the gradient of f (equation (2.10)).

(b) Since $\nabla^2 f(\mathbf{x}) = 2\mathbf{A} \succeq \mathbf{0}$, it follows by Theorem 2.38 that the global minimum points are exactly the stationary points, which combined with part (a) implies the result.

(c) When $\mathbf{A} \succ \mathbf{0}$, the vector $\mathbf{x} = -\mathbf{A}^{-1}\mathbf{b}$ is the unique solution to $\mathbf{A}\mathbf{x} = -\mathbf{b}$, and hence by parts (a) and (b), it is the unique global minimum point of f. □

We note that when $\mathbf{A} \succ \mathbf{0}$, the global minimizer of f is $\mathbf{x}^* = -\mathbf{A}^{-1}\mathbf{b}$, and consequently the minimal value of the function is

$$\begin{aligned} f(\mathbf{x}^*) &= (\mathbf{x}^*)^T \mathbf{A}\mathbf{x}^* + 2\mathbf{b}^T\mathbf{x}^* + c \\ &= (-\mathbf{A}^{-1}\mathbf{b})^T \mathbf{A}(-\mathbf{A}^{-1}\mathbf{b}) - 2\mathbf{b}^T\mathbf{A}^{-1}\mathbf{b} + c \\ &= c - \mathbf{b}^T \mathbf{A}^{-1}\mathbf{b}. \end{aligned}$$

Another useful property of quadratic functions is that they are coercive if and only if the associated matrix $\mathbf{A}$ is positive definite.

Lemma 2.42 (coerciveness of quadratic functions). *Let* $f(\mathbf{x}) = \mathbf{x}^T \mathbf{A}\mathbf{x} + 2\mathbf{b}^T\mathbf{x} + c$, *where* $\mathbf{A} \in \mathbb{R}^{n \times n}$ *is symmetric,* $\mathbf{b} \in \mathbb{R}^n$, *and* $c \in \mathbb{R}$. *Then* f *is coercive if and only if* $\mathbf{A} \succ \mathbf{0}$.

Proof. If $\mathbf{A} \succ \mathbf{0}$, then by Lemma 1.11, $\mathbf{x}^T \mathbf{A} \mathbf{x} \geq \alpha \|\mathbf{x}\|^2$ for $\alpha = \lambda_{\min}(\mathbf{A}) > 0$. We can thus write

$$f(\mathbf{x}) \geq \alpha \|\mathbf{x}\|^2 + 2\mathbf{b}^T\mathbf{x} + c \geq \alpha \|\mathbf{x}\|^2 - 2\|\mathbf{b}\| \cdot \|\mathbf{x}\| + c = \alpha \|\mathbf{x}\| \left(\|\mathbf{x}\| - 2\frac{\|\mathbf{b}\|}{\alpha} \right) + c,$$

where we also used the Cauchy–Schwarz inequality. Since obviously $\alpha \|\mathbf{x}\|(\|\mathbf{x}\| - 2\frac{\|\mathbf{b}\|}{\alpha}) + c \to \infty$ as $\|\mathbf{x}\| \to \infty$, it follows that $f(\mathbf{x}) \to \infty$ as $\|\mathbf{x}\| \to \infty$, establishing the coerciveness of f. Now assume that f is coercive. We need to prove that $\mathbf{A}$ is positive definite, or in other words that all its eigenvalues are positive. We begin by showing that there does not exist a negative eigenvalue. Suppose in contradiction that there exists such a negative eigenvalue; that is, there exists a nonzero vector $\mathbf{v} \in \mathbb{R}^n$ and $\lambda < 0$ such that $\mathbf{A}\mathbf{v} = \lambda \mathbf{v}$. Then

$$f(\alpha \mathbf{v}) = \lambda \|\mathbf{v}\|^2 \alpha^2 + 2(\mathbf{b}^T \mathbf{v})\alpha + c \to -\infty$$

as α tends to ∞, thus contradicting the assumption that f is coercive. We thus conclude that all the eigenvalues of $\mathbf{A}$ are nonnegative. We will show that 0 cannot be an eigenvalue of $\mathbf{A}$. By contradiction assume that there exists $\mathbf{v} \neq \mathbf{0}$ such that $\mathbf{A}\mathbf{v} = \mathbf{0}$. Then for any $\alpha \in \mathbb{R}$

$$f(\alpha \mathbf{v}) = 2(\mathbf{b}^T \mathbf{v})\alpha + c.$$

Then if $\mathbf{b}^T \mathbf{v} = 0$, we have $f(\alpha \mathbf{v}) \to c$ as $\alpha \to \infty$. If $\mathbf{b}^T \mathbf{v} > 0$, then $f(\alpha \mathbf{v}) \to -\infty$ as $\alpha \to -\infty$, and if $\mathbf{b}^T \mathbf{v} < 0$, then $f(\alpha \mathbf{v}) \to -\infty$ as $\alpha \to \infty$, contradicting the coerciveness of the function. We have thus proven that $\mathbf{A}$ is positive definite. $\square$

The last result describes an important characterization of the property that a quadratic function is nonnegative over the entire space. It is a generalization of the property that $\mathbf{A} \succeq \mathbf{0}$ if and only if $\mathbf{x}^T \mathbf{A} \mathbf{x} \geq 0$ for any $\mathbf{x} \in \mathbb{R}^n$.

Theorem 2.43 (characterization of the nonnegativity of quadratic functions). *Let $f(\mathbf{x}) = \mathbf{x}^T \mathbf{A} \mathbf{x} + 2\mathbf{b}^T \mathbf{x} + c$, where $\mathbf{A} \in \mathbb{R}^{n \times n}$ is symmetric, $\mathbf{b} \in \mathbb{R}^n$, and $c \in \mathbb{R}$. Then the following two claims are equivalent:*

(a) $f(\mathbf{x}) \equiv \mathbf{x}^T \mathbf{A} \mathbf{x} + 2\mathbf{b}^T \mathbf{x} + c \geq 0$ *for all* $\mathbf{x} \in \mathbb{R}^n$.

(b) $\begin{pmatrix} \mathbf{A} & \mathbf{b} \\ \mathbf{b}^T & c \end{pmatrix} \succeq \mathbf{0}$.

Proof. Suppose that (b) holds. Then in particular for any $\mathbf{x} \in \mathbb{R}^n$ the inequality

$$\begin{pmatrix} \mathbf{x} \\ 1 \end{pmatrix}^T \begin{pmatrix} \mathbf{A} & \mathbf{b} \\ \mathbf{b}^T & c \end{pmatrix} \begin{pmatrix} \mathbf{x} \\ 1 \end{pmatrix} \geq 0$$

holds, which is the same as the inequality $\mathbf{x}^T \mathbf{A} \mathbf{x} + 2\mathbf{b}^T \mathbf{x} + c \geq 0$, proving the validity of (a). Now, assume that (a) holds. We begin by showing that $\mathbf{A} \succeq \mathbf{0}$. Suppose in contradiction that $\mathbf{A}$ is not positive semidefinite. Then there exists an eigenvector $\mathbf{v}$ corresponding to a negative eigenvalue $\lambda < 0$ of $\mathbf{A}$:

$$\mathbf{A}\mathbf{v} = \lambda \mathbf{v}.$$

Thus, for any $\alpha \in \mathbb{R}$

$$f(\alpha \mathbf{v}) = \lambda \|\mathbf{v}\|^2 \alpha^2 + 2(\mathbf{b}^T \mathbf{v})\alpha + c \to -\infty$$

as $\alpha \to -\infty$, contradicting the nonnegativity of f. Our objective is to prove (b); that is, we want to show that for any $\mathbf{y} \in \mathbb{R}^n$ and $t \in \mathbb{R}$

$$\begin{pmatrix} \mathbf{y} \\ t \end{pmatrix}^T \begin{pmatrix} \mathbf{A} & \mathbf{b} \\ \mathbf{b}^T & c \end{pmatrix} \begin{pmatrix} \mathbf{y} \\ t \end{pmatrix} \geq 0,$$

which is equivalent to

$$\mathbf{y}^T \mathbf{A} \mathbf{y} + 2t \mathbf{b}^T \mathbf{y} + c t^2 \geq 0. \tag{2.12}$$

To show the validity of (2.12) for any $\mathbf{y} \in \mathbb{R}^n$ and $t \in \mathbb{R}$, we consider two cases. If $t = 0$, then (2.12) reads as $\mathbf{y}^T \mathbf{A} \mathbf{y} \geq 0$, which is a valid inequality since we have shown that $\mathbf{A} \succeq 0$. The second case is when $t \neq 0$. To show that (2.12) holds in this case, note that (2.12) is the same as the inequality

$$t^2 f\left(\frac{\mathbf{y}}{t}\right) = t^2 \left[\left(\frac{\mathbf{y}}{t}\right)^T \mathbf{A} \left(\frac{\mathbf{y}}{t}\right) + 2 \mathbf{b}^T \left(\frac{\mathbf{y}}{t}\right) + c \right] \geq 0,$$

which holds true by the nonnegativity of f. $\square$

Exercises

2.1. Find the global minimum and maximum points of the function $f(x, y) = x^2 + y^2 + 2x - 3y$ over the unit ball $S = B[\mathbf{0}, 1] = \{(x, y) : x^2 + y^2 \leq 1\}$.

2.2. Let $\mathbf{a} \in \mathbb{R}^n$ be a nonzero vector. Show that the maximum of $\mathbf{a}^T \mathbf{x}$ over $B[\mathbf{0}, 1] = \{\mathbf{x} \in \mathbb{R}^n : \|\mathbf{x}\| \leq 1\}$ is attained at $\mathbf{x}^* = \frac{\mathbf{a}}{\|\mathbf{a}\|}$ and that the maximal value is $\|\mathbf{a}\|$.

2.3. Find the global minimum and maximum points of the function $f(x, y) = 2x - 3y$ over the set $S = \{(x, y) : 2x^2 + 5y^2 \leq 1\}$.

2.4. Show that if $\mathbf{A}, \mathbf{B}$ are $n \times n$ positive semidefinite matrices, then their sum $\mathbf{A} + \mathbf{B}$ is also positive semidefinite.

2.5. Let $\mathbf{A} \in \mathbb{R}^{n \times n}$ and $\mathbf{B} \in \mathbb{R}^{m \times m}$ be two symmetric matrices. Prove that the following two claims are equivalent:

   (i) $\mathbf{A}$ and $\mathbf{B}$ are positive semidefinite.

   (ii) $\begin{pmatrix} \mathbf{A} & \mathbf{0}_{n \times m} \\ \mathbf{0}_{m \times n} & \mathbf{B} \end{pmatrix}$ is positive semidefinite.

2.6. Let $\mathbf{B} \in \mathbb{R}^{n \times k}$ and let $\mathbf{A} = \mathbf{B}\mathbf{B}^T$.

   (i) Prove $\mathbf{A}$ is positive semidefinite.

   (ii) Prove that $\mathbf{A}$ is positive definite if and only if $\mathbf{B}$ has a full row rank.

2.7. (i) Let $\mathbf{A}$ be an $n \times n$ symmetric matrix. Show that $\mathbf{A}$ is positive semidefinite if and only if there exists a matrix $\mathbf{B} \in \mathbb{R}^{n \times n}$ such that $\mathbf{A} = \mathbf{B}\mathbf{B}^T$.

   (ii) Let $\mathbf{x} \in \mathbb{R}^n$ and let $\mathbf{A}$ be defined as

   $$A_{ij} = x_i x_j, \quad i, j = 1, 2, \ldots, n.$$

   Show that $\mathbf{A}$ is positive semidefinite and that it is *not* a positive definite matrix when $n > 1$.

2.8. Let $\mathbf{Q} \in \mathbb{R}^{n \times n}$ be a positive definite matrix. Show that the "Q-norm" defined by

$$\|\mathbf{x}\|_Q = \sqrt{\mathbf{x}^T \mathbf{Q} \mathbf{x}}$$

is indeed a norm.

2.9. Let $\mathbf{A}$ be an $n \times n$ positive semidefinite matrix.

(i) Show that for any $i \neq j$

$$A_{ii} A_{jj} \geq A_{ij}^2.$$

(ii) Show that if for some $i \in \{1, 2, \ldots, n\}$ $A_{ii} = 0$, then the ith row of $\mathbf{A}$ consists of zeros.

2.10. Let $\mathbf{A}^\alpha$ be the $n \times n$ matrix ($n > 1$) defined by

$$A_{ij}^\alpha = \begin{cases} \alpha, & i = j, \\ 1, & i \neq j. \end{cases}$$

Show that $\mathbf{A}^\alpha$ is positive semidefinite if and only if $\alpha \geq 1$.

2.11. Let $\mathbf{d} \in \Delta_n$ (Δ_n being the unit-simplex). Show that the $n \times n$ matrix $\mathbf{A}$ defined by

$$A_{ij} = \begin{cases} d_i - d_i^2, & i = j, \\ -d_i d_j, & i \neq j, \end{cases}$$

is positive semidefinite.

2.12. Prove that a 2×2 matrix $\mathbf{A}$ is negative semidefinite if and only if $\text{Tr}(\mathbf{A}) \leq 0$ and $\det(\mathbf{A}) \geq 0$.

2.13. For each of the following matrices determine whether they are positive/negative semidefinite/definite or indefinite:

(i) $\mathbf{A} = \begin{pmatrix} 2 & 2 & 0 & 0 \\ 2 & 2 & 0 & 0 \\ 0 & 0 & 3 & 1 \\ 0 & 0 & 1 & 3 \end{pmatrix}.$

(ii) $\mathbf{B} = \begin{pmatrix} 2 & 2 & 2 \\ 2 & 3 & 3 \\ 2 & 3 & 3 \end{pmatrix}.$

(iii) $\mathbf{C} = \begin{pmatrix} 2 & 1 & 3 \\ 1 & 2 & 1 \\ 3 & 1 & 2 \end{pmatrix}.$

(iv) $\mathbf{D} = \begin{pmatrix} -5 & 1 & 1 \\ 1 & -7 & 1 \\ 1 & 1 & -5 \end{pmatrix}.$

2.14. (Schur complement lemma) Let

$$\mathbf{D} = \begin{pmatrix} \mathbf{A} & \mathbf{b} \\ \mathbf{b}^T & c \end{pmatrix},$$

where $\mathbf{A} \in \mathbb{R}^{n \times n}, \mathbf{b} \in \mathbb{R}^n, c \in \mathbb{R}$. Suppose that $\mathbf{A} \succ 0$. Prove that $\mathbf{D} \succeq 0$ if and only if $c - \mathbf{b}^T \mathbf{A}^{-1} \mathbf{b} \geq 0$.

2.15. For each of the following functions, determine whether it is coercive or not:
   (i) $f(x_1,x_2) = x_1^4 + x_2^4$.
   (ii) $f(x_1,x_2) = e^{x_1^2} + e^{x_2^2} - x_1^{200} - x_2^{200}$.
   (iii) $f(x_1,x_2) = 2x_1^2 - 8x_1x_2 + x_2^2$.
   (iv) $f(x_1,x_2) = 4x_1^2 + 2x_1x_2 + 2x_2^2$.
   (v) $f(x_1,x_2,x_3) = x_1^3 + x_2^3 + x_3^3$.
   (vi) $f(x_1,x_2) = x_1^2 - 2x_1x_2^2 + x_2^4$.
   (vii) $f(\mathbf{x}) = \frac{\mathbf{x}^T \mathbf{A} \mathbf{x}}{\|\mathbf{x}\|+1}$, where $\mathbf{A} \in \mathbb{R}^{n \times n}$ is positive definite.

2.16. Find a function $f : \mathbb{R}^2 \to \mathbb{R}$ which is not coercive and satisfies that for any $\alpha \in \mathbb{R}$
$$\lim_{|x_1| \to \infty} f(x_1, \alpha x_1) = \lim_{|x_2| \to \infty} f(\alpha x_2, x_2) = \infty.$$

2.17. For each of the following functions, find all the stationary points and classify them according to whether they are saddle points, strict/nonstrict local/global minimum/maximum points:
   (i) $f(x_1,x_2) = (4x_1^2 - x_2)^2$.
   (ii) $f(x_1,x_2,x_3) = x_1^4 - 2x_1^2 + x_2^2 + 2x_2x_3 + 2x_3^2$.
   (iii) $f(x_1,x_2) = 2x_2^3 - 6x_2^2 + 3x_1^2 x_2$.
   (iv) $f(x_1,x_2) = x_1^4 + 2x_1^2 x_2 + x_2^2 - 4x_1^2 - 8x_1 - 8x_2$.
   (v) $f(x_1,x_2) = (x_1 - 2x_2)^4 + 64 x_1 x_2$.
   (vi) $f(x_1,x_2) = 2x_1^2 + 3x_2^2 - 2x_1 x_2 + 2x_1 - 3x_2$.
   (vii) $f(x_1,x_2) = x_1^2 + 4x_1x_2 + x_2^2 + x_1 - x_2$.

2.18. Let f be twice continuously differentiable function over $\mathbb{R}^n$. Suppose that $\nabla^2 f(\mathbf{x}) \succ \mathbf{0}$ for any $\mathbf{x} \in \mathbb{R}^n$. Prove that a stationary point of f is necessarily a strict global minimum point.

2.19. Let $f(\mathbf{x}) = \mathbf{x}^T \mathbf{A} \mathbf{x} + 2\mathbf{b}^T \mathbf{x} + c$, where $\mathbf{A} \in \mathbb{R}^{n \times n}$ is symmetric, $\mathbf{b} \in \mathbb{R}^n$, and $c \in \mathbb{R}$. Suppose that $\mathbf{A} \succeq \mathbf{0}$. Show that f is bounded below[1] over $\mathbb{R}^n$ if and only if $\mathbf{b} \in \text{Range}(\mathbf{A}) = \{\mathbf{A}\mathbf{y} : \mathbf{y} \in \mathbb{R}^n\}$.

---

[1] A function f is bounded below over a set C if there exists a constant α such that $f(\mathbf{x}) \geq \alpha$ for any $\mathbf{x} \in C$.

Chapter 3

Least Squares

3.1 ▪ "Solution" of Overdetermined Systems

Suppose that we are given a linear system of the form

$$\mathbf{Ax} = \mathbf{b},$$

where $\mathbf{A} \in \mathbb{R}^{m \times n}$ and $\mathbf{b} \in \mathbb{R}^m$. Assume that the system is *overdetermined*, meaning that $m > n$. In addition, we assume that $\mathbf{A}$ has a full column rank; that is, $\text{rank}(\mathbf{A}) = n$. In this setting, the system is usually *inconsistent* (has no solution) and a common approach for finding an approximate solution is to pick the solution resulting with the minimal squared norm of the residual $\mathbf{r} = \mathbf{Ax} - \mathbf{b}$:

$$\text{(LS)} \quad \min_{\mathbf{x} \in \mathbb{R}^n} \|\mathbf{Ax} - \mathbf{b}\|^2.$$

Problem (LS) is a problem of minimizing a quadratic function over the entire space. The quadratic objective function is given by

$$f(\mathbf{x}) = \mathbf{x}^T \mathbf{A}^T \mathbf{A} \mathbf{x} - 2 \mathbf{b}^T \mathbf{A} \mathbf{x} + \|\mathbf{b}\|^2.$$

Since $\mathbf{A}$ is of full column rank, it follows that for any $\mathbf{x} \in \mathbb{R}^n$ it holds that $\nabla^2 f(\mathbf{x}) = 2\mathbf{A}^T\mathbf{A} \succ \mathbf{0}$. (Otherwise, if $\mathbf{A}$ is not of full column, only positive semidefiniteness can be guaranteed.) Hence, by Lemma 2.41 the unique stationary point

$$\mathbf{x}_{\text{LS}} = (\mathbf{A}^T \mathbf{A})^{-1} \mathbf{A}^T \mathbf{b} \tag{3.1}$$

is the optimal solution of problem (LS). The vector $\mathbf{x}_{\text{LS}}$ is called *the least squares solution* or *the least squares estimate* of the system $\mathbf{Ax} = \mathbf{b}$. It is quite common not to write the explicit expression for $\mathbf{x}_{\text{LS}}$ but instead to write the associated system of equations that defines it:

$$(\mathbf{A}^T \mathbf{A}) \mathbf{x}_{\text{LS}} = \mathbf{A}^T \mathbf{b}.$$

The above system of equations is called *the normal system*. We can actually omit the assumption that the system is overdetermined and just keep the assumption that $\mathbf{A}$ is of full column rank. Under this assumption $m \geq n$, and in the case when $m = n$, the matrix $\mathbf{A}$ is nonsingular and the least squares solution is actually the solution of the linear system, that is, $\mathbf{A}^{-1}\mathbf{b}$.

Example 3.1. Consider the inconsistent linear system

$$x_1 + 2x_2 = 0,$$
$$2x_1 + x_2 = 1,$$
$$3x_1 + 2x_2 = 1.$$

We will denote by $\mathbf{A}$ and $\mathbf{b}$ the coefficients matrix and right-hand-side vector of the system. The least squares problem can be explicitly written as

$$\min_{x_1,x_2}(x_1+2x_2)^2 + (2x_1+x_2-1)^2 + (3x_1+2x_2-1)^2.$$

Essentially, the solution to the above problem is the vector that yields the minimal sum of squares of the errors corresponding to the three equations. To find the least squares solution, we will solve the normal equations:

$$\begin{pmatrix} 1 & 2 \\ 2 & 1 \\ 3 & 2 \end{pmatrix}^T \begin{pmatrix} 1 & 2 \\ 2 & 1 \\ 3 & 2 \end{pmatrix} \begin{pmatrix} x_1 \\ x_2 \end{pmatrix} = \begin{pmatrix} 1 & 2 \\ 2 & 1 \\ 3 & 2 \end{pmatrix}^T \begin{pmatrix} 0 \\ 1 \\ 1 \end{pmatrix},$$

which are the same as

$$\begin{pmatrix} 14 & 10 \\ 10 & 9 \end{pmatrix} \begin{pmatrix} x_1 \\ x_2 \end{pmatrix} = \begin{pmatrix} 5 \\ 3 \end{pmatrix}.$$

The solution of the above system is the least squares estimate:

$$\mathbf{x}_{LS} = \begin{pmatrix} 15/26 \\ -8/26 \end{pmatrix}.$$

Note that

$$\mathbf{A}\mathbf{x}_{LS} = \begin{pmatrix} -0.038 \\ 0.846 \\ 1.115 \end{pmatrix},$$

so that the residual vector containing the errors in each of the equations is

$$\mathbf{A}\mathbf{x}_{LS} - \mathbf{b} = \begin{pmatrix} -0.038 \\ -0.154 \\ 0.115 \end{pmatrix}.$$

The total sum of squares of the errors, which is the optimal value of the least squares problem is $(-0.038)^2 + (-0.154)^2 + 0.115^2 = 0.038$. ∎

In MATLAB, finding the least squares solution is a very easy task.

MATLAB Implementation

To find the least squares solution of an overdetermined linear system $\mathbf{A}\mathbf{x} = \mathbf{b}$ in MATLAB, the backslash operator \ should be used. Therefore, Example 3.1 can be solved by the following commands

```
>> A =[1,2;2,1;3,2];
>> b=[0;1;1];
>> format rational;
>> A\b
ans =
      15/26
      -4/13
```

3.2 • Data Fitting

One area in which least squares is being frequently used is data fitting. We begin by describing the problem of linear fitting. Suppose that we are given a set of data points $(s_i, t_i), i = 1, 2, \ldots, m$, where $s_i \in \mathbb{R}^n$ and $t_i \in \mathbb{R}$, and assume that a linear relation of the form

$$t_i = s_i^T x, \quad i = 1, 2, \ldots, m,$$

approximately holds. In the least squares approach the objective is to find the parameters vector $x \in \mathbb{R}^n$ that solves the problem

$$\min_{x \in \mathbb{R}^n} \sum_{i=1}^{m} (s_i^T x - t_i)^2.$$

We can alternatively write the problem as

$$\min_{x \in \mathbb{R}^n} \|Sx - t\|^2,$$

where

$$S = \begin{pmatrix} -s_1^T- \\ -s_2^T- \\ \vdots \\ -s_m^T- \end{pmatrix}, \quad t = \begin{pmatrix} t_1 \\ t_2 \\ \vdots \\ t_m \end{pmatrix}.$$

Example 3.2. Consider the 30 points in $\mathbb{R}^2$ described in the left image of Figure 3.1. The 30 x-coordinates are $x_i = (i-1)/29, i = 1, 2, \ldots, 30$, and the corresponding y-coordinates are defined by $y_i = 2x_i + 1 + \varepsilon_i$, where for every i, ε_i is randomly generated from a standard normal distribution with zero mean and standard deviation of 0.1. The MATLAB commands that generated the points and plotted them are

```
randn('seed',319);
d=linspace(0,1,30)';
e=2*d+1+0.1*randn(30,1);
plot(d,e,'*')
```

Note that we have used the command `randn('seed',sd)` in order to control the random number generator. In future versions of MATLAB, it is possible that this command will not be supported anymore, and will be replaced by the command `rng(sd,'v4')`. Given the 30 points, the objective is to find a line of the form $y = ax + b$ that best fits them. The corresponding linear system that needs to be "solved" is

$$\underbrace{\begin{pmatrix} x_1 & 1 \\ x_2 & 1 \\ \vdots & \vdots \\ x_{30} & 1 \end{pmatrix}}_{X} \begin{pmatrix} a \\ b \end{pmatrix} = \underbrace{\begin{pmatrix} y_1 \\ y_2 \\ \vdots \\ y_{30} \end{pmatrix}}_{y}.$$

The least squares solution of the above system is $(X^T X)^{-1} X^T y$. In MATLAB, the parameters a and b can be extracted via the commands

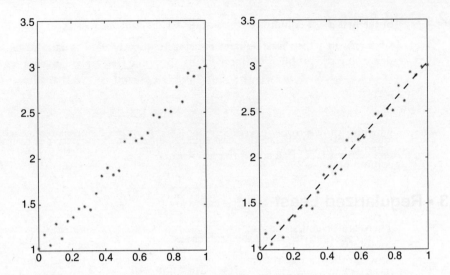

Figure 3.1. *Left image: 30 points in the plane. Right image: the points and the corresponding least squares line.*

```
>> u=[d,ones(30,1)]\e;
>> a=u(1),b=u(2)
a =
    2.0616
b =
    0.9725
```

Note that the obtained estimates of a and b are very close to the "true" a and b (2 and 1, respectively) that were used to generate the data. The least squares line as well as the 30 points is described in the right image of Figure 3.1. ∎

The least squares approach can be used also in nonlinear fitting. Suppose, for example, that we are given a set of points in $\mathbb{R}^2$: $(u_i, y_i), i = 1, 2, \ldots, m$, and that we know a priori that these points are approximately related via a polynomial of degree at most d; i.e., there exists $a_0, \ldots, a_d$ such that

$$\sum_{j=0}^{d} a_j u_i^j \approx y_i, \quad i = 1, \ldots, m.$$

The least squares approach to this problem seeks $a_0, a_1, \ldots, a_d$ that are the least squares solution to the linear system

$$\underbrace{\begin{pmatrix} 1 & u_1 & u_1^2 & \cdots & u_1^d \\ 1 & u_2 & u_2^2 & \cdots & u_2^d \\ \vdots & \vdots & \vdots & & \vdots \\ 1 & u_m & u_m^2 & \cdots & u_m^d \end{pmatrix}}_{\mathbf{U}} \begin{pmatrix} a_0 \\ a_1 \\ \vdots \\ a_d \end{pmatrix} = \begin{pmatrix} y_0 \\ y_1 \\ \vdots \\ y_m \end{pmatrix}.$$

The least squares solution is of course well-defined if the $m \times (d+1)$ matrix is of a full column rank. This of course suggests in particular that $m \geq d+1$. The matrix $\mathbf{U}$ consists of the first $d+1$ columns of the so-called *Vandermonde* matrix,

$$\begin{pmatrix} 1 & u_1 & u_1^2 & \cdots & u_1^{m-1} \\ 1 & u_2 & u_2^2 & \cdots & u_2^{m-1} \\ \vdots & \vdots & \vdots & & \vdots \\ 1 & u_m & u_m^2 & \cdots & u_m^{m-1} \end{pmatrix},$$

which is known to be invertible when all the u_is are different from each other. Thus, when $m \geq d+1$, and all the u_is are different from each other, the matrix $\mathbf{U}$ is of a full column rank.

3.3 • Regularized Least Squares

There are several situations in which the least squares solution does not give rise to a good estimate of the "true" vector $\mathbf{x}$. For example, when $\mathbf{A}$ is underdetermined, that is, when there are fewer equations than variables, there are several optimal solutions to the least squares problem, and it is unclear which of these optimal solutions is the one that should be considered. In these cases, some type of prior information on $\mathbf{x}$ should be incorporated into the optimization model. One way to do this is to consider a penalized problem in which a *regularization function* $R(\cdot)$ is added to the objective function. The regularized least squares (RLS) problem has the form

$$\text{(RLS)} \quad \min_{\mathbf{x}} \|\mathbf{A}\mathbf{x} - \mathbf{b}\|^2 + \lambda R(\mathbf{x}). \tag{3.2}$$

The positive constant λ is the *regularization parameter*. As λ gets larger, more weight is given to the regularization function.

In many cases, the regularization is taken to be quadratic. In particular, $R(\mathbf{x}) = \|\mathbf{D}\mathbf{x}\|^2$ where $\mathbf{D} \in \mathbb{R}^{p \times n}$ is a given matrix. The quadratic regularization function aims to control the norm of $\mathbf{D}\mathbf{x}$ and is formulated as follows:

$$\min_{\mathbf{x}} \|\mathbf{A}\mathbf{x} - \mathbf{b}\|^2 + \lambda \|\mathbf{D}\mathbf{x}\|^2.$$

To find the optimal solution of this problem, note that it can be equivalently written as

$$\min_{\mathbf{x}} \{f_{\text{RLS}}(\mathbf{x}) \equiv \mathbf{x}^T(\mathbf{A}^T\mathbf{A} + \lambda \mathbf{D}^T\mathbf{D})\mathbf{x} - 2\mathbf{b}^T\mathbf{A}\mathbf{x} + \|\mathbf{b}\|^2\}.$$

Since the Hessian of the objective function is $\nabla^2 f_{\text{RLS}}(\mathbf{x}) = 2(\mathbf{A}^T\mathbf{A} + \lambda \mathbf{D}^T\mathbf{D}) \succeq \mathbf{0}$, it follows by Lemma 2.41 that any stationary point is a global minimum point. The stationary points are those satisfying $\nabla f_{\text{RLS}}(\mathbf{x}) = \mathbf{0}$, that is,

$$(\mathbf{A}^T\mathbf{A} + \lambda \mathbf{D}^T\mathbf{D})\mathbf{x} = \mathbf{A}^T\mathbf{b}.$$

Therefore, if $\mathbf{A}^T\mathbf{A} + \lambda \mathbf{D}^T\mathbf{D} \succ \mathbf{0}$, then the RLS solution is given by

$$\mathbf{x}_{\text{RLS}} = (\mathbf{A}^T\mathbf{A} + \lambda \mathbf{D}^T\mathbf{D})^{-1} \mathbf{A}^T \mathbf{b}. \tag{3.3}$$

Example 3.3. Let $\mathbf{A} \in \mathbb{R}^{3 \times 3}$ be given by

$$\mathbf{A} = \begin{pmatrix} 2 + 10^{-3} & 3 & 4 \\ 3 & 5 + 10^{-3} & 7 \\ 4 & 7 & 10 + 10^{-3} \end{pmatrix}.$$

The matrix was constructed via the MATLAB code

```
B=[1,1,1;1,2,3];
A=B'*B+0.001*eye(3);
```

The "true" vector was chosen to be $\mathbf{x}_{\text{true}} = (1,2,3)^T$, and $\mathbf{b}$ is a noisy measurement of $\mathbf{A}\mathbf{x}_{\text{true}}$:

```
>> x_true=[1;2;3];
>> randn('seed',315);
>> b=A*x_true+0.01*randn(3,1)
b =
   20.0019
   34.0004
   48.0202
```

The matrix $\mathbf{A}$ is in fact of a full column rank since its eigenvalues are all positive (which can be checked, for example, by the MATLAB command `eig(A)`), and the least squares solution is given by $\mathbf{x}_{\text{LS}}$, whose value can be computed by

```
A\b
ans =
    4.5446
   -5.1295
    6.5742
```

$\mathbf{x}_{\text{LS}}$ is rather far from the true vector $\mathbf{x}_{\text{true}}$. One difference between the solutions is that the squared norm $\|\mathbf{x}_{\text{LS}}\|^2 = 90.1855$ is much larger then the correct squared norm $\|\mathbf{x}_{\text{true}}\|^2 = 14$. In order to control the norm of the solution we will add the quadratic regularization function $\|\mathbf{x}\|^2$. The regularized solution will thus have the form (see (3.3))

$$\mathbf{x}_{\text{RLS}} = (\mathbf{A}^T\mathbf{A} + \lambda\mathbf{I})^{-1}\mathbf{A}^T\mathbf{b}.$$

Picking the regularization parameter as $\lambda = 1$, the RLS solution becomes

```
>> x_rls=(A'*A+eye(3))\(A'*b)
x_rls =
    1.1763
    2.0318
    2.8872
```

which is a much better estimate for $\mathbf{x}_{\text{true}}$ than $\mathbf{x}_{\text{LS}}$. ∎

3.4 ▪ Denoising

One application area in which regularization is commonly used is *denoising*. Suppose that a noisy measurement of a signal $\mathbf{x} \in \mathbb{R}^n$ is given:

$$\mathbf{b} = \mathbf{x} + \mathbf{w}.$$

Here $\mathbf{x}$ is an unknown signal, $\mathbf{w}$ is an unknown noise vector, and $\mathbf{b}$ is the known measurements vector. The denoising problem is the following: Given $\mathbf{b}$, find a "good" estimate of $\mathbf{x}$. The least squares problem associated with the approximate equations $\mathbf{x} \approx \mathbf{b}$ is

$$\min \|\mathbf{x} - \mathbf{b}\|^2.$$

However, the optimal solution of this problem is obviously $\mathbf{x} = \mathbf{b}$, which is meaningless. This is a case in which the least squares solution is not informative even though the associated matrix—the identity matrix—is of a full column rank. To find a more relevant

3.4. Denoising

problem, we will add a regularization term. For that, we need to exploit some a priori information on the signal. For example, we might know in advance that the signal is smooth in some sense. In that case, it is very natural to add a quadratic penalty, which is the sum of the squares of the differences of consecutive components of the vector; that is, the regularization function is

$$R(\mathbf{x}) = \sum_{i=1}^{n-1}(x_i - x_{i+1})^2.$$

This quadratic function can also be written as $R(\mathbf{x}) = \|\mathbf{L}\mathbf{x}\|^2$, where $\mathbf{L} \in \mathbb{R}^{(n-1)\times n}$ is given by

$$\mathbf{L} = \begin{pmatrix} 1 & -1 & 0 & 0 & \cdots & 0 & 0 \\ 0 & 1 & -1 & 0 & \cdots & 0 & 0 \\ 0 & 0 & 1 & -1 & \cdots & 0 & 0 \\ \vdots & \vdots & \vdots & \vdots & & \vdots & \vdots \\ 0 & 0 & 0 & 0 & \cdots & 1 & -1 \end{pmatrix}.$$

The resulting regularized least squares problem is (with λ a given regularization parameter)

$$\min_{\mathbf{x}} \|\mathbf{x} - \mathbf{b}\|^2 + \lambda \|\mathbf{L}\mathbf{x}\|^2,$$

and its optimal solution is given by

$$\mathbf{x}_{\text{RLS}}(\lambda) = (\mathbf{I} + \lambda \mathbf{L}^T \mathbf{L})^{-1} \mathbf{b}. \tag{3.4}$$

Example 3.4. Consider the signal $\mathbf{x} \in \mathbb{R}^{300}$ constructed by the following MATLAB commands:

```
t=linspace(0,4,300)';
x=sin(t)+t.*(cos(t).^2);
```

Essentially, this is the signal given by $x_i = \sin(4\frac{i-1}{299}) + (4\frac{i-1}{299})\cos^2(4\frac{i-1}{299}), i = 1, 2, \ldots, 300$. A normally distributed noise with zero mean and standard deviation of 0.05 was added to each of the components:

```
randn('seed',314);
b=x+0.05*randn(300,1);
```

The true and noisy signals are given in Figure 3.2, which was constructed by the MATLAB commands.

```
subplot(1,2,1);
plot(1:300,x,'LineWidth',2);
subplot(1,2,2);
plot(1:300,b,'LineWidth',2);
```

In order to denoise the signal $\mathbf{b}$, we look at the optimal solution of the RLS problem given by (3.4) for four different values of the regularization parameter: $\lambda = 1, 10, 100, 1000$. The original true signal is denoted by a dotted line. As can be seen in Figure 3.3, as λ gets larger, the RLS solution becomes smoother. For $\lambda = 1$ the RLS solution $\mathbf{x}_{\text{RLS}}(1)$ is not

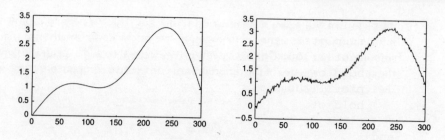

Figure 3.2. *A signal (left image) and its noisy version (right image).*

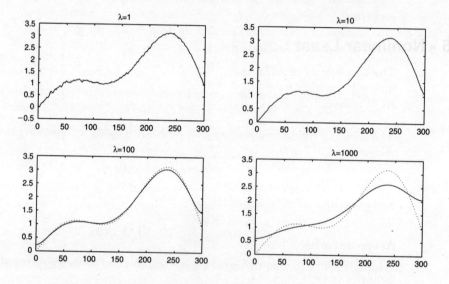

Figure 3.3. *Four reconstructions of a noisy signal by RLS solutions.*

smooth enough and is very close to the noisy signal **b**. For $\lambda = 10$ the RLS solution is a rather good estimate of the original vector **x**. For $\lambda = 100$ we get a smoother RLS signal, but evidently it is less accurate than $\mathbf{x}_{RLS}(10)$, especially near the boundaries. The RLS solution for $\lambda = 1000$ is very smooth, but it is a rather poor estimate of the original signal. In any case, it is evident that the parameter λ is chosen via a trade off between data fidelity (closeness of **x** to **b**) and smoothness (size of **Lx**). The four plots where produced by the MATLAB commands

```
L=zeros(299,300);
for i=1:299
    L(i,i)=1;
    L(i,i+1)=-1;
end

x_rls=(eye(300)+1*L'*L)\b;
x_rls=[x_rls,(eye(300)+10*L'*L)\b];
x_rls=[x_rls,(eye(300)+100*L'*L)\b];
x_rls=[x_rls,(eye(300)+1000*L'*L)\b];
figure(2)
```

```
for j=1:4
    subplot(2,2,j);
    plot(1:300,x_rls(:,j),'LineWidth',2);
    hold on
    plot(1:300,x,':r','LineWidth',2);
    hold off
    title(['\lambda=',num2str(10^(j-1))]);
end
```

∎

3.5 • Nonlinear Least Squares

The least squares problem considered so far is also referred to as "linear least squares" since it is a method for finding a solution to a set of approximate linear equalities. There are of course situations in which we are given a system of *nonlinear* equations

$$f_i(\mathbf{x}) \approx c_i, \quad i = 1, 2, \ldots, m.$$

In this case, the appropriate problem is the *nonlinear least squares (NLS) problem*, which is formulated as

$$\min \sum_{i=1}^{m} (f_i(\mathbf{x}) - c_i)^2. \tag{3.5}$$

As opposed to linear least squares, there is no easy way to solve NLS problems. In Section 4.5 we will describe the Gauss–Newton method which is specifically devised to solve NLS problems of the form (3.5), but the method is not guaranteed to converge to the global optimal solution of (3.5) but rather to a stationary point.

3.6 • Circle Fitting

Suppose that we are given m points $\mathbf{a}_1, \mathbf{a}_2, \ldots, \mathbf{a}_m \in \mathbb{R}^n$. The *circle fitting problem* seeks to find a circle

$$C(\mathbf{x}, r) = \{\mathbf{y} \in \mathbb{R}^n : \|\mathbf{y} - \mathbf{x}\| = r\}$$

that best fits the m points. Note that we use the term "circle," although this terminology is usually used in the plane ($n = 2$), and here we consider the general n-dimensional space $\mathbb{R}^n$. Additionally, note that $C(\mathbf{x}, r)$ is the boundary set of the corresponding ball $B(\mathbf{x}, r)$. An illustration of such a fit is given in Figure 3.4. The circle fitting problem has applications in many areas, such as archaeology, computer graphics, coordinate metrology, petroleum engineering, statistics, and more. The nonlinear (approximate) equations associated with the problem are

$$\|\mathbf{x} - \mathbf{a}_i\| \approx r, \quad i = 1, 2, \ldots, m.$$

Since we wish to deal with differentiable functions, and the norm function is not differentiable, we will consider the squared version of the latter:

$$\|\mathbf{x} - \mathbf{a}_i\|^2 \approx r^2, \quad i = 1, 2, \ldots, m.$$

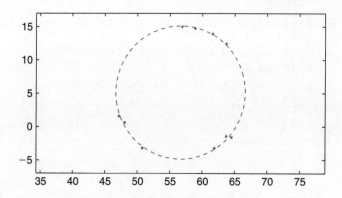

Figure 3.4. *The best circle fit (the optimal solution of problem* (3.6)*) of 10 points denoted by asterisks.*

The NLS problem associated with these equations is

$$\min_{\mathbf{x}\in\mathbb{R}^n, r\in\mathbb{R}_+} \sum_{i=1}^{m} (\|\mathbf{x}-\mathbf{a}_i\|^2 - r^2)^2. \tag{3.6}$$

From a first glance, problem (3.6) seems to be a standard NLS problem, but in this case we can show that it is in fact *equivalent* to a linear least squares problem, and therefore the global optimal solution can be easily obtained. We begin by noting that problem (3.6) is the same as

$$\min_{\mathbf{x},r} \left\{ \sum_{i=1}^{m} (-2\mathbf{a}_i^T \mathbf{x} + \|\mathbf{x}\|^2 - r^2 + \|\mathbf{a}_i\|^2)^2 : \mathbf{x}\in\mathbb{R}^n, r\in\mathbb{R} \right\}. \tag{3.7}$$

Making the change of variables $R = \|\mathbf{x}\|^2 - r^2$, the above problem reduces to

$$\min_{\mathbf{x}\in\mathbb{R}^n, R\in\mathbb{R}} \left\{ f(\mathbf{x},R) \equiv \sum_{i=1}^{m} (-2\mathbf{a}_i^T \mathbf{x} + R + \|\mathbf{a}_i\|^2)^2 : \|\mathbf{x}\|^2 \geq R \right\}. \tag{3.8}$$

Note that the change of variables imposed an additional relation between the variables that is given by the constraint $\|\mathbf{x}\|^2 \geq R$. We will show that in fact this constraint can be dropped; that is, problem (3.8) is equivalent to the linear least squares problem

$$\min_{\mathbf{x},R} \left\{ \sum_{i=1}^{m} (-2\mathbf{a}_i^T \mathbf{x} + R + \|\mathbf{a}_i\|^2)^2 : \mathbf{x}\in\mathbb{R}^n, R\in\mathbb{R} \right\}. \tag{3.9}$$

Indeed, any optimal solution $(\hat{\mathbf{x}}, \hat{R})$ of (3.9) automatically satisfies $\|\hat{\mathbf{x}}\|^2 \geq \hat{R}$ since otherwise, if $\|\hat{\mathbf{x}}\|^2 < \hat{R}$, we would have

$$-2\mathbf{a}_i^T \hat{\mathbf{x}} + \hat{R} + \|\mathbf{a}_i\|^2 > -2\mathbf{a}_i^T \hat{\mathbf{x}} + \|\hat{\mathbf{x}}\|^2 + \|\mathbf{a}_i\|^2 = \|\hat{\mathbf{x}} - \mathbf{a}_i\|^2 \geq 0, \qquad i=1,\ldots,m.$$

Squaring both sides of the first inequality in the above equation and summing over i yield

$$f(\hat{\mathbf{x}}, \hat{R}) = \sum_{i=1}^{m}\left(-2\mathbf{a}_i^T\hat{\mathbf{x}} + \hat{R} + \|\mathbf{a}_i\|^2\right)^2 > \sum_{i=1}^{m}\left(-2\mathbf{a}_i^T\hat{\mathbf{x}} + \|\hat{\mathbf{x}}\|^2 + \|\mathbf{a}_i\|^2\right)^2 = f(\hat{\mathbf{x}}, \|\hat{\mathbf{x}}\|^2),$$

showing that $(\hat{\mathbf{x}}, \|\hat{\mathbf{x}}\|^2)$ gives a lower function value than $(\hat{\mathbf{x}}, \hat{R})$, in contradiction to the optimality of $(\hat{\mathbf{x}}, \hat{R})$. To conclude, problem (3.6) is equivalent to the least squares problem (3.9), which can also be written as

$$\min_{\mathbf{y}\in\mathbb{R}^{n+1}} \|\tilde{\mathbf{A}}\mathbf{y} - \mathbf{b}\|^2, \qquad (3.10)$$

where $\mathbf{y} = \begin{pmatrix}\mathbf{x}\\R\end{pmatrix}$ and

$$\tilde{\mathbf{A}} = \begin{pmatrix} 2\mathbf{a}_1^T & -1 \\ 2\mathbf{a}_2^T & -1 \\ \vdots & \vdots \\ 2\mathbf{a}_m^T & -1 \end{pmatrix}, \quad \mathbf{b} = \begin{pmatrix} \|\mathbf{a}_1\|^2 \\ \|\mathbf{a}_2\|^2 \\ \vdots \\ \|\mathbf{a}_m\|^2 \end{pmatrix}. \qquad (3.11)$$

If $\tilde{\mathbf{A}}$ is of full column rank, then the unique solution of the linear least squares problem (3.10) is

$$\mathbf{y} = (\tilde{\mathbf{A}}^T\tilde{\mathbf{A}})^{-1}\tilde{\mathbf{A}}^T\mathbf{b}.$$

The optimal $\mathbf{x}$ is given by the first n components of $\mathbf{y}$ and the radius r is given by

$$r = \sqrt{\|\mathbf{x}\|^2 - R},$$

where R is the last (i.e., $(n+1)$th) component of $\mathbf{y}$. We summarize the above discussion in the following lemma.

Lemma 3.5. *Let $\mathbf{y} = (\tilde{\mathbf{A}}^T\tilde{\mathbf{A}})^{-1}\tilde{\mathbf{A}}^T\mathbf{b}$, where $\tilde{\mathbf{A}}$ and $\mathbf{b}$ are given in (3.11). Then the optimal solution of problem (3.6) is given by $(\hat{\mathbf{x}}, \hat{r})$, where $\hat{\mathbf{x}}$ consists of the first n components of $\mathbf{y}$ and $\hat{r} = \sqrt{\|\hat{\mathbf{x}}\|^2 - y_{n+1}}$.*

Exercises

3.1. Let $\mathbf{A} \in \mathbb{R}^{m\times n}, \mathbf{b} \in \mathbb{R}^n, \mathbf{L} \in \mathbb{R}^{p\times n}$, and $\lambda \in \mathbb{R}_{++}$. Consider the regularized least squares problem

$$(\text{RLS}) \quad \min_{\mathbf{x}\in\mathbb{R}^n} \|\mathbf{A}\mathbf{x} - \mathbf{b}\|^2 + \lambda\|\mathbf{L}\mathbf{x}\|^2.$$

Show that (RLS) has a unique solution if and only if Null($\mathbf{A}$) $\cap$ Null($\mathbf{L}$) = $\{\mathbf{0}\}$, where here for a matrix $\mathbf{B}$, Null($\mathbf{B}$) is the null space of $\mathbf{B}$ given by $\{\mathbf{x} : \mathbf{B}\mathbf{x} = \mathbf{0}\}$.

3.2. Generate thirty points $(x_i, y_i), i = 1, 2, \ldots, 30$, by the MATLAB code

```
randn('seed',314);
x=linspace(0,1,30)';
y=2*x.^2-3*x+1+0.05*randn(size(x));
```

Find the quadratic function $y = ax^2 + bx + c$ that best fits the points in the least squares sense. Indicate what are the parameters a, b, c found by the least squares

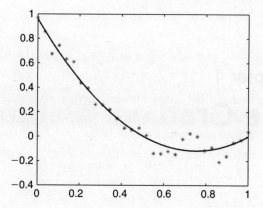

Figure 3.5. 30 *points and their best quadratic least squares fit.*

solution, and plot the points along with the derived quadratic function. The resulting plot should look like the one in Figure 3.5.

3.3. Write a MATLAB function `circle_fit` whose input is an $n \times m$ matrix **A**; the columns of **A** are the m vectors in $\mathbb{R}^n$ to which a circle should be fitted. The call to the function will be of the form

   `[x,r]=circle_fit(A)`

The output $(\mathbf{x}, r)$ is the optimal solution of (3.6). Use the code in order to find the best circle fit in the sense of (3.6) of the 5 points

$$\mathbf{a}_1 = \begin{pmatrix} 0 \\ 0 \end{pmatrix}, \quad \mathbf{a}_2 = \begin{pmatrix} 0.5 \\ 0 \end{pmatrix}, \quad \mathbf{a}_3 = \begin{pmatrix} 1 \\ 0 \end{pmatrix}, \quad \mathbf{a}_4 = \begin{pmatrix} 1 \\ 1 \end{pmatrix}, \quad \mathbf{a}_5 = \begin{pmatrix} 0 \\ 1 \end{pmatrix}.$$

Chapter 4
The Gradient Method

4.1 ▪ Descent Directions Methods

In this chapter we consider the unconstrained minimization problem

$$\min\{f(\mathbf{x}) : \mathbf{x} \in \mathbb{R}^n\}.$$

We assume that the objective function is continuously differentiable over $\mathbb{R}^n$. We have already seen in Chapter 2 that a first order necessary optimality condition is that the gradient vanishes at optimal points, so in principle the optimal solution of the problem can be obtained by finding among all the stationary points of f the one with the minimal function value. In Chapter 2 several examples were presented in which such an approach can lead to the detection of the unconstrained global minimum of f, but unfortunately these were exceptional examples. In the majority of problems such an approach is not implementable for the following reasons: (i) it might be a very difficult task to solve the set of (usually nonlinear) equations $\nabla f(\mathbf{x}) = \mathbf{0}$; (ii) even if it is possible to find all the stationary points, it might be that there are infinite number of stationary points and the task of finding the one corresponding to the minimal function value is an optimization problem which by itself might be as difficult as the original problem. For these reasons, instead of trying to find an analytic solution to the stationarity condition, we will consider an iterative algorithm for finding stationary points.

The iterative algorithms that we will consider in this chapter take the form

$$\mathbf{x}_{k+1} = \mathbf{x}_k + t_k \mathbf{d}_k, \quad k = 0, 1, 2, \ldots,$$

where $\mathbf{d}_k$ is the so-called *direction* and t_k is the *stepsize*. We will limit ourselves to "descent directions," whose definition is now given.

Definition 4.1 (descent direction). *Let $f : \mathbb{R}^n \to \mathbb{R}$ be a continuously differentiable function over $\mathbb{R}^n$. A vector $\mathbf{0} \neq \mathbf{d} \in \mathbb{R}^n$ is called* **a descent direction** *of f at $\mathbf{x}$ if the directional derivative $f'(\mathbf{x}; \mathbf{d})$ is negative, meaning that*

$$f'(\mathbf{x}; \mathbf{d}) = \nabla f(\mathbf{x})^T \mathbf{d} < 0.$$

The most important property of descent directions is that taking small enough steps along these directions lead to a decrease of the objective function.

Lemma 4.2 (descent property of descent directions). *Let f be a continuously differentiable function over an open set U, and let $\mathbf{x} \in U$. Suppose that $\mathbf{d}$ is a descent direction of f at $\mathbf{x}$. Then there exists $\varepsilon > 0$ such that*

$$f(\mathbf{x} + t\mathbf{d}) < f(\mathbf{x})$$

for any $t \in (0, \varepsilon]$.

Proof. Since $f'(\mathbf{x}; \mathbf{d}) < 0$, it follows from the definition of the directional derivative that

$$\lim_{t \to 0^+} \frac{f(\mathbf{x} + t\mathbf{d}) - f(\mathbf{x})}{t} = f'(\mathbf{x}; \mathbf{d}) < 0.$$

Therefore, there exists an $\varepsilon > 0$ such that

$$\frac{f(\mathbf{x} + t\mathbf{d}) - f(\mathbf{x})}{t} < 0$$

for any $t \in (0, \varepsilon]$, which readily implies the desired result. $\square$

We are now ready to write in a schematic way a general descent directions method.

Schematic Descent Directions Method

Initialization: Pick $\mathbf{x}_0 \in \mathbb{R}^n$ arbitrarily.
General step: For any $k = 0, 1, 2, \ldots$ set

(a) Pick a descent direction $\mathbf{d}_k$.

(b) Find a stepsize t_k satisfying $f(\mathbf{x}_k + t_k \mathbf{d}_k) < f(\mathbf{x}_k)$.

(c) Set $\mathbf{x}_{k+1} = \mathbf{x}_k + t_k \mathbf{d}_k$.

(d) If a stopping criterion is satisfied, then STOP and $\mathbf{x}_{k+1}$ is the output.

Of course, many details are missing in the above description of the schematic algorithm:

- What is the starting point?
- How to choose the descent direction?
- What stepsize should be taken?
- What is the stopping criteria?

Without specification of these missing details, the descent direction method remains "conceptual" and cannot be implemented. The initial starting point can be chosen arbitrarily (in the absence of an educated guess for the optimal solution). The main difference between different methods is the choice of the descent direction, and in this chapter we will elaborate on one of these choices. An example of a popular stopping criteria is $\|\nabla f(\mathbf{x}_{k+1})\| \leq \varepsilon$. We will assume that the stepsize is chosen in such a way such that $f(\mathbf{x}_{k+1}) < f(\mathbf{x}_k)$. This means that the method is assumed to be a *descent method*, that is, a method in which the function values decrease from iteration to iteration. The process of finding the stepsize t_k is called *line search*, since it is essentially a minimization

4.1. Descent Directions Methods

procedure on the one-dimensional function $g(t) = f(\mathbf{x}_k + t\mathbf{d}_k)$. There are many choices for stepsize selection rules. We describe here three popular choices:

- **constant stepsize** $t_k = \bar{t}$ for any k.
- **exact line search** t_k is a minimizer of f along the ray $\mathbf{x}_k + t\mathbf{d}_k$:
$$t_k \in \mathrm{argmin}_{t \geq 0} f(\mathbf{x}_k + t\mathbf{d}_k).$$
- **backtracking** The method requires three parameters: $s > 0, \alpha \in (0,1), \beta \in (0,1)$. The choice of t_k is done by the following procedure. First, t_k is set to be equal to the initial guess s. Then, while
$$f(\mathbf{x}_k) - f(\mathbf{x}_k + t_k \mathbf{d}_k) < -\alpha t_k \nabla f(\mathbf{x}_k)^T \mathbf{d}_k,$$
we set $t_k \leftarrow \beta t_k$. In other words, the stepsize is chosen as $t_k = s\beta^{i_k}$, where i_k is the smallest nonnegative integer for which the condition
$$f(\mathbf{x}_k) - f(\mathbf{x}_k + s\beta^{i_k}\mathbf{d}_k) \geq -\alpha s\beta^{i_k} \nabla f(\mathbf{x}_k)^T \mathbf{d}_k$$
is satisfied.

The main advantage of the constant stepsize strategy is of course its simplicity, but at this point it is unclear how to choose the constant. A large constant might cause the algorithm to be nondecreasing, and a small constant can cause slow convergence of the method. The option of exact line search seems more attractive from a first glance, but it is not always possible to actually find the exact minimizer. The third option is in a sense a compromise between the latter two approaches. It does not perform an exact line search procedure, but it does find a good enough stepsize, where the meaning of "good enough" is that it satisfies the following sufficient decrease condition:

$$f(\mathbf{x}_k) - f(\mathbf{x}_k + t_k \mathbf{d}_k) \geq -\alpha t_k \nabla f(\mathbf{x}_k)^T \mathbf{d}_k. \tag{4.1}$$

The next result shows that the sufficient decrease condition (4.1) is always satisfied for small enough t_k.

Lemma 4.3 (validity of the sufficient decrease condition). *Let f be a continuously differentiable function over $\mathbb{R}^n$, and let $\mathbf{x} \in \mathbb{R}^n$. Suppose that $0 \neq \mathbf{d} \in \mathbb{R}^n$ is a descent direction of f at $\mathbf{x}$ and let $\alpha \in (0,1)$. Then there exists $\varepsilon > 0$ such that the inequality*
$$f(\mathbf{x}) - f(\mathbf{x} + t\mathbf{d}) \geq -\alpha t \nabla f(\mathbf{x})^T \mathbf{d}$$
holds for all $t \in [0, \varepsilon]$.

Proof. Since f is continuously differentiable it follows that (see Proposition 1.23)
$$f(\mathbf{x} + t\mathbf{d}) = f(\mathbf{x}) + t\nabla f(\mathbf{x})^T \mathbf{d} + o(t\|\mathbf{d}\|),$$
and hence
$$f(\mathbf{x}) - f(\mathbf{x} + t\mathbf{d}) = -\alpha t \nabla f(\mathbf{x})^T \mathbf{d} - (1-\alpha)t\nabla f(\mathbf{x})^T \mathbf{d} - o(t\|\mathbf{d}\|). \tag{4.2}$$

Since $\mathbf{d}$ is a descent direction of f at $\mathbf{x}$ we have
$$\lim_{t \to 0^+} \frac{(1-\alpha)t\nabla f(\mathbf{x})^T \mathbf{d} + o(t\|\mathbf{d}\|)}{t} = (1-\alpha)\nabla f(\mathbf{x})^T \mathbf{d} < 0.$$

Hence, there exists $\varepsilon > 0$ such that for all $t \in (0, \varepsilon]$ the inequality
$$(1-\alpha)t\nabla f(\mathbf{x})^T \mathbf{d} + o(t\|\mathbf{d}\|) < 0$$
holds, which combined with (4.2) implies the desired result. □

We will now show that for a quadratic function, the exact line stepsize can be easily computed.

Example 4.4 (exact line search for quadratic functions). Let $f(\mathbf{x}) = \mathbf{x}^T \mathbf{A}\mathbf{x} + 2\mathbf{b}^T \mathbf{x} + c$, where $\mathbf{A}$ is an $n \times n$ positive definite matrix, $\mathbf{b} \in \mathbb{R}^n$, and $c \in \mathbb{R}$. Let $\mathbf{x} \in \mathbb{R}^n$ and let $\mathbf{d} \in \mathbb{R}^n$ be a descent direction of f at $\mathbf{x}$. We will find an explicit formula for the stepsize generated by exact line search, that is, an expression for the solution of
$$\min_{t \geq 0} f(\mathbf{x} + t\mathbf{d}).$$
We have
$$g(t) = f(\mathbf{x} + t\mathbf{d}) = (\mathbf{x} + t\mathbf{d})^T \mathbf{A}(\mathbf{x} + t\mathbf{d}) + 2\mathbf{b}^T(\mathbf{x} + t\mathbf{d}) + c$$
$$= (\mathbf{d}^T \mathbf{A}\mathbf{d})t^2 + 2(\mathbf{d}^T \mathbf{A}\mathbf{x} + \mathbf{d}^T \mathbf{b})t + \mathbf{x}^T \mathbf{A}\mathbf{x} + 2\mathbf{b}^T \mathbf{x} + c$$
$$= (\mathbf{d}^T \mathbf{A}\mathbf{d})t^2 + 2(\mathbf{d}^T \mathbf{A}\mathbf{x} + \mathbf{d}^T \mathbf{b})t + f(\mathbf{x}).$$
Since $g'(t) = 2(\mathbf{d}^T \mathbf{A}\mathbf{d})t + 2\mathbf{d}^T(\mathbf{A}\mathbf{x} + \mathbf{b})$ and $\nabla f(\mathbf{x}) = 2(\mathbf{A}\mathbf{x} + \mathbf{b})$, it follows that $g'(t) = 0$ if and only if
$$t = \bar{t} \equiv -\frac{\mathbf{d}^T \nabla f(\mathbf{x})}{2\mathbf{d}^T \mathbf{A}\mathbf{d}}.$$
Since $\mathbf{d}$ is a descent direction of f at $\mathbf{x}$, it follows that $\mathbf{d}^T \nabla f(\mathbf{x}) < 0$ and hence $\bar{t} > 0$, which implies that the stepsize dictated by the exact line search rule is
$$\bar{t} = -\frac{\mathbf{d}^T \nabla f(\mathbf{x})}{2\mathbf{d}^T \mathbf{A}\mathbf{d}}. \tag{4.3}$$
Note that we have implicitly used in the above analysis the fact that the second order derivative of g is always positive. ∎

4.2 • The Gradient Method

In the gradient method the descent direction is chosen to be the minus of the gradient at the current point: $\mathbf{d}_k = -\nabla f(\mathbf{x}_k)$. The fact that this is a descent direction whenever $\nabla f(\mathbf{x}_k) \neq 0$ can be easily shown—just note that
$$f'(\mathbf{x}_k; -\nabla f(\mathbf{x}_k)) = -\nabla f(\mathbf{x}_k)^T \nabla f(\mathbf{x}_k) = -\|\nabla f(\mathbf{x}_k)\|^2 < 0.$$
In addition for being a descent direction, minus the gradient is also the steepest direction method. This means that the normalized direction $-\nabla f(\mathbf{x}_k)/\|\nabla f(\mathbf{x}_k)\|$ corresponds to the minimal directional derivative among all normalized directions. A formal statement and proof is given in the following result.

Lemma 4.5. *Let f be a continuously differentiable function, and let $\mathbf{x} \in \mathbb{R}^n$ be a nonstationary point ($\nabla f(\mathbf{x}) \neq 0$). Then an optimal solution of*
$$\min_{\mathbf{d} \in \mathbb{R}^n} \{f'(\mathbf{x}; \mathbf{d}) : \|\mathbf{d}\| = 1\} \tag{4.4}$$
is $\mathbf{d} = -\frac{\nabla f(\mathbf{x})}{\|\nabla f(\mathbf{x})\|}$.

4.2. The Gradient Method

Proof. Since $f'(\mathbf{x};\mathbf{d}) = \nabla f(\mathbf{x})^T \mathbf{d}$, problem (4.4) is the same as

$$\min_{\mathbf{d} \in \mathbb{R}^n} \{\nabla f(\mathbf{x})^T \mathbf{d} : \|\mathbf{d}\| = 1\}.$$

By the Cauchy–Schwarz inequality we have

$$\nabla f(\mathbf{x})^T \mathbf{d} \geq -\|\nabla f(\mathbf{x})\| \cdot \|\mathbf{d}\| = -\|\nabla f(\mathbf{x})\|.$$

Thus, $-\|\nabla f(\mathbf{x})\|$ is a lower bound on the optimal value of (4.4). On the other hand, plugging $\mathbf{d} = -\frac{\nabla f(\mathbf{x})}{\|\nabla f(\mathbf{x})\|}$ in the objective function of (4.4) we obtain that

$$f'\left(\mathbf{x}, -\frac{\nabla f(\mathbf{x})}{\|\nabla f(\mathbf{x})\|}\right) = -\nabla f(\mathbf{x})^T \left(\frac{\nabla f(\mathbf{x})}{\|\nabla f(\mathbf{x})\|}\right) = -\|\nabla f(\mathbf{x})\|,$$

and we thus come to the conclusion that the lower bound $-\|\nabla f(\mathbf{x})\|$ is attained at $\mathbf{d} = -\frac{\nabla f(\mathbf{x})}{\|\nabla f(\mathbf{x})\|}$, which readily implies that this is an optimal solution of (4.4). □

We will now present the gradient method with the standard stopping criteria, which is the condition $\|\nabla f(\mathbf{x}_{k+1})\| \leq \varepsilon$.

The Gradient Method

Input: $\varepsilon > 0$ - tolerance parameter.

Initialization: Pick $\mathbf{x}_0 \in \mathbb{R}^n$ arbitrarily.
General step: For any $k = 0, 1, 2, \ldots$ execute the following steps:

(a) Pick a stepsize t_k by a line search procedure on the function

$$g(t) = f(\mathbf{x}_k - t \nabla f(\mathbf{x}_k)).$$

(b) Set $\mathbf{x}_{k+1} = \mathbf{x}_k - t_k \nabla f(\mathbf{x}_k)$.

(c) If $\|\nabla f(\mathbf{x}_{k+1})\| \leq \varepsilon$, then STOP and $\mathbf{x}_{k+1}$ is the output.

Example 4.6 (exact line search). Consider the two-dimensional minimization problem

$$\min_{x,y} x^2 + 2y^2 \tag{4.5}$$

whose optimal solution is $(x,y) = (0,0)$ with corresponding optimal value 0. To invoke the gradient method with stepsize chosen by exact line search, we constructed a MATLAB function, called `gradient_method_quadratic`, that finds (up to a tolerance) the optimal solution of a quadratic problem of the form

$$\min_{\mathbf{x} \in \mathbb{R}^n} \{\mathbf{x}^T \mathbf{A} \mathbf{x} + 2\mathbf{b}^T \mathbf{x}\},$$

where $\mathbf{A} \in \mathbb{R}^{n \times n}$ positive definite and $\mathbf{b} \in \mathbb{R}^n$. The function invokes the gradient method with exact line search, which by Example 4.4 corresponds to the stepsize $t_k = \frac{\|\nabla f(\mathbf{x}_k)\|^2}{2\nabla f(\mathbf{x}_k)^T \mathbf{A} \nabla f(\mathbf{x}_k)}$. The MATLAB function is described below.

```
function [x,fun_val]=gradient_method_quadratic(A,b,x0,epsilon)
% INPUT
% =======================
% A ....... the positive definite matrix associated with the
%           objective function
% b ....... a column vector associated with the linear part of the
%           objective function
% x0 ...... starting point of the method
% epsilon . tolerance parameter
% OUTPUT
% =======================
% x ....... an optimal solution (up to a tolerance) of
%           min(x^T A x+2 b^T x)
% fun_val . the optimal function value up to a tolerance


x=x0;
iter=0;
grad=2*(A*x+b);
while (norm(grad)>epsilon)
    iter=iter+1;
    t=norm(grad)^2/(2*grad'*A*grad);
    x=x-t*grad;
    grad=2*(A*x+b);
    fun_val=x'*A*x+2*b'*x;
    fprintf('iter_number = %3d norm_grad = %2.6f fun_val = %2.6f\n',...
    iter,norm(grad),fun_val);
end
```

In order to solve problem (4.5) using the gradient method with exact line search, tolerance parameter $\varepsilon = 10^{-5}$, and initial vector $\mathbf{x}_0 = (2,1)^T$, the following MATLAB command was executed:

```
[x,fun_val]=gradient_method_quadratic([1,0;0,2],[0;0],[2;1],1e-5)
```

The output is

```
iter_number =    1 norm_grad = 1.885618 fun_val = 0.666667
iter_number =    2 norm_grad = 0.628539 fun_val = 0.074074
iter_number =    3 norm_grad = 0.209513 fun_val = 0.008230
iter_number =    4 norm_grad = 0.069838 fun_val = 0.000914
iter_number =    5 norm_grad = 0.023279 fun_val = 0.000102
iter_number =    6 norm_grad = 0.007760 fun_val = 0.000011
iter_number =    7 norm_grad = 0.002587 fun_val = 0.000001
iter_number =    8 norm_grad = 0.000862 fun_val = 0.000000
iter_number =    9 norm_grad = 0.000287 fun_val = 0.000000
iter_number =   10 norm_grad = 0.000096 fun_val = 0.000000
iter_number =   11 norm_grad = 0.000032 fun_val = 0.000000
iter_number =   12 norm_grad = 0.000011 fun_val = 0.000000
iter_number =   13 norm_grad = 0.000004 fun_val = 0.000000
```

The method therefore stopped after 13 iterations with a solution which is pretty close to the optimal value:

```
x =
 1.0e-005 *
   0.1254
  -0.0627
```

4.2. The Gradient Method

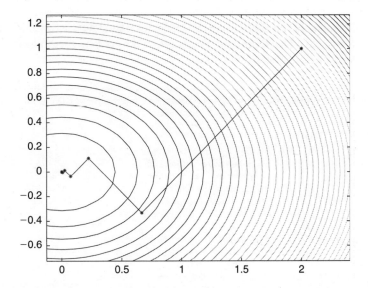

Figure 4.1. *The iterates of the gradient method along with the contour lines of the objective function.*

It is also very informative to visually look at the progress of the iterates. The iterates and the contour plots of the objective function are given in Figure 4.1. ∎

An evident behavior of the gradient method as illustrated in Figure 4.1 is the "zig-zag" effect, meaning that the direction found at the kth iteration $\mathbf{x}_{k+1} - \mathbf{x}_k$ is orthogonal to the direction found at the $(k+1)$th iteration $\mathbf{x}_{k+2} - \mathbf{x}_{k+1}$. This is a general property whose proof will be given now.

Lemma 4.7. *Let $\{\mathbf{x}_k\}_{k \geq 0}$ be the sequence generated by the gradient method with exact line search for solving a problem of minimizing a continuously differentiable function f. Then for any $k = 0, 1, 2, \ldots$*

$$(\mathbf{x}_{k+2} - \mathbf{x}_{k+1})^T (\mathbf{x}_{k+1} - \mathbf{x}_k) = 0.$$

Proof. By the definition of the gradient method we have that $\mathbf{x}_{k+1} - \mathbf{x}_k = -t_k \nabla f(\mathbf{x}_k)$ and $\mathbf{x}_{k+2} - \mathbf{x}_{k+1} = -t_{k+1} \nabla f(\mathbf{x}_{k+1})$. Therefore, we wish to prove that $\nabla f(\mathbf{x}_k)^T \nabla f(\mathbf{x}_{k+1}) = 0$. Since

$$t_k \in \operatorname{argmin}_{t \geq 0} \{g(t) \equiv f(\mathbf{x}_k - t \nabla f(\mathbf{x}_k))\},$$

and the optimal solution is not $t_k = 0$, it follows that $g'(t_k) = 0$. Hence,

$$-\nabla f(\mathbf{x}_k)^T \nabla f(\mathbf{x}_k - t_k \nabla f(\mathbf{x}_k)) = 0,$$

meaning that the desired result $\nabla f(\mathbf{x}_k)^T \nabla f(\mathbf{x}_{k+1}) = 0$ holds. □

Let us now consider an example with a constant stepsize.

Example 4.8 (constant stepsize). Consider the same optimization problem given in Example 4.6

$$\min_{x,y} x^2 + 2y^2.$$

To solve this problem using the gradient method with a constant stepsize, we use the following MATLAB function that employs the gradient method with a constant stepsize for an arbitrary objective function.

```
function [x,fun_val]=gradient_method_constant(f,g,x0,t,epsilon)
% Gradient method with constant stepsize
%
% INPUT
%=======================================
% f ......... objective function
% g ........ gradient of the objective function
% x0......... initial point
% t ......... constant stepsize
% epsilon ... tolerance parameter
% OUTPUT
%=======================================
% x ........ optimal solution (up to a tolerance)
%            of min f(x)
% fun_val ... optimal function value
x=x0;
grad=g(x);
iter=0;
while (norm(grad)>epsilon)
    iter=iter+1;
    x=x-t*grad;
    fun_val=f(x);
    grad=g(x);
    fprintf('iter_number = %3d norm_grad = %2.6f fun_val = %2.6f \n',...
    iter,norm(grad),fun_val);
end
```

We can employ the gradient method with constant stepsize $t_k = 0.1$ and initial vector $\mathbf{x}_0 = (2,1)^T$ by executing the MATLAB commands

```
A=[1,0;0,2];
[x,fun_val]=gradient_method_constant(@(x)x'*A*x,@(x)2*A*x,[2;1],0.1,1e-5);
```

and the long output is

```
iter_number =   1 norm_grad = 4.000000 fun_val = 3.280000
iter_number =   2 norm_grad = 2.937210 fun_val = 1.897600
iter_number =   3 norm_grad = 2.222791 fun_val = 1.141888
       :              :                       :
iter_number =  56 norm_grad = 0.000015 fun_val = 0.000000
iter_number =  57 norm_grad = 0.000012 fun_val = 0.000000
iter_number =  58 norm_grad = 0.000010 fun_val = 0.000000
```

The excessive number of iterations is due to the fact that the stepsize was chosen to be too small. However, taking a stepsize which is large might lead to divergence of the iterates. For example, taking the constant stepsize to be 100 results in a divergent sequence.

```
>> A=[1,0;0,2];
>> [x,fun_val]=gradient_method_constant(@(x)x'*A*x,@(x)2*A*x,[2;1],100,1e-5);
iter_number =   1 norm_grad = 1783.488716 fun_val = 476806.000000
iter_number =   2 norm_grad = 656209.693339 fun_val = 56962873606.000000
iter_number =   3 norm_grad = 256032703.004797 fun_val = 8318300807190406.0
       :              :                       :
iter_number = 119 norm_grad = NaN fun_val = NaN
```

4.2. The Gradient Method

The important question is therefore how to choose the constant stepsize so that it will not be too large (to ensure convergence) and not be too small (to ensure that the convergence will not be too slow). We will consider again the theoretical issue of choosing the constant stepsize in Section 4.7. ∎

Let us now consider an example with a backtracking stepsize selection rule.

Example 4.9 (stepsize selection by backtracking). The following MATLAB function implements the gradient method with a backtracking stepsize selection rule.

```
function [x,fun_val]=gradient_method_backtracking(f,g,x0,s,alpha,...
   beta,epsilon)
% Gradient method with backtracking stepsize rule
%
% INPUT
%=========================================
% f ......... objective function
% g ......... gradient of the objective function
% x0......... initial point
% s ......... initial choice of stepsize
% alpha ..... tolerance parameter for the stepsize selection
% beta ...... the constant in which the stepsize is multiplied
%             at each backtracking step (0<beta<1)
% epsilon ... tolerance parameter for stopping rule
% OUTPUT
%=========================================
% x ........ optimal solution (up to a tolerance)
%            of min f(x)
% fun_val ... optimal function value
x=x0;
grad=g(x);
fun_val=f(x);
iter=0;
while (norm(grad)>epsilon)
    iter=iter+1;
    t=s;
    while (fun_val-f(x-t*grad)<alpha*t*norm(grad)^2)
        t=beta*t;
    end
    x=x-t*grad;
    fun_val=f(x);
    grad=g(x);
    fprintf('iter_number = %3d norm_grad = %2.6f fun_val = %2.6f \n',...
    iter,norm(grad),fun_val);
end
```

As in the previous examples, we will consider the problem of minimizing the function x^2+2y^2. Employing the gradient method with backtracking stepsize selection rule, a starting vector $\mathbf{x}_0=(2,1)^T$ and parameters $\varepsilon=10^{-5}, s=2, \alpha=\frac{1}{4}, \beta=\frac{1}{2}$ results in the following output:

```
>> A=[1,0;0,2];
>> [x,fun_val]=gradient_method_backtracking(@(x)x'*A*x,@(x)2*A*x,...
[2;1],2,0.25,0.5,1e-5);
iter_number =   1 norm_grad = 2.000000 fun_val = 1.000000
iter_number =   2 norm_grad = 0.000000 fun_val = 0.000000
```

That is, the gradient method with backtracking terminated after only 2 iterations. In fact, in this case (and this is probably a result of pure luck), it converged to the *exact* optimal solution. In this example there was no advantage in performing an exact line search procedure, and in fact better results were obtained by the nonexact/backtracking line search. Computational experience teaches us that in practice backtracking does not have real disadvantages in comparison to exact line search.

The gradient method can behave quite badly. As an example, consider the minimization problem

$$\min_{x,y} x^2 + \frac{1}{100}y^2,$$

and suppose that we employ the backtracking gradient method with initial vector $(\frac{1}{100}, 1)^T$:

```
>> A=[1,0;0,0.01];
>> [x,fun_val]=gradient_method_backtracking(@(x)x'*A*x,@(x)2*A*x,...
[0.01;1],2,0.25,0.5,1e-5);
iter_number =    1 norm_grad = 0.028003 fun_val = 0.009704
iter_number =    2 norm_grad = 0.027730 fun_val = 0.009324
iter_number =    3 norm_grad = 0.027465 fun_val = 0.008958
        :              :                      :
iter_number =  201 norm_grad = 0.000010 fun_val = 0.000000
```

∎

Clearly, 201 iterations is a large number of steps in order to obtain convergence for a two-dimensional problem. The main question that arises is whether we can find a quantity that can predict in some sense the number of iterations required by the gradient method for convergence; this measure should quantify in some sense the "hardness" of the problem. This is an important issue that actually does not have a full answer, but a partial answer can be found in the notion of *condition number*.

4.3 ▪ The Condition Number

Consider the quadratic minimization problem

$$\min_{\mathbf{x} \in \mathbb{R}^n} \{f(\mathbf{x}) \equiv \mathbf{x}^T \mathbf{A} \mathbf{x}\}, \qquad (4.6)$$

where $\mathbf{A} \succ \mathbf{0}$. The optimal solution is obviously $\mathbf{x}^* = \mathbf{0}$. The gradient method with exact line search takes the form

$$\mathbf{x}_{k+1} = \mathbf{x}_k - t_k \mathbf{d}_k,$$

where $\mathbf{d}_k = 2\mathbf{A}\mathbf{x}_k$ is the gradient of f at $\mathbf{x}_k$ and the stepsize t_k chosen by the exact minimization rule is (see formula (4.3))

$$t_k = \frac{\mathbf{d}_k^T \mathbf{d}_k}{2\mathbf{d}_k^T \mathbf{A} \mathbf{d}_k}. \qquad (4.7)$$

It holds that

$$\begin{aligned}
f(\mathbf{x}_{k+1}) &= \mathbf{x}_{k+1}^T \mathbf{A} \mathbf{x}_{k+1} \\
&= (\mathbf{x}_k - t_k \mathbf{d}_k)^T \mathbf{A} (\mathbf{x}_k - t_k \mathbf{d}_k) \\
&= \mathbf{x}_k^T \mathbf{A} \mathbf{x}_k - 2t_k \mathbf{d}_k^T \mathbf{A} \mathbf{x}_k + t_k^2 \mathbf{d}_k^T \mathbf{A} \mathbf{d}_k \\
&= \mathbf{x}_k^T \mathbf{A} \mathbf{x}_k - t_k \mathbf{d}_k^T \mathbf{d}_k + t_k^2 \mathbf{d}_k^T \mathbf{A} \mathbf{d}_k.
\end{aligned}$$

4.3. The Condition Number

Plugging in the expression for t_k given in (4.7) into the last equation we obtain that

$$f(\mathbf{x}_{k+1}) = \mathbf{x}_k^T \mathbf{A} \mathbf{x}_k - \frac{1}{4}\frac{(\mathbf{d}_k^T \mathbf{d}_k)^2}{\mathbf{d}_k^T \mathbf{A} \mathbf{d}_k} = \mathbf{x}_k^T \mathbf{A} \mathbf{x}_k \left(1 - \frac{1}{4}\frac{(\mathbf{d}_k^T \mathbf{d}_k)^2}{(\mathbf{d}_k^T \mathbf{A} \mathbf{d}_k)(\mathbf{x}_k^T \mathbf{A} \mathbf{A}^{-1} \mathbf{A} \mathbf{x}_k)}\right)$$

$$= \left(1 - \frac{(\mathbf{d}_k^T \mathbf{d}_k)^2}{(\mathbf{d}_k^T \mathbf{A} \mathbf{d}_k)(\mathbf{d}_k^T \mathbf{A}^{-1} \mathbf{d}_k)}\right) f(\mathbf{x}_k). \tag{4.8}$$

We will now use the following well-known result, also known as the *Kantorovich inequality*.

Lemma 4.10 (Kantorovich inequality). *Let $\mathbf{A}$ be a positive definite $n \times n$ matrix. Then for any $0 \neq \mathbf{x} \in \mathbb{R}^n$ the inequality*

$$\frac{(\mathbf{x}^T \mathbf{x})^2}{(\mathbf{x}^T \mathbf{A} \mathbf{x})(\mathbf{x}^T \mathbf{A}^{-1} \mathbf{x})} \geq \frac{4\lambda_{\max}(\mathbf{A})\lambda_{\min}(\mathbf{A})}{(\lambda_{\max}(\mathbf{A}) + \lambda_{\min}(\mathbf{A}))^2} \tag{4.9}$$

holds.

Proof. Denote $m = \lambda_{\min}(\mathbf{A})$ and $M = \lambda_{\max}(\mathbf{A})$. The eigenvalues of the matrix $\mathbf{A} + Mm\mathbf{A}^{-1}$ are $\lambda_i(\mathbf{A}) + \frac{Mm}{\lambda_i(\mathbf{A})}, i = 1, \ldots, n$. It is easy to show that the maximum of the one-dimensional function $\varphi(t) = t + \frac{Mm}{t}$ over $[m, M]$ is attained at the endpoints m and M with a corresponding value of $M + m$, and therefore, since $m \leq \lambda_i(\mathbf{A}) \leq M$, it follows that the eigenvalues of $\mathbf{A} + Mm\mathbf{A}^{-1}$ are smaller than $(M + m)$. Thus,

$$\mathbf{A} + Mm\mathbf{A}^{-1} \preceq (M + m)\mathbf{I}.$$

Consequently, for any $\mathbf{x} \in \mathbb{R}^n$,

$$\mathbf{x}^T \mathbf{A} \mathbf{x} + Mm(\mathbf{x}^T \mathbf{A}^{-1} \mathbf{x}) \leq (M + m)(\mathbf{x}^T \mathbf{x}),$$

which combined with the simple inequality $\alpha\beta \leq \frac{1}{4}(\alpha + \beta)^2$ (for any two real numbers α, β) yields

$$(\mathbf{x}^T \mathbf{A} \mathbf{x})[Mm(\mathbf{x}^T \mathbf{A}^{-1} \mathbf{x})] \leq \frac{1}{4}\left[(\mathbf{x}^T \mathbf{A} \mathbf{x}) + Mm(\mathbf{x}^T \mathbf{A}^{-1} \mathbf{x})\right]^2 \leq \frac{(M + m)^2}{4}(\mathbf{x}^T \mathbf{x})^2,$$

which after some simple rearrangement of terms establishes the desired result. $\square$

Coming back to the convergence rate analysis of the gradient method on problem (4.6), it follows by using the Kantorovich inequality that (4.8) yields

$$f(\mathbf{x}_{k+1}) \leq \left(1 - \frac{4Mm}{(M+m)^2}\right) f(\mathbf{x}_k) = \left(\frac{M-m}{M+m}\right)^2 f(\mathbf{x}_k),$$

where $M = \lambda_{\max}(\mathbf{A}), m = \lambda_{\min}(\mathbf{A})$. We summarize the above discussion in the following lemma.

Lemma 4.11. *Let $\{\mathbf{x}_k\}_{k \geq 0}$ be the sequence generated by the gradient descent method with exact line search for solving problem (4.6). Then for any $k = 0, 1, \ldots$*

$$f(\mathbf{x}_{k+1}) \leq \left(\frac{M-m}{M+m}\right)^2 f(\mathbf{x}_k), \tag{4.10}$$

where $M = \lambda_{\max}(\mathbf{A}), m = \lambda_{\min}(\mathbf{A})$.

Inequality (4.10) implies
$$f(\mathbf{x}_k) \leq c^k f(\mathbf{x}_0),$$
where $c = (\frac{M-m}{M+m})^2$. That is, the sequence of function values is bounded above by a decreasing geometric sequence. In this case we say that the sequence of function values converges at a *linear rate* to the optimal value. The speed of the convergence depends on c; as c gets larger, the convergence speed becomes slower. The quantity c can also be written as
$$c = \left(\frac{x-1}{x+1}\right)^2,$$
where $x = \frac{M}{m} = \frac{\lambda_{\max}(\mathbf{A})}{\lambda_{\min}(\mathbf{A})}$. Since c is an increasing function of x, it follows that the behavior of the gradient method depends on the ratio between the maximal and minimal eigenvalues of $\mathbf{A}$; this number is called the *condition number*. Although the condition number can be defined for general matrices, we will restrict ourselves to positive definite matrices.

Definition 4.12 (condition number). *Let $\mathbf{A}$ be an $n \times n$ positive definite matrix. Then the* **condition number** *of $\mathbf{A}$ is defined by*
$$x(\mathbf{A}) = \frac{\lambda_{\max}(\mathbf{A})}{\lambda_{\min}(\mathbf{A})}.$$

We have already found one illustration in Example 4.9 that the gradient method applied to problems with large condition number might require a large number of iterations and vice versa, the gradient method employed on problems with small condition number is likely to converge within a small number of steps. Indeed, the condition number of the matrix associated with the function $x^2 + 0.01y^2$ is $\frac{1}{0.01} = 100$, which is relatively large, is the cause for the 201 iterations that were required for convergence, and the small condition number of the matrix associated with the function $x^2 + 2y^2$ ($x = 2$) is the reason for the small number of required steps. Matrices with large condition number are called *ill-conditioned*, and matrices with small condition number are called *well-conditioned*. Of course, the entire discussion until now was on the restrictive class of quadratic objective functions, where the Hessian matrix is constant, but the notion of condition number also appears in the context of nonquadratic objective functions. In that case, it is well known that the rate of convergence of $\mathbf{x}_k$ to a given stationary point $\mathbf{x}^*$ depends on the condition number of $x(\nabla^2 f(\mathbf{x}^*))$. We will not focus on these theoretical results, but will illustrate it on a well-known ill-conditioned problem.

Example 4.13 (Rosenbrock function). The *Rosenbrock function* is the following function:
$$f(x_1, x_2) = 100(x_2 - x_1^2)^2 + (1 - x_1)^2.$$
The optimal solution is obviously $(x_1, x_2) = (1, 1)$ with corresponding optimal value 0. The Rosenbrock function is extremely ill-conditioned at the optimal solution. Indeed,
$$\nabla f(\mathbf{x}) = \begin{pmatrix} -400x_1(x_2 - x_1^2) - 2(1 - x_1) \\ 200(x_2 - x_1^2) \end{pmatrix},$$
$$\nabla^2 f(\mathbf{x}) = \begin{pmatrix} -400x_2 + 1200x_1^2 + 2 & -400x_1 \\ -400x_1 & 200 \end{pmatrix}.$$

It is not difficult to show that $(x_1, x_2) = (1, 1)$ is the unique stationary point. In addition,
$$\nabla^2 f(1,1) = \begin{pmatrix} 802 & -400 \\ -400 & 200 \end{pmatrix}$$

4.3. The Condition Number

and hence the condition number is

```
>> A=[802,-400;-400,200];
>> cond(A)
ans =
    2.5080e+003
```

A condition number of more than 2500 should have severe effects on the convergence speed of the gradient method. Let us then employ the gradient method with backtracking on the Rosenbrock function with starting vector $\mathbf{x}_0 = (2,5)^T$:

```
>> f=@(x)100*(x(2)-x(1)^2)^2+(1-x(1))^2;
>> g=@(x)[-400*(x(2)-x(1)^2)*x(1)-2*(1-x(1));200*(x(2)-x(1)^2)];
>> [x,fun_val]=gradient_method_backtracking(f,g,[2;5],2,0.25,0.5,1e-5);
iter_number =    1 norm_grad = 118.254478 fun_val = 3.221022
iter_number =    2 norm_grad = 0.723051 fun_val = 1.496586
    :            :                :
iter_number = 6889 norm_grad = 0.000019 fun_val = 0.000000
iter_number = 6890 norm_grad = 0.000009 fun_val = 0.000000
```

This run required the huge amount of 6890 iterations, so the ill-conditioning effect has a significant impact. To better understand the nature of this pessimistic run, consider the contour plots of the Rosenbrock function along with the iterates as illustrated in Figure 4.2. Note that the function has banana-shaped contour lines surrounding the unique stationary point $(1,1)$. ∎

Sensitivity of Solutions to Linear Systems

The condition number has an important role in the study of the sensitivity of linear systems of equations to perturbation in the right-hand-side vector. Specifically, suppose that

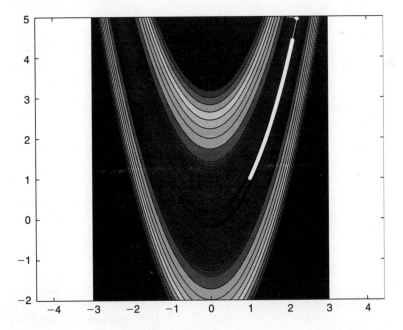

Figure 4.2. *Contour lines of the Rosenbrock function along with thousands of iterations of the gradient method.*

we are given a linear system $\mathbf{Ax} = \mathbf{b}$, and for the sake of simplicity, let us assume that $\mathbf{A}$ is positive definite. The solution of the linear system is of course $\mathbf{x} = \mathbf{A}^{-1}\mathbf{b}$. Now, suppose that instead of $\mathbf{b}$ in the right-hand side, we consider a perturbation $\mathbf{b} + \Delta\mathbf{b}$. Let us denote the solution of the new system by $\mathbf{x} + \Delta\mathbf{x}$, that is, $\mathbf{A}(\mathbf{x} + \Delta\mathbf{x}) = \mathbf{b} + \Delta\mathbf{b}$. We have

$$\mathbf{x} + \Delta\mathbf{x} = \mathbf{A}^{-1}(\mathbf{b} + \Delta\mathbf{b}) = \mathbf{x} + \mathbf{A}^{-1}\Delta\mathbf{b},$$

so that the change in the solution is $\Delta\mathbf{x} = \mathbf{A}^{-1}\Delta\mathbf{b}$. Our purpose is to find a bound on the relative error $\frac{\|\Delta\mathbf{x}\|}{\|\mathbf{x}\|}$ in terms of the relative error of the right-hand-side perturbation $\frac{\|\Delta\mathbf{b}\|}{\|\mathbf{b}\|}$:

$$\frac{\|\Delta\mathbf{x}\|}{\|\mathbf{x}\|} = \frac{\|\mathbf{A}^{-1}\Delta\mathbf{b}\|}{\|\mathbf{x}\|} \leq \frac{\|\mathbf{A}^{-1}\| \cdot \|\Delta\mathbf{b}\|}{\|\mathbf{x}\|} = \frac{\lambda_{\max}(\mathbf{A}^{-1})\|\Delta\mathbf{b}\|}{\|\mathbf{x}\|},$$

where the last equality follows from the fact that the spectral norm of a positive definite matrix $\mathbf{D}$ is $\|\mathbf{D}\| = \lambda_{\max}(\mathbf{D})$. By the positive definiteness of $\mathbf{A}$, it follows that $\lambda_{\max}(\mathbf{A}^{-1}) = \frac{1}{\lambda_{\min}(\mathbf{A})}$, and we can therefore continue the chain of equalities and inequalities:

$$\begin{aligned}
\frac{\|\Delta\mathbf{x}\|}{\|\mathbf{x}\|} &\leq \frac{1}{\lambda_{\min}(\mathbf{A})} \frac{\|\Delta\mathbf{b}\|}{\|\mathbf{x}\|} = \frac{1}{\lambda_{\min}(\mathbf{A})} \frac{\|\Delta\mathbf{b}\|}{\|\mathbf{A}^{-1}\mathbf{b}\|} \\
&\leq \frac{1}{\lambda_{\min}(\mathbf{A})} \cdot \frac{\|\Delta\mathbf{b}\|}{\lambda_{\min}(\mathbf{A}^{-1})\|\mathbf{b}\|} \\
&= \frac{\lambda_{\max}(\mathbf{A})}{\lambda_{\min}(\mathbf{A})} \frac{\|\Delta\mathbf{b}\|}{\|\mathbf{b}\|} \\
&= \varkappa(\mathbf{A}) \frac{\|\Delta\mathbf{b}\|}{\|\mathbf{b}\|},
\end{aligned} \quad (4.11)$$

where inequality (4.11) follows from the fact that for a positive definite matrix $\mathbf{A}$, the inequality $\|\mathbf{A}^{-1}\mathbf{b}\| \geq \lambda_{\min}(\mathbf{A}^{-1})\|\mathbf{b}\|$ holds. Indeed,

$$\|\mathbf{A}^{-1}\mathbf{b}\| = \sqrt{\mathbf{b}^T \mathbf{A}^{-2} \mathbf{b}} \geq \sqrt{\lambda_{\min}(\mathbf{A}^{-2})\|\mathbf{b}\|^2} = \lambda_{\min}(\mathbf{A}^{-1})\|\mathbf{b}\|.$$

We can therefore deduce that the sensitivity of the solution of the linear system to right-hand-side perturbations depends on the condition number of the coefficients matrix.

Example 4.14. As an example, consider the matrix

$$\mathbf{A} = \begin{pmatrix} 1 + 10^{-5} & 1 \\ 1 & 1 + 10^{-5} \end{pmatrix},$$

whose condition number is larger than 200000:

```
>> format long
>> A=[1+1e-5,1;1,1+1e-5];
>> cond(A)
ans =
    2.000009999998795e+005
```

The solution of the system

$$\mathbf{Ax} = \begin{pmatrix} 1 \\ 1 \end{pmatrix}$$

is

```
>> A\[1;1]
ans =
   0.499997500018278
   0.499997500006722
```

That is, approximately $(0.5, 0.5)^T$. However, if we make a small change to the right-hand-side vector: $(1.1, 1)^T$ instead of $(1,1)^T$ (relative error of $\frac{\|(0.1,0)^T\|}{\|(1,1)^T\|} = 0.0707$), then the perturbed solution is very much different from $(0.5, 0.5)^T$:

```
>> A\[1.1;1]
ans =
   1.0e+003 *
   5.000524997400047
  -4.999475002650021
```

∎

4.4 • Diagonal Scaling

The problem of ill-conditioned problems is a major one, and many methods have been developed in order to circumvent it. One of the most popular approaches is to "condition" the problem by making an appropriate linear transformation of the decision variables. More precisely, consider the unconstrained minimization problem

$$\min\{f(\mathbf{x}) : \mathbf{x} \in \mathbb{R}^n\}.$$

For a given nonsingular matrix $\mathbf{S} \in \mathbb{R}^{n \times n}$, we make the linear transformation $\mathbf{x} = \mathbf{S}\mathbf{y}$ and obtain the equivalent problem

$$\min\{g(\mathbf{y}) \equiv f(\mathbf{S}\mathbf{y}) : \mathbf{y} \in \mathbb{R}^n\}.$$

Since $\nabla g(\mathbf{y}) = \mathbf{S}^T \nabla f(\mathbf{S}\mathbf{y}) = \mathbf{S}^T \nabla f(\mathbf{x})$, it follows that the gradient method applied to the transformed problem takes the form

$$\mathbf{y}_{k+1} = \mathbf{y}_k - t_k \mathbf{S}^T \nabla f(\mathbf{S}\mathbf{y}_k).$$

Multiplying the last equality by $\mathbf{S}$ from the left, and using the notation $\mathbf{x}_k = \mathbf{S}\mathbf{y}_k$, we obtain the recursive formula

$$\mathbf{x}_{k+1} = \mathbf{x}_k - t_k \mathbf{S}\mathbf{S}^T \nabla f(\mathbf{x}_k).$$

Defining $\mathbf{D} = \mathbf{S}\mathbf{S}^T$, we obtain the following version of the gradient method, which we call *the scaled gradient method* with scaling matrix $\mathbf{D}$:

$$\mathbf{x}_{k+1} = \mathbf{x}_k - t_k \mathbf{D} \nabla f(\mathbf{x}_k).$$

By its definition, the matrix $\mathbf{D}$ is positive definite. The direction $-\mathbf{D}\nabla f(\mathbf{x}_k)$ is a descent direction of f at $\mathbf{x}_k$ when $\nabla f(\mathbf{x}_k) \neq 0$ since

$$f'(\mathbf{x}_k; -\mathbf{D}\nabla f(\mathbf{x}_k)) = -\nabla f(\mathbf{x}_k)^T \mathbf{D} \nabla f(\mathbf{x}_k) < 0,$$

where the latter strict inequality follows from the positive definiteness of the matrix $\mathbf{D}$. To summarize the above discussion, we have shown that the scaled gradient method with scaling matrix $\mathbf{D}$ is equivalent to the gradient method employed on the function $g(\mathbf{y}) = f(\mathbf{D}^{1/2}\mathbf{y})$. We note that the gradient and Hessian of g are given by

$$\nabla g(\mathbf{y}) = \mathbf{D}^{1/2}\nabla f(\mathbf{D}^{1/2}\mathbf{y}) = \mathbf{D}^{1/2}\nabla f(\mathbf{x}),$$
$$\nabla^2 g(\mathbf{y}) = \mathbf{D}^{1/2}\nabla^2 f(\mathbf{D}^{1/2}\mathbf{y})\mathbf{D}^{1/2} = \mathbf{D}^{1/2}\nabla^2 f(\mathbf{x})\mathbf{D}^{1/2},$$

where $\mathbf{x} = \mathbf{D}^{1/2}\mathbf{y}$.

The stepsize in the scaled gradient method can be chosen by each of the three options described in Section 4.1. It is often beneficial to choose the scaling matrix $\mathbf{D}$ differently at each iteration, and we describe this version explicitly below.

Scaled Gradient Method

Input: ε - tolerance parameter.

Initialization: Pick $\mathbf{x}_0 \in \mathbb{R}^n$ arbitrarily.
General step: For any $k = 0, 1, 2, \ldots$ execute the following steps:

(a) Pick a scaling matrix $\mathbf{D}_k \succ 0$.

(b) Pick a stepsize t_k by a line search procedure on the function

$$g(t) = f(\mathbf{x}_k - t\mathbf{D}_k \nabla f(\mathbf{x}_k)).$$

(c) Set $\mathbf{x}_{k+1} = \mathbf{x}_k - t_k \mathbf{D}_k \nabla f(\mathbf{x}_k)$.

(d) If $\|\nabla f(\mathbf{x}_{k+1})\| \leq \varepsilon$, then STOP, and $\mathbf{x}_{k+1}$ is the output.

Note that the stopping criteria was chosen to be $\|\nabla f(\mathbf{x}_{k+1})\| \leq \varepsilon$. A different and perhaps more appropriate stopping criteria might involve bounding the norm of $\mathbf{D}_k^{1/2}\nabla f(\mathbf{x}_{k+1})$.

The main question that arises is of course how to choose the scaling matrix $\mathbf{D}_k$. To accelerate the rate of convergence of the generated sequence, which depends on the condition number of the scaled Hessian $\mathbf{D}_k^{1/2}\nabla^2 f(\mathbf{x}_k)\mathbf{D}_k^{1/2}$, the scaling matrix is often chosen to make this scaled Hessian to be as close as possible to the identity matrix. When $\nabla^2 f(\mathbf{x}_k) \succ 0$, we can actually choose $\mathbf{D}_k = (\nabla^2 f(\mathbf{x}_k))^{-1}$ and the scaled Hessian becomes the identity matrix. The resulting method

$$\mathbf{x}_{k+1} = \mathbf{x}_k - t_k \nabla^2 f(\mathbf{x}_k)^{-1}\nabla f(\mathbf{x}_k)$$

is the celebrated *Newton's method*, which will be the topic of Chapter 5. One difficulty associated with Newton's method is that it requires full knowledge of the Hessian and in addition, the term $\nabla^2 f(\mathbf{x}_k)^{-1}\nabla f(\mathbf{x}_k)$ suggests that a linear system of the form

4.4. Diagonal Scaling

$\nabla^2 f(\mathbf{x}_k)\mathbf{d} = \nabla f(\mathbf{x}_k)$ needs to be solved at each iteration, which might be costly from a computational point of view. This is why simpler scaling matrices are suggested in the literature. The simplest of all scaling matrices are diagonal matrices. Diagonal scaling is in fact a natural idea since the ill-conditioning of optimization problems often arises as a result of a large variety of magnitudes of the decision variables. For example, if one variable is given in kilometers and the other in millimeters, then the first variables is 6 orders of magnitude larger than the second. This is a problem that could have been solved at the initial stage of the formulation of the problem, but when it is not, it can be resolved via diagonal rescaling. A natural choice for diagonal elements is

$$D_{ii} = (\nabla^2 f(\mathbf{x}_k))_{ii}^{-1}.$$

With the above choice, the diagonal elements of $\mathbf{D}^{1/2}\nabla^2 f(\mathbf{x}_k)\mathbf{D}^{1/2}$ are all one. Of course, this choice can be made only when the diagonal of the Hessian is positive.

Example 4.15. Consider the problem

$$\min\{1000x_1^2 + 40x_1x_2 + x_2^2\}.$$

We begin by employing the gradient method with exact stepsize, initial point $(1,1000)^T$, and tolerance parameter $\varepsilon = 10^{-5}$ using the MATLAB function `gradient_method_quadratic` that uses an exact line search.

```
>> A=[1000,20;20,1];
>> gradient_method_quadratic(A,[0;0],[1;1000],1e-5)

iter_number =    1 norm_grad = 1199.023961 fun_val = 598776.964973
iter_number =    2 norm_grad = 24186.628410 fun_val = 344412.923902
iter_number =    3 norm_grad = 689.671401 fun_val = 198104.25098
       :                  :                      :
iter_number =   68 norm_grad = 0.000287 fun_val = 0.000000
iter_number =   69 norm_grad = 0.000008 fun_val = 0.000000
ans =
   1.0e-005 *
   -0.0136
    0.6812
```

The excessive number of iterations is not surprising since the condition number of the associated matrix is large:

```
>> cond(A)
ans =
   1.6680e+003
```

The scaled gradient method with diagonal scaling matrix

$$\mathbf{D} = \begin{pmatrix} \frac{1}{1000} & 0 \\ 0 & 1 \end{pmatrix}$$

should converge faster since the condition number of the scaled matrix $\mathbf{D}^{1/2}\mathbf{A}\mathbf{D}^{1/2}$ is substantially smaller:

```
>> D=diag(1./diag(A));
>> sqrtm(D)*A*sqrtm(D)
```

```
ans =
    1.0000    0.6325
    0.6325    1.0000
>> cond(ans)
ans =
    4.4415
```

To check the performance of the scaled gradient method, we will use a slight modification of the MATLAB function `gradient_method_quadratic`, which we call `gradient_scaled_quadratic`.

```
function [x,fun_val]=gradient_scaled_quadratic(A,b,D,x0,epsilon)
% INPUT
% =======================
% A ....... the positive definite matrix associated
%           with the objective function
% b ....... a column vector associated with the linear part
%           of the objective function
% D ....... scaling matrix
% x0 ...... starting point of the method
% epsilon . tolerance parameter
% OUTPUT
% =======================
% x ....... an optimal solution (up to a tolerance)...
%           of min(x^T A x+2 b^T x)
% fun_val . the optimal function value up to a tolerance

x=x0;
iter=0;
grad=2*(A*x+b);
while (norm(grad)>epsilon)
    iter=iter+1;
    t=grad'*D*grad/(2*(grad'*D')*A*(D*grad));
    x=x-t*D*grad;
    grad=2*(A*x+b);
    fun_val=x'*A*x+2*b'*x;
    fprintf('iter_number = %3d norm_grad = %2.6f fun_val = %2.6f \n',...
    iter,norm(grad),fun_val);
end
```

The running of this code requires only 19 iterations:

```
>> gradient_scaled_quadratic(A,[0;0],D,[1;1000],1e-5)
iter_number =    1 norm_grad = 10461.338850 fun_val = 102437.875289
iter_number =    2 norm_grad = 4137.812524 fun_val = 10080.228908
      :                    :                       :
iter_number =   18 norm_grad = 0.000036 fun_val = 0.000000
iter_number =   19 norm_grad = 0.000009 fun_val = 0.000000
ans =
  1.0e-006 *
   -0.0106
    0.3061
```

4.5 • The Gauss–Newton Method

In Section 3.5 we considered the nonlinear least squares problem

$$\min_{\mathbf{x}\in\mathbb{R}^n}\left\{g(\mathbf{x})\equiv\sum_{i=1}^{m}(f_i(\mathbf{x})-c_i)^2\right\}. \tag{4.12}$$

We will assume here that $f_1,\ldots,f_m$ are continuously differentiable over $\mathbb{R}^n$ for all $i=1,2,\ldots,m$ and that $c_1,\ldots,c_m\in\mathbb{R}$. The problem is sometimes also written in the terms of the vector-valued function

$$F(\mathbf{x})=\begin{pmatrix}f_1(\mathbf{x})-c_1\\f_2(\mathbf{x})-c_2\\\vdots\\f_m(\mathbf{x})-c_m\end{pmatrix},$$

and then it takes the form

$$\min\|F(\mathbf{x})\|^2.$$

The general step of the Gauss–Newton method goes as follows: given the kth iterate $\mathbf{x}_k$, the next iterate is chosen to minimize the sum of squares of the linear approximations of f_i at $\mathbf{x}_k$, that is,

$$\mathbf{x}_{k+1}=\operatorname{argmin}_{\mathbf{x}\in\mathbb{R}^n}\left\{\sum_{i=1}^{m}\left[f_i(\mathbf{x}_k)+\nabla f_i(\mathbf{x}_k)^T(\mathbf{x}-\mathbf{x}_k)-c_i\right]^2\right\}. \tag{4.13}$$

The minimization problem above is essentially a linear least squares problem

$$\min_{\mathbf{x}\in\mathbb{R}^n}\|\mathbf{A}_k\mathbf{x}-\mathbf{b}_k\|^2,$$

where

$$\mathbf{A}_k=\begin{pmatrix}\nabla f_1(\mathbf{x}_k)^T\\\nabla f_2(\mathbf{x}_k)^T\\\vdots\\\nabla f_m(\mathbf{x}_k)^T\end{pmatrix}=J(\mathbf{x}_k)$$

is the so-called *Jacobian* matrix and

$$\mathbf{b}_k=\begin{pmatrix}\nabla f_1(\mathbf{x}_k)^T\mathbf{x}_k-f_1(\mathbf{x}_k)+c_1\\\nabla f_2(\mathbf{x}_k)^T\mathbf{x}_k-f_2(\mathbf{x}_k)+c_2\\\vdots\\\nabla f_m(\mathbf{x}_k)^T\mathbf{x}_k-f_m(\mathbf{x}_k)+c_m\end{pmatrix}=J(\mathbf{x}_k)\mathbf{x}_k-F(\mathbf{x}_k).$$

The underlying assumption is of course that $J(\mathbf{x}_k)$ is of a full column rank; otherwise the minimization in (4.13) will not produce a unique minimizer. In that case, we can also write an explicit expression for the Gauss–Newton iterates (see formula (3.1) in Chapter 3):

$$\mathbf{x}_{k+1}=(J(\mathbf{x}_k)^TJ(\mathbf{x}_k))^{-1}J(\mathbf{x}_k)^T\mathbf{b}_k.$$

Note that the method can also be written as

$$\mathbf{x}_{k+1}=(J(\mathbf{x}_k)^TJ(\mathbf{x}_k))^{-1}J(\mathbf{x}_k)^T(J(\mathbf{x}_k)\mathbf{x}_k-F(\mathbf{x}_k))$$
$$=\mathbf{x}_k-(J(\mathbf{x}_k)^TJ(\mathbf{x}_k))^{-1}J(\mathbf{x}_k)^TF(\mathbf{x}_k).$$

The Gauss–Newton direction is therefore $\mathbf{d}_k = (J(\mathbf{x}_k)^T J(\mathbf{x}_k))^{-1} J(\mathbf{x}_k)^T F(\mathbf{x}_k)$. Noting that $\nabla g(\mathbf{x}) = 2J(\mathbf{x})^T F(\mathbf{x})$, we can conclude that

$$\mathbf{d}_k = \frac{1}{2}(J(\mathbf{x}_k)^T J(\mathbf{x}_k))^{-1} \nabla g(\mathbf{x}_k),$$

meaning that the Gauss–Newton method is essentially a scaled gradient method with the following positive definite scaling matrix

$$\mathbf{D}_k = \frac{1}{2}(J(\mathbf{x}_k)^T J(\mathbf{x}_k))^{-1}.$$

This fact also explains why the Gauss–Newton method is a descent direction method. The method described so far is also called *the pure Gauss–Newton method* since no stepsize is involved. To transform this method into a practical algorithm, a stepsize is introduced, leading to the *damped Gauss–Newton* method.

Damped Gauss–Newton Method

Input: $\varepsilon > 0$ - tolerance parameter.

Initialization: Pick $\mathbf{x}_0 \in \mathbb{R}^n$ arbitrarily.
General step: For any $k = 0, 1, 2, \ldots$ execute the following steps:

(a) Set $\mathbf{d}_k = (J(\mathbf{x}_k)^T J(\mathbf{x}_k))^{-1} J(\mathbf{x}_k)^T F(\mathbf{x}_k)$.

(b) Set t_k by a line search procedure on the function

$$h(t) = g(\mathbf{x}_k - t\mathbf{d}_k).$$

(c) Set $\mathbf{x}_{k+1} = \mathbf{x}_k - t_k \mathbf{d}_k$.

(c) If $\|\nabla g(\mathbf{x}_{k+1})\| \leq \varepsilon$, then STOP, and $\mathbf{x}_{k+1}$ is the output.

4.6 ▪ The Fermat–Weber Problem

The gradient method is the basis for many other methods that might seem at first glance to be unrelated to it. One interesting example is the Fermat–Weber problem. In the 17th century Pierre de Fermat posed the following problem: "Given three distinct points in the plane, find the point having the minimal sum of distances to these three points." The Italian physicist Torricelli solved this problem and defined a construction by ruler and compass for finding it (the point is thus called "the Torricelli point" or "the Torricelli–Fermat point"). Later on, it was generalized by the German economist Weber to a problem in the space $\mathbb{R}^n$ and with an arbitrary number of points. The problem known today as "the Fermat–Weber problem" is the following: given m points in $\mathbb{R}^n: \mathbf{a}_1, \ldots, \mathbf{a}_m$—also called the "anchor points"—and m weights $\omega_1, \omega_2, \ldots, \omega_m > 0$, find a point $\mathbf{x} \in \mathbb{R}^n$ that minimizes the weighted distance of $\mathbf{x}$ to each of the points $\mathbf{a}_1, \ldots, \mathbf{a}_m$. Mathematically, this problem can be cast as the minimization problem

$$\min_{\mathbf{x} \in \mathbb{R}^n} \left\{ f(\mathbf{x}) \equiv \sum_{i=1}^m \omega_i \|\mathbf{x} - \mathbf{a}_i\| \right\}.$$

4.6. The Fermat–Weber Problem

Note that the objective function is not differentiable at the anchor points $\mathbf{a}_1,\ldots,\mathbf{a}_m$. This problem is one of the fundamental localization problems, and it is an instance of a *facility location* problem. For example, $\mathbf{a}_1, \mathbf{a}_2, \ldots, \mathbf{a}_m$ can represent locations of cities, and $\mathbf{x}$ will be the location of a new airport or hospital (or any other facility that serves the cities); the weights might be proportional to the size of population at each of the cities. One popular approach for solving the problem was introduced by Weiszfeld in 1937. The starting point is the first order optimality condition:

$$\nabla f(\mathbf{x}) = 0.$$

Note that we implicitly assume here that $\mathbf{x}$ is not an anchor point. The latter equality can be explicitly written as

$$\sum_{i=1}^m \omega_i \frac{\mathbf{x} - \mathbf{a}_i}{\|\mathbf{x} - \mathbf{a}_i\|} = 0.$$

After some algebraic manipulation, the latter relation can be written as

$$\left(\sum_{i=1}^m \frac{\omega_i}{\|\mathbf{x} - \mathbf{a}_i\|}\right)\mathbf{x} = \sum_{i=1}^m \frac{\omega_i \mathbf{a}_i}{\|\mathbf{x} - \mathbf{a}_i\|},$$

which is the same as

$$\mathbf{x} = \frac{1}{\sum_{i=1}^m \frac{\omega_i}{\|\mathbf{x} - \mathbf{a}_i\|}} \sum_{i=1}^m \frac{\omega_i \mathbf{a}_i}{\|\mathbf{x} - \mathbf{a}_i\|}.$$

We can thus reformulate the optimality condition as $\mathbf{x} = T(\mathbf{x})$, where T is the operator

$$T(\mathbf{x}) \equiv \frac{1}{\sum_{i=1}^m \frac{\omega_i}{\|\mathbf{x} - \mathbf{a}_i\|}} \sum_{i=1}^m \frac{\omega_i \mathbf{a}_i}{\|\mathbf{x} - \mathbf{a}_i\|}.$$

Thus, the problem of finding a stationary point of f can be recast as the problem of finding a fixed point of T. Therefore, a natural approach for solving the problem is via a fixed point method, that is, by the iterations

$$\mathbf{x}_{k+1} = T(\mathbf{x}_k).$$

We can now write explicitly Weiszfeld's theorem for solving the Fermat–Weber problem.

Weiszfeld's Method

Initialization: Pick $\mathbf{x}_0 \in \mathbb{R}^n$ such that $\mathbf{x}_0 \neq \mathbf{a}_1, \mathbf{a}_2, \ldots, \mathbf{a}_m$.
General step: For any $k = 0, 1, 2, \ldots$ compute

$$\mathbf{x}_{k+1} = T(\mathbf{x}_k) = \frac{1}{\sum_{i=1}^m \frac{\omega_i}{\|\mathbf{x}_k - \mathbf{a}_i\|}} \sum_{i=1}^m \frac{\omega_i \mathbf{a}_i}{\|\mathbf{x}_k - \mathbf{a}_i\|}. \tag{4.14}$$

Note that the algorithm is defined only when the iterates $\mathbf{x}_k$ are all different from $\mathbf{a}_1,\ldots,\mathbf{a}_m$. Although the algorithm was initially presented as a fixed point method, the surprising fact is that it is basically a gradient method. Indeed,

$$\mathbf{x}_{k+1} = \frac{1}{\sum_{i=1}^{m} \frac{\omega_i}{\|\mathbf{x}_k - \mathbf{a}_i\|}} \sum_{i=1}^{m} \frac{\omega_i \mathbf{a}_i}{\|\mathbf{x}_k - \mathbf{a}_i\|}$$

$$= \mathbf{x}_k - \frac{1}{\sum_{i=1}^{m} \frac{\omega_i}{\|\mathbf{x}_k - \mathbf{a}_i\|}} \sum_{i=1}^{m} \omega_i \frac{\mathbf{x}_k - \mathbf{a}_i}{\|\mathbf{x}_k - \mathbf{a}_i\|}$$

$$= \mathbf{x}_k - \frac{1}{\sum_{i=1}^{m} \frac{\omega_i}{\|\mathbf{x}_k - \mathbf{a}_i\|}} \nabla f(\mathbf{x}_k).$$

Therefore, Weiszfeld's method is essentially the gradient method with a special choice of stepsize:

$$t_k = \frac{1}{\sum_{i=1}^{m} \frac{\omega_i}{\|\mathbf{x}_k - \mathbf{a}_i\|}}.$$

We are left of course with several questions. Is the method well-defined? That is, can we guarantee that none of the iterates $\mathbf{x}_k$ is equal to any of the points $\mathbf{a}_1, \ldots, \mathbf{a}_m$? Is the sequence of objective function values decreases? Does the sequence $\{\mathbf{x}_k\}_{k \geq 0}$ converge to a global optimal solution? We will answer at least part of these questions in this section.

We would like to show that the generated sequence of function values is nonincreasing. For that, we define the auxiliary function

$$h(\mathbf{y}, \mathbf{x}) \equiv \sum_{i=1}^{m} \omega_i \frac{\|\mathbf{y} - \mathbf{a}_i\|^2}{\|\mathbf{x} - \mathbf{a}_i\|}, \qquad \mathbf{y} \in \mathbb{R}^n, \quad \mathbf{x} \in \mathbb{R}^n \setminus \mathcal{A},$$

where $\mathcal{A} \equiv \{\mathbf{a}_1, \mathbf{a}_2, \ldots, \mathbf{a}_m\}$. The function $h(\cdot, \cdot)$ has several important properties. First of all, the operator T can be computed on a vector $\mathbf{x} \notin \mathcal{A}$ by minimizing the function $h(\mathbf{y}, \mathbf{x})$ over all $\mathbf{y} \in \mathbb{R}^n$.

Lemma 4.16. *For any $\mathbf{x} \in \mathbb{R}^n \setminus \mathcal{A}$, one has*

$$T(\mathbf{x}) = \mathrm{argmin}_{\mathbf{y}} \{h(\mathbf{y}, \mathbf{x}) : \mathbf{y} \in \mathbb{R}^n\}. \tag{4.15}$$

Proof. The function $h(\cdot, \mathbf{x})$ is a quadratic function whose associated matrix is positive definite. In fact, the associated matrix is $(\sum_{i=1}^{m} \frac{\omega_i}{\|\mathbf{x} - \mathbf{a}_i\|})\mathbf{I}$. Therefore, by Lemma 2.41, the unique global minimum of (4.15), which we denote by $\mathbf{y}^*$, is the unique stationary point of $h(\cdot, \mathbf{x})$, that is, the point for which the gradient vanishes:

$$\nabla_{\mathbf{y}} h(\mathbf{y}^*, \mathbf{x}) = \mathbf{0}.$$

Thus,

$$2 \sum_{i=1}^{m} \omega_i \frac{\mathbf{y}^* - \mathbf{a}_i}{\|\mathbf{x} - \mathbf{a}_i\|} = \mathbf{0}.$$

Extracting $\mathbf{y}^*$ from the last equation yields $\mathbf{y}^* = T(\mathbf{x})$, and the result is established. $\square$

Lemma 4.16 basically shows that the update formula of Weiszfeld's method can be written as

$$\mathbf{x}_{k+1} = \mathrm{argmin}\{h(\mathbf{x}, \mathbf{x}_k) : \mathbf{x} \in \mathbb{R}^n\}.$$

We are now ready to prove other important properties of h, which will be crucial for showing that the sequence $\{f(\mathbf{x}_k)\}_{k \geq 0}$ is nonincreasing.

4.6. The Fermat–Weber Problem

Lemma 4.17. *If* $\mathbf{x} \in \mathbb{R}^n \setminus \mathcal{A}$, *then*

(a) $h(\mathbf{x},\mathbf{x}) = f(\mathbf{x})$,

(b) $h(\mathbf{y},\mathbf{x}) \geq 2f(\mathbf{y}) - f(\mathbf{x})$ *for any* $\mathbf{y} \in \mathbb{R}^n$,

(c) $f(T(\mathbf{x})) \leq f(\mathbf{x})$ *and* $f(T(\mathbf{x})) = f(\mathbf{x})$ *if and only if* $\mathbf{x} = T(\mathbf{x})$,

(d) $\mathbf{x} = T(\mathbf{x})$ *if and only if* $\nabla f(\mathbf{x}) = \mathbf{0}$.

Proof. (a) $h(\mathbf{x},\mathbf{x}) = \sum_{i=1}^{m} \omega_i \frac{\|\mathbf{x}-\mathbf{a}_i\|^2}{\|\mathbf{x}-\mathbf{a}_i\|} = \sum_{i=1}^{m} \omega_i \|\mathbf{x}-\mathbf{a}_i\| = f(\mathbf{x})$.

(b) For any nonnegative number a and positive number b, the inequality

$$\frac{a^2}{b} \geq 2a - b$$

holds. Substituting $a = \|\mathbf{y} - \mathbf{a}_i\|$ and $b = \|\mathbf{x} - \mathbf{a}_i\|$, it follows that for any $i = 1, 2, \ldots, m$

$$\frac{\|\mathbf{y}-\mathbf{a}_i\|^2}{\|\mathbf{x}-\mathbf{a}_i\|} \geq 2\|\mathbf{y}-\mathbf{a}_i\| - \|\mathbf{x}-\mathbf{a}_i\|.$$

Multiplying the above inequality by ω_i and summing over $i = 1, 2, \ldots, m$, we obtain that

$$\sum_{i=1}^{m} \omega_i \frac{\|\mathbf{y}-\mathbf{a}_i\|^2}{\|\mathbf{x}-\mathbf{a}_i\|} \geq 2\sum_{i=1}^{m} \omega_i \|\mathbf{y}-\mathbf{a}_i\| - \sum_{i=1}^{m} \omega_i \|\mathbf{x}-\mathbf{a}_i\|.$$

Therefore,

$$h(\mathbf{y},\mathbf{x}) \geq 2f(\mathbf{y}) - f(\mathbf{x}). \tag{4.16}$$

(c) Since $T(\mathbf{x}) = \mathrm{argmin}_{\mathbf{y} \in \mathbb{R}^n} h(\mathbf{y},\mathbf{x})$, it follows that

$$h(T(\mathbf{x}),\mathbf{x}) \leq h(\mathbf{x},\mathbf{x}) = f(\mathbf{x}), \tag{4.17}$$

where the last equality follows from part (a). Part (b) of the lemma yields

$$h(T(\mathbf{x}),\mathbf{x}) \geq 2f(T(\mathbf{x})) - f(\mathbf{x}),$$

which combined with (4.17) implies

$$f(\mathbf{x}) \geq h(T(\mathbf{x}),\mathbf{x}) \geq 2f(T(\mathbf{x})) - f(\mathbf{x}), \tag{4.18}$$

establishing the fact that $f(T(\mathbf{x})) \leq f(\mathbf{x})$. To complete the proof we need to show that $f(T(\mathbf{x})) = f(\mathbf{x})$ if and only if $T(\mathbf{x}) = \mathbf{x}$. Of course, if $T(\mathbf{x}) = \mathbf{x}$, then $f(T(\mathbf{x})) = f(\mathbf{x})$. To show the reverse implication, let us assume that $f(T(\mathbf{x})) = f(\mathbf{x})$. By the chain of inequalities (4.18) it follows that $h(\mathbf{x},\mathbf{x}) = f(\mathbf{x}) = h(T(\mathbf{x}),\mathbf{x})$. Since the unique minimizer of $h(\cdot,\mathbf{x})$ is $T(\mathbf{x})$, it follows that $\mathbf{x} = T(\mathbf{x})$.

(d) The proof of (d) follows by simple algebraic manipulation. □

We are now ready to prove the descent property of the sequence $\{f(\mathbf{x}_k)\}_{k \geq 0}$ under the assumption that all the iterates are not anchor points.

Lemma 4.18. *Let* $\{\mathbf{x}_k\}_{k\geq 0}$ *be the sequence generated by Weiszfeld's method* (4.14), *where we assume that* $\mathbf{x}_k \notin \mathcal{A}$ *for all* $k \geq 0$. *Then we have the following:*

(a) *The sequence* $\{f(\mathbf{x}_k)\}$ *is nonincreasing: for any* $k \geq 0$ *the inequality* $f(\mathbf{x}_{k+1}) \leq f(\mathbf{x}_k)$ *holds.*

(b) *For any* k, $f(\mathbf{x}_k) = f(\mathbf{x}_{k+1})$ *if and only if* $\nabla f(\mathbf{x}_k) = \mathbf{0}$.

Proof. (a) Since $\mathbf{x}_k \notin \mathcal{A}$ for all k, this result follows by substituting $\mathbf{x} = \mathbf{x}_k$ in part (c) of Lemma 4.17.

(b) By part (c) of Lemma 4.17 it follows that $f(\mathbf{x}_k) = f(\mathbf{x}_{k+1}) = f(T(\mathbf{x}_k))$ if and only if $\mathbf{x}_k = \mathbf{x}_{k+1} = T(\mathbf{x}_k)$. By part (d) of Lemma 4.17 the latter is equivalent to $\nabla f(\mathbf{x}_k) = \mathbf{0}$. □

We have thus shown that sequence $\{f(\mathbf{x}_k)\}_{k\geq 0}$ is strictly decreasing as long as we are not stuck at a stationary point. The underlying assumption that $\mathbf{x}_k \notin \mathcal{A}$ is problematic in the sense that it cannot be easily guaranteed. One approach to make sure that the sequence generated by the method does not contain anchor points is to choose the starting point $\mathbf{x}_0$ so that its value is strictly smaller than the values of the anchor points:

$$f(\mathbf{x}_0) < \min\{f(\mathbf{x}_0), f(\mathbf{x}_1), \ldots, f(\mathbf{a}_m)\}.$$

This assumption, combined with the monotonicity of the function values of the sequence generated by the method, implies that the iterates do not include anchor points. We will prove that under this assumption any convergent subsequence of $\{\mathbf{x}_k\}_{k\geq 0}$ converges to a stationary point.

Theorem 4.19. *Let* $\{\mathbf{x}_k\}_{k\geq 0}$ *be the sequence generated by Weiszfeld's method and assume that* $f(\mathbf{x}_0) < \min\{f(\mathbf{a}_1), f(\mathbf{a}_2), \ldots, f(\mathbf{a}_m)\}$. *Then all the limit points of* $\{\mathbf{x}_k\}_{k\geq 0}$ *are stationary points of* f.

Proof. Let $\{\mathbf{x}_{k_n}\}_{n\geq 0}$ be a subsequence of $\{\mathbf{x}_k\}_{k\geq 0}$ that converges to a point $\mathbf{x}^*$. By the monotonicity of the method and the continuity of the objective function we have

$$f(\mathbf{x}^*) \leq f(\mathbf{x}_0) < \min\{f(\mathbf{a}_1), f(\mathbf{a}_2), \ldots, f(\mathbf{a}_m)\}.$$

Therefore, $\mathbf{x}^* \notin \mathcal{A}$, and hence $\nabla f(\mathbf{x}^*)$ is defined. We will show that $\nabla f(\mathbf{x}^*) = \mathbf{0}$. By the continuity of the operator T at $\mathbf{x}^*$, it follows that the sequence $\mathbf{x}_{k_n+1} = T(\mathbf{x}_{k_n}) \to T(\mathbf{x}^*)$ as $n \to \infty$. The sequence of function values $\{f(\mathbf{x}_k)\}_{k\geq 0}$ is nonincreasing and bounded below by 0 and thus converges to a value which we denote by f^*. Obviously, both $\{f(\mathbf{x}_{k_n})\}_{n\geq 0}$ and $\{f(\mathbf{x}_{k_n+1})\}_{n\geq 0}$ converge to f^*. By the continuity of f, we thus obtain that $f(T(\mathbf{x}^*)) = f(\mathbf{x}^*) = f^*$, which by parts (c) and (d) of Lemma 4.17 implies that $\nabla f(\mathbf{x}^*) = \mathbf{0}$. □

It can be shown that for the Fermat–Weber problem, stationary points are global optimal solutions, and hence the latter theorem shows that all limit points of the sequence generated by the method are global optimal solutions. In fact, it is also possible to show that the entire sequence converges to a global optimal solution, but this analysis is beyond the scope of the book.

4.7 • Convergence Analysis of the Gradient Method

4.7.1 • Lipschitz Property of the Gradient

In this section we will present a convergence analysis of the gradient method employed on the unconstrained minimization problem

$$\min\{f(\mathbf{x}) : \mathbf{x} \in \mathbb{R}^n\}.$$

We will assume that the objective function f is continuously differentiable and that its gradient ∇f is Lipschitz continuous over $\mathbb{R}^n$, meaning that there exists $L > 0$ for which

$$\|\nabla f(\mathbf{x}) - \nabla f(\mathbf{y})\| \leq L\|\mathbf{x} - \mathbf{y}\| \text{ for any } \mathbf{x}, \mathbf{y} \in \mathbb{R}^n.$$

Note that if ∇f is Lipschitz with constant L, then it is also Lipschitz with constant $\tilde{L}$ for all $\tilde{L} \geq L$. Therefore, there are essentially infinite number of Lipschitz constants for a function with Lipschitz gradient. Frequently, we are interested in the smallest possible Lipschitz constant. The class of functions with Lipschitz gradient with constant L is denoted by $C_L^{1,1}(\mathbb{R}^n)$ or just $C_L^{1,1}$. Occasionally, when the exact value of the Lipschitz constant is unimportant, we will omit it and denote the class by $C^{1,1}$. For a given $D \subseteq \mathbb{R}^n$, the set of all continuously differentiable functions over D whose gradient satisfies the above Lipschitz condition for any $\mathbf{x}, \mathbf{y} \in D$ is denoted by $C_L^{1,1}(D)$. The following are some simple examples of $C^{1,1}$ functions:

- **Linear functions** Given $\mathbf{a} \in \mathbb{R}^n$, the function $f(\mathbf{x}) = \mathbf{a}^T\mathbf{x}$ is in $C_0^{1,1}$.

- **Quadratic functions** Let $\mathbf{A}$ be an $n \times n$ symmetric matrix, $\mathbf{b} \in \mathbb{R}^n$, and $c \in \mathbb{R}$. Then the function $f(\mathbf{x}) = \mathbf{x}^T\mathbf{A}\mathbf{x} + 2\mathbf{b}^T\mathbf{x} + c$ is a $C^{1,1}$ function.

To compute a Lipschitz constant of the quadratic function $f(\mathbf{x}) = \mathbf{x}^T\mathbf{A}\mathbf{x} + 2\mathbf{b}^T\mathbf{x} + c$, we can use the definition of Lipschitz continuity:

$$\|\nabla f(\mathbf{x}) - \nabla f(\mathbf{y})\| = 2\|(\mathbf{A}\mathbf{x} + \mathbf{b}) - (\mathbf{A}\mathbf{y} + \mathbf{b})\| = 2\|\mathbf{A}\mathbf{x} - \mathbf{A}\mathbf{y}\| = 2\|\mathbf{A}(\mathbf{x} - \mathbf{y})\| \leq 2\|\mathbf{A}\| \cdot \|\mathbf{x} - \mathbf{y}\|.$$

We thus conclude that a Lipschitz constant of ∇f is $2\|\mathbf{A}\|$.

A useful fact, stated in the following result, is that for twice continuously differentiable functions, Lipschitz continuity of the gradient is equivalent to boundedness of the Hessian.

Theorem 4.20. *Let f be a twice continuously differentiable function over $\mathbb{R}^n$. Then the following two claims are equivalent:*

(a) $f \in C_L^{1,1}(\mathbb{R}^n)$.

(b) $\|\nabla^2 f(\mathbf{x})\| \leq L$ *for any* $\mathbf{x} \in \mathbb{R}^n$.

Proof. (b) $\Rightarrow$ (a). Suppose that $\|\nabla^2 f(\mathbf{x})\| \leq L$ for any $\mathbf{x} \in \mathbb{R}^n$. Then by the fundamental theorem of calculus we have for all $\mathbf{x}, \mathbf{y} \in \mathbb{R}^n$

$$\nabla f(\mathbf{y}) = \nabla f(\mathbf{x}) + \int_0^1 \nabla^2 f(\mathbf{x} + t(\mathbf{y} - \mathbf{x}))(\mathbf{y} - \mathbf{x}) dt$$

$$= \nabla f(\mathbf{x}) + \left(\int_0^1 \nabla^2 f(\mathbf{x} + t(\mathbf{y} - \mathbf{x})) dt\right) \cdot (\mathbf{y} - \mathbf{x}),$$

Thus,

$$\|\nabla f(\mathbf{y}) - \nabla f(\mathbf{x})\| = \left\|\left(\int_0^1 \nabla^2 f(\mathbf{x}+t(\mathbf{y}-\mathbf{x}))dt\right)\cdot(\mathbf{y}-\mathbf{x})\right\|$$

$$\leq \left\|\int_0^1 \nabla^2 f(\mathbf{x}+t(\mathbf{y}-\mathbf{x}))dt\right\| \|\mathbf{y}-\mathbf{x}\|$$

$$\leq \left(\int_0^1 \|\nabla^2 f(\mathbf{x}+t(\mathbf{y}-\mathbf{x}))\|dt\right)\|\mathbf{y}-\mathbf{x}\|$$

$$\leq L\|\mathbf{y}-\mathbf{x}\|,$$

establishing the desired result $f \in C_L^{1,1}$.

(a) $\Rightarrow$ (b). Suppose now that $f \in C_L^{1,1}$. Then by the fundamental theorem of calculus for any $\mathbf{d} \in \mathbb{R}^n$ and $\alpha > 0$ we have

$$\nabla f(\mathbf{x}+\alpha\mathbf{d}) - \nabla f(\mathbf{x}) = \int_0^\alpha \nabla^2 f(\mathbf{x}+t\mathbf{d})\mathbf{d}dt.$$

Thus,

$$\left\|\left(\int_0^\alpha \nabla^2 f(\mathbf{x}+t\mathbf{d})dt\right)\mathbf{d}\right\| = \|\nabla f(\mathbf{x}+\alpha\mathbf{d}) - \nabla f(\mathbf{x})\| \leq \alpha L\|\mathbf{d}\|.$$

Dividing by α and taking the limit $\alpha \to 0^+$, we obtain

$$\|\nabla^2 f(\mathbf{x})\mathbf{d}\| \leq L\|\mathbf{d}\|,$$

implying that $\|\nabla^2 f(\mathbf{x})\| \leq L$. $\square$

Example 4.21. Let $f : \mathbb{R} \to \mathbb{R}$ be given by $f(x) = \sqrt{1+x^2}$. Then

$$0 \leq f''(x) = \frac{1}{(1+x^2)^{3/2}} \leq 1$$

for any $x \in \mathbb{R}$, and thus $f \in C_1^{1,1}$. ∎

4.7.2 ▪ The Descent Lemma

An important result for $C^{1,1}$ functions is that they can be bounded above by a quadratic function over the entire space. This result, known as "the descent lemma," is fundamental in convergence proofs of gradient-based methods.

Lemma 4.22 (descent lemma). *Let $D \subseteq \mathbb{R}^n$ and $f \in C_L^{1,1}(D)$ for some $L > 0$. Then for any $\mathbf{x}, \mathbf{y} \in D$ satisfying $[\mathbf{x}, \mathbf{y}] \subseteq D$ it holds that*

$$f(\mathbf{y}) \leq f(\mathbf{x}) + \nabla f(\mathbf{x})^T(\mathbf{y}-\mathbf{x}) + \frac{L}{2}\|\mathbf{x}-\mathbf{y}\|^2.$$

Proof. By the fundamental theorem of calculus,

$$f(\mathbf{y}) - f(\mathbf{x}) = \int_0^1 \langle \nabla f(\mathbf{x}+t(\mathbf{y}-\mathbf{x})), \mathbf{y}-\mathbf{x}\rangle dt.$$

4.7. Convergence Analysis of the Gradient Method

Therefore,

$$f(\mathbf{y})-f(\mathbf{x}) = \langle \nabla f(\mathbf{x}), \mathbf{y}-\mathbf{x}\rangle + \int_0^1 \langle \nabla f(\mathbf{x}+t(\mathbf{y}-\mathbf{x}))-\nabla f(\mathbf{x}), \mathbf{y}-\mathbf{x}\rangle dt.$$

Thus,

$$\begin{aligned}
|f(\mathbf{y})-f(\mathbf{x})-\langle \nabla f(\mathbf{x}),\mathbf{y}-\mathbf{x}\rangle| &= \left|\int_0^1 \langle \nabla f(\mathbf{x}+t(\mathbf{y}-\mathbf{x}))-\nabla f(\mathbf{x}), \mathbf{y}-\mathbf{x}\rangle dt\right| \\
&\leq \int_0^1 |\langle \nabla f(\mathbf{x}+t(\mathbf{y}-\mathbf{x}))-\nabla f(\mathbf{x}), \mathbf{y}-\mathbf{x}\rangle| dt \\
&\leq \int_0^1 \|\nabla f(\mathbf{x}+t(\mathbf{y}-\mathbf{x}))-\nabla f(\mathbf{x})\|\cdot\|\mathbf{y}-\mathbf{x}\| dt \\
&\leq \int_0^1 tL\|\mathbf{y}-\mathbf{x}\|^2 dt \\
&= \frac{L}{2}\|\mathbf{y}-\mathbf{x}\|^2. \quad \square
\end{aligned}$$

Note that the proof of the descent lemma actually shows both upper and lower bounds on the function:

$$f(\mathbf{x})+\nabla f(\mathbf{x})^T(\mathbf{y}-\mathbf{x})-\frac{L}{2}\|\mathbf{x}-\mathbf{y}\|^2 \leq f(\mathbf{y}) \leq f(\mathbf{x})+\nabla f(\mathbf{x})^T(\mathbf{y}-\mathbf{x})+\frac{L}{2}\|\mathbf{x}-\mathbf{y}\|^2.$$

4.7.3 ▪ Convergence of the Gradient Method

Equipped with the descent lemma, we are now ready to prove the convergence of the gradient method for $C^{1,1}$ functions. Of course, we cannot guarantee convergence to a global optimal solution, but we can show the convergence to stationary points in the sense that the gradient converges to zero. We begin by the following result showing a "sufficient decrease" of the gradient method at each iteration. More precisely, we show that at each iteration the decrease in the function value is at least a constant times the squared norm of the gradient.

Lemma 4.23 (sufficient decrease lemma). *Suppose that $f \in C_L^{1,1}(\mathbb{R}^n)$. Then for any $\mathbf{x} \in \mathbb{R}^n$ and $t > 0$*

$$f(\mathbf{x})-f(\mathbf{x}-t\nabla f(\mathbf{x})) \geq t\left(1-\frac{Lt}{2}\right)\|\nabla f(\mathbf{x})\|^2. \tag{4.19}$$

Proof. By the descent lemma we have

$$\begin{aligned}
f(\mathbf{x}-t\nabla f(\mathbf{x})) &\leq f(\mathbf{x})-t\|\nabla f(\mathbf{x})\|^2 + \frac{Lt^2}{2}\|\nabla f(\mathbf{x})\|^2 \\
&= f(\mathbf{x})-t\left(1-\frac{Lt}{2}\right)\|\nabla f(\mathbf{x})\|^2.
\end{aligned}$$

The result then follows by simple rearrangement of terms. $\square$

Our goal now is to show that a sufficient decrease property occurs in each of the stepsize selection strategies: constant, exact line search, and backtracking. In the constant stepsize setting, we assume that $t_k = \bar{t} \in \left(0, \frac{2}{L}\right)$. Substituting $\mathbf{x} = \mathbf{x}_k, t = \bar{t}$ in (4.19) yields the inequality

$$f(\mathbf{x}_k) - f(\mathbf{x}_{k+1}) \geq \bar{t}\left(1 - \frac{L\bar{t}}{2}\right)\|\nabla f(\mathbf{x}_k)\|^2. \qquad (4.20)$$

Note that the guaranteed descent in the gradient method per iteration is

$$\bar{t}\left(1 - \frac{\bar{t}L}{2}\right)\|\nabla f(\mathbf{x}_k)\|^2.$$

If we wish to obtain the largest guaranteed bound on the decrease, then we seek the maximum of $\bar{t}(1 - \frac{\bar{t}L}{2})$ over $(0, \frac{2}{L})$. This maximum is attained at $\bar{t} = \frac{1}{L}$, and thus a popular choice for a stepsize is $\frac{1}{L}$. In this case we have

$$f(\mathbf{x}_k) - f(\mathbf{x}_{k+1}) = f(\mathbf{x}_k) - f\left(\mathbf{x}_k - \frac{1}{L}\nabla f(\mathbf{x}_k)\right) \geq \frac{1}{2L}\|\nabla f(\mathbf{x}_k)\|^2. \qquad (4.21)$$

In the exact line search setting, the update formula of the algorithm is

$$\mathbf{x}_{k+1} = \mathbf{x}_k - t_k \nabla f(\mathbf{x}_k),$$

where $t_k \in \operatorname{argmin}_{t \geq 0} f(\mathbf{x}_k - t\nabla f(\mathbf{x}_k))$. By the definition of t_k we know that

$$f(\mathbf{x}_k - t_k \nabla f(\mathbf{x}_k)) \leq f\left(\mathbf{x}_k - \frac{1}{L}\nabla f(\mathbf{x}_k)\right),$$

and thus we have

$$f(\mathbf{x}_k) - f(\mathbf{x}_{k+1}) \geq f(\mathbf{x}_k) - f\left(\mathbf{x}_k - \frac{1}{L}\nabla f(\mathbf{x}_k)\right) \geq \frac{1}{2L}\|\nabla f(\mathbf{x}_k)\|^2, \qquad (4.22)$$

where the last inequality was shown in (4.21).

In the backtracking setting we seek a small enough stepsize t_k for which

$$f(\mathbf{x}_k) - f(\mathbf{x}_k - t_k \nabla f(\mathbf{x}_k)) \geq \alpha t_k \|\nabla f(\mathbf{x}_k)\|^2, \qquad (4.23)$$

where $\alpha \in (0, 1)$. We would like to find a lower bound on t_k. There are two options. Either $t_k = s$ (the initial value of the stepsize) or t_k is determined by the backtracking procedure, meaning that the stepsize $\tilde{t}_k = t_k/\beta$ is not acceptable and does not satisfy (4.23):

$$f(\mathbf{x}_k) - f\left(\mathbf{x}_k - \frac{t_k}{\beta}\nabla f(\mathbf{x}_k)\right) < \alpha \frac{t_k}{\beta}\|\nabla f(\mathbf{x}_k)\|^2, \qquad (4.24)$$

Substituting $\mathbf{x} = \mathbf{x}_k, t = \frac{t_k}{\beta}$ in (4.19) we obtain that

$$f(\mathbf{x}_k) - f\left(\mathbf{x}_k - \frac{t_k}{\beta}\nabla f(\mathbf{x}_k)\right) \geq \frac{t_k}{\beta}\left(1 - \frac{Lt_k}{2\beta}\right)\|\nabla f(\mathbf{x}_k)\|^2,$$

which combined with (4.24) implies that

$$\frac{t_k}{\beta}\left(1 - \frac{Lt_k}{2\beta}\right) < \alpha \frac{t_k}{\beta},$$

4.7. Convergence Analysis of the Gradient Method

which is the same as $t_k > \frac{2(1-\alpha)\beta}{L}$. Overall, we obtained that in the backtracking setting we have

$$t_k \geq \min\left\{s, \frac{2(1-\alpha)\beta}{L}\right\},$$

which combined with (4.23) implies that

$$f(\mathbf{x}_k) - f(\mathbf{x}_k - t_k \nabla f(\mathbf{x}_k)) \geq \alpha \min\left\{s, \frac{2(1-\alpha)\beta}{L}\right\} \|\nabla f(\mathbf{x}_k)\|^2. \qquad (4.25)$$

We summarize the above discussion, or, better said, the three inequalities (4.20), (4.22), and (4.25), in the following result.

Lemma 4.24 (sufficient decrease of the gradient method). *Let $f \in C_L^{1,1}(\mathbb{R}^n)$. Let $\{\mathbf{x}_k\}_{k \geq 0}$ be the sequence generated by the gradient method for solving*

$$\min_{\mathbf{x} \in \mathbb{R}^n} f(\mathbf{x})$$

with one of the following stepsize strategies:

- *constant stepsize $\bar{t} \in (0, \frac{2}{L})$,*

- *exact line search,*

- *backtracking procedure with parameters $s \in \mathbb{R}_{++}, \alpha \in (0,1)$, and $\beta \in (0,1)$.*

Then

$$f(\mathbf{x}_k) - f(\mathbf{x}_{k+1}) \geq M \|\nabla f(\mathbf{x}_k)\|^2, \qquad (4.26)$$

where

$$M = \begin{cases} \bar{t}\left(1 - \frac{\bar{t}L}{2}\right) & \text{constant stepsize,} \\ \frac{1}{2L} & \text{exact line search,} \\ \alpha \min\left\{s, \frac{2(1-\alpha)\beta}{L}\right\} & \text{backtracking.} \end{cases}$$

We will now show the convergence of the norms of the gradients $\|\nabla f(\mathbf{x}_k)\|$ to zero.

Theorem 4.25 (convergence of the gradient method). *Let $f \in C_L^{1,1}(\mathbb{R}^n)$, and let $\{\mathbf{x}_k\}_{k \geq 0}$ be the sequence generated by the gradient method for solving*

$$\min_{\mathbf{x} \in \mathbb{R}^n} f(\mathbf{x})$$

with one of the following stepsize strategies:

- *constant stepsize $\bar{t} \in (0, \frac{2}{L})$,*

- *exact line search,*

- *backtracking procedure with parameters $s \in \mathbb{R}_{++}, \alpha \in (0,1)$, and $\beta \in (0,1)$.*

Assume that f is bounded below over $\mathbb{R}^n$, that is, there exists $m \in \mathbb{R}$ such that $f(\mathbf{x}) > m$ for all $\mathbf{x} \in \mathbb{R}^n$.

Then we have the following:

(a) The sequence $\{f(\mathbf{x}_k)\}_{k \geq 0}$ is nonincreasing. In addition, for any $k \geq 0$, $f(\mathbf{x}_{k+1}) < f(\mathbf{x}_k)$ unless $\nabla f(\mathbf{x}_k) = \mathbf{0}$.

(b) $\nabla f(\mathbf{x}_k) \to \mathbf{0}$ as $k \to \infty$.

Proof. (a) By (4.26) we have that

$$f(\mathbf{x}_k) - f(\mathbf{x}_{k+1}) \geq M \|\nabla f(\mathbf{x}_k)\|^2 \geq 0 \qquad (4.27)$$

for some constant $M > 0$, and hence the equality $f(\mathbf{x}_k) = f(\mathbf{x}_{k+1})$ can hold only when $\nabla f(\mathbf{x}_k) = \mathbf{0}$.

(b) Since the sequence $\{f(\mathbf{x}_k)\}_{k \geq 0}$ is nonincreasing and bounded below, it converges. Thus, in particular $f(\mathbf{x}_k) - f(\mathbf{x}_{k+1}) \to 0$ as $k \to \infty$, which combined with (4.27) implies that $\|\nabla f(\mathbf{x}_k)\| \to 0$ as $k \to \infty$. $\square$

We can also get an estimate for the rate of convergence of the gradient method, or more precisely of the norm of the gradients. This is done in the following result.

Theorem 4.26 (rate of convergence of gradient norms). *Under the setting of Theorem 4.25, let f^* be the limit of the convergent sequence $\{f(\mathbf{x}_k)\}_{k \geq 0}$. Then for any $n = 0, 1, 2, \ldots$*

$$\min_{k=0,1,\ldots,n} \|\nabla f(\mathbf{x}_k)\| \leq \sqrt{\frac{f(\mathbf{x}_0) - f^*}{M(n+1)}},$$

where

$$M = \begin{cases} \bar{t}\left(1 - \frac{\bar{t}L}{2}\right) & \text{constant stepsize,} \\ \frac{1}{2L} & \text{exact line search,} \\ \alpha \min\left\{s, \frac{2\beta(1-\alpha)}{L}\right\} & \text{backtracking.} \end{cases}$$

Proof. Summing the inequality (4.26) over $k = 0, 1, \ldots, n$, we obtain

$$f(\mathbf{x}_0) - f(\mathbf{x}_{n+1}) \geq M \sum_{k=0}^{n} \|\nabla f(\mathbf{x}_k)\|^2.$$

Since $f(\mathbf{x}_{n+1}) \geq f^*$, we can thus conclude that

$$f(\mathbf{x}_0) - f^* \geq M \sum_{k=0}^{n} \|\nabla f(\mathbf{x}_k)\|^2.$$

Finally, using the latter inequality along with the fact that for every $k = 0, 1, \ldots, n$ the obvious inequality $\|\nabla f(\mathbf{x}_k)\|^2 \geq \min_{k=0,1,\ldots,n} \|\nabla f(\mathbf{x}_k)\|^2$ holds, it follows that

$$f(\mathbf{x}_0) - f^* \geq M(n+1) \min_{k=0,1,\ldots,n} \|\nabla f(\mathbf{x}_k)\|^2,$$

implying the desired result. □

Exercises

4.1. Let $f \in C_L^{1,1}(\mathbb{R}^n)$ and let $\{\mathbf{x}_k\}_{k \geq 0}$ be the sequence generated by the gradient method with a constant stepsize $t_k = \frac{1}{L}$. Assume that $\mathbf{x}_k \to \mathbf{x}^*$. Show that if $\nabla f(\mathbf{x}_k) \neq 0$ for all $k \geq 0$, then $\mathbf{x}^*$ is *not* a local maximum point.

4.2. [9, Exercise 1.3.3] Consider the minimization problem

$$\min\{\mathbf{x}^T \mathbf{Q} \mathbf{x} : \mathbf{x} \in \mathbb{R}^2\},$$

where $\mathbf{Q}$ is a positive definite 2×2 matrix. Suppose we use the diagonal scaling matrix

$$\mathbf{D} = \begin{pmatrix} Q_{11}^{-1} & 0 \\ 0 & Q_{22}^{-1} \end{pmatrix}.$$

Show that the above scaling matrix improves the condition number of $\mathbf{Q}$ in the sense that

$$\varkappa(\mathbf{D}^{1/2} \mathbf{Q} \mathbf{D}^{1/2}) \leq \varkappa(\mathbf{Q}).$$

4.3. Consider the quadratic minimization problem

$$\min\{\mathbf{x}^T \mathbf{A} \mathbf{x} : \mathbf{x} \in \mathbb{R}^5\},$$

where $\mathbf{A}$ is the 5×5 Hilbert matrix defined by

$$A_{i,j} = \frac{1}{i+j-1}, \quad i, j = 1, 2, 3, 4, 5.$$

The matrix can be constructed via the MATLAB command `A = hilb(5)`. Run the following methods and compare the number of iterations required by each of the methods when the initial vector is $\mathbf{x}_0 = (1, 2, 3, 4, 5)^T$ to obtain a solution $\mathbf{x}$ with $\|\nabla f(\mathbf{x})\| \leq 10^{-4}$:

- gradient method with backtracking stepsize rule and parameters $\alpha = 0.5, \beta = 0.5, s = 1$;
- gradient method with backtracking stepsize rule and parameters $\alpha = 0.1, \beta = 0.5, s = 1$;
- gradient method with exact line search;
- diagonally scaled gradient method with diagonal elements $D_{ii} = \frac{1}{A_{ii}}, i = 1, 2, 3, 4, 5$ and exact line search;
- diagonally scaled gradient method with diagonal elements $D_{ii} = \frac{1}{A_{ii}}, i = 1, 2, 3, 4, 5$ and backtracking line search with parameters $\alpha = 0.1, \beta = 0.5, s = 1$.

4.4. Consider the Fermat–Weber problem

$$\min_{\mathbf{x} \in \mathbb{R}^n} \left\{ f(\mathbf{x}) \equiv \sum_{i=1}^m \omega_i \|\mathbf{x} - \mathbf{a}_i\| \right\},$$

where $\omega_1, \ldots, \omega_m > 0$ and $\mathbf{a}_1, \ldots, \mathbf{a}_m \in \mathbb{R}^n$ are m different points. Let

$$p \in \operatorname{argmin}_{i=1,2,\ldots,m} f(\mathbf{a}_i).$$

Suppose that

$$\left\| \sum_{i \neq p} \omega_i \frac{\mathbf{a}_p - \mathbf{a}_i}{\|\mathbf{a}_p - \mathbf{a}_i\|} \right\| > \omega_p.$$

(i) Show that there exists a direction $\mathbf{d} \in \mathbb{R}^n$ such that $f'(\mathbf{a}_p; \mathbf{d}) < 0$.

(ii) Show that there exists $\mathbf{x}_0 \in \mathbb{R}^n$ satisfying $f(\mathbf{x}_0) < \min\{f(\mathbf{a}_1), f(\mathbf{a}_2), \ldots, f(\mathbf{a}_p)\}$. Explain how to compute such a vector.

4.5. In the "source localization problem" we are given m locations of sensors $\mathbf{a}_1, \mathbf{a}_2, \ldots, \mathbf{a}_m \in \mathbb{R}^n$ and approximate distances between the sensors and an unknown "source" located at $\mathbf{x} \in \mathbb{R}^n$:

$$d_i \approx \|\mathbf{x} - \mathbf{a}_i\|.$$

The problem is to find and estimate $\mathbf{x}$ given the locations $\mathbf{a}_1, \mathbf{a}_2, \ldots, \mathbf{a}_m$ and the approximate distances $d_1, d_2, \ldots, d_m$. A natural formulation as an optimization problem is to consider the nonlinear least squares problem

$$\text{(SL)} \quad \min \left\{ f(\mathbf{x}) \equiv \sum_{i=1}^m (\|\mathbf{x} - \mathbf{a}_i\| - d_i)^2 \right\}.$$

We will denote the set of sensors by $\mathcal{A} \equiv \{\mathbf{a}_1, \mathbf{a}_2, \ldots, \mathbf{a}_m\}$.

(i) Show that the optimality condition $\nabla f(\mathbf{x}) = \mathbf{0}$ ($\mathbf{x} \notin \mathcal{A}$) is the same as

$$\mathbf{x} = \frac{1}{m} \left\{ \sum_{i=1}^m \mathbf{a}_i + \sum_{i=1}^m d_i \frac{\mathbf{x} - \mathbf{a}_i}{\|\mathbf{x} - \mathbf{a}_i\|} \right\}.$$

(ii) Show that the corresponding fixed point method

$$\mathbf{x}_{k+1} = \frac{1}{m} \left\{ \sum_{i=1}^m \mathbf{a}_i + \sum_{i=1}^m d_i \frac{\mathbf{x}_k - \mathbf{a}_i}{\|\mathbf{x}_k - \mathbf{a}_i\|} \right\}$$

is a gradient method, assuming that $\mathbf{x}_k \notin \mathcal{A}$ for all $k \geq 0$. What is the stepsize?

4.6. Another formulation of the source localization problem consists of minimizing the following objective function:

$$\text{(SL2)} \quad \min_{\mathbf{x} \in \mathbb{R}^n} \left\{ f(\mathbf{x}) \equiv \sum_{i=1}^m (\|\mathbf{x} - \mathbf{a}_i\|^2 - d_i^2)^2 \right\}.$$

This is of course a nonlinear least squares problem, and thus the Gauss–Newton method can be employed in order to solve it. We will assume that $n = 2$.

(i) Show that as long as all the points $\mathbf{a}_1, \mathbf{a}_2, \ldots, \mathbf{a}_m$ do not reside on the same line in the plane, the method is well-defined, meaning that the linear least squares problem solved at each iteration has a unique solution.

(ii) Write a MATLAB function that implements the damped Gauss–Newton method employed on problem (SL2) with a backtracking line search strategy with parameters $s = 1, \alpha = \beta = 0.5, \varepsilon = 10^{-4}$. Run the function on the two-dimensional problem ($n = 2$) with 5 anchors ($m = 5$) and data generated by the MATLAB commands

```
randn('seed',317);
A=randn(2,5);
x=randn(2,1);
d=sqrt(sum((A-x*ones(1,5)).^2))+0.05*randn(1,5);
d=d';
```

The columns of the 2×5 matrix $\mathbf{A}$ are the locations of the five sensors, $\mathbf{x}$ is the "true" location of the source, and $\mathbf{d}$ is the vector of noisy measurements between the source and the sensors. Compare your results (e.g., number of iterations) to the gradient method with backtracking and parameters $s = 1, \alpha = \beta = 0.5, \varepsilon = 10^{-4}$. Start both methods with the initial vector $(1000, -500)^T$.

4.7. Let $f(\mathbf{x}) = \mathbf{x}^T \mathbf{A} \mathbf{x} + 2\mathbf{b}^T \mathbf{x} + c$, where $\mathbf{A}$ is a symmetric $n \times n$ matrix, $\mathbf{b} \in \mathbb{R}^n$, and $c \in \mathbb{R}$. Show that the *smallest* Lipschitz constant of ∇f is $2\|\mathbf{A}\|$.

4.8. Let $f : \mathbb{R}^n \to \mathbb{R}$ be given by $f(\mathbf{x}) = \sqrt{1 + \|\mathbf{x}\|^2}$. Show that $f \in C_1^{1,1}$.

4.9. Let $f \in C_L^{1,1}(\mathbb{R}^m)$, and let $\mathbf{A} \in \mathbb{R}^{m \times n}, \mathbf{b} \in \mathbb{R}^m$. Show that the function $g : \mathbb{R}^n \to \mathbb{R}$ defined by $g(\mathbf{x}) = f(\mathbf{A}\mathbf{x} + \mathbf{b})$ satisfies $g \in C_{\tilde{L}}^{1,1}(\mathbb{R}^n)$, where $\tilde{L} = \|\mathbf{A}\|^2 L$.

4.10. Give an example of a function $f \in C_L^{1,1}(\mathbb{R})$ and a starting point $x_0 \in \mathbb{R}$ such that the problem $\min f(x)$ has an optimal solution and the gradient method with constant stepsize $t = \frac{2}{L}$ diverges.

4.11. Suppose that $f \in C_L^{1,1}(\mathbb{R}^n)$ and assume that $\nabla^2 f(\mathbf{x}) \succeq 0$ for any $\mathbf{x} \in \mathbb{R}^n$. Suppose that the optimal value of the problem $\min_{\mathbf{x} \in \mathbb{R}^n} f(\mathbf{x})$ is f^*. Let $\{\mathbf{x}_k\}_{k \geq 0}$ be the sequence generated by the gradient method with constant stepsize $\frac{1}{L}$. Show that if $\{\mathbf{x}_k\}_{k \geq 0}$ is bounded, then $f(\mathbf{x}_k) \to f^*$ as $k \to \infty$.

Chapter 5
Newton's Method

5.1 • Pure Newton's Method

In the previous chapter we considered the unconstrained minimization problem

$$\min\{f(\mathbf{x}): \mathbf{x} \in \mathbb{R}^n\},$$

where we assumed that f is continuously differentiable. We studied the gradient method which only uses first order information, namely information on the function values and gradients. In this chapter we assume that f is twice continuously differentiable, and we will present a *second order* method, namely a method that uses, in addition to the information on function values and gradients, evaluations of the Hessian matrices. We will concentrate on Newton's method. The main idea of Newton's method is the following. Given an iterate $\mathbf{x}_k$, the next iterate $\mathbf{x}_{k+1}$ is chosen to minimize the quadratic approximation of the function around $\mathbf{x}_k$:

$$\mathbf{x}_{k+1} = \mathrm{argmin}_{\mathbf{x} \in \mathbb{R}^n} \left\{ f(\mathbf{x}_k) + \nabla f(\mathbf{x}_k)^T (\mathbf{x} - \mathbf{x}_k) + \frac{1}{2}(\mathbf{x} - \mathbf{x}_k)^T \nabla^2 f(\mathbf{x}_k)(\mathbf{x} - \mathbf{x}_k) \right\}. \quad (5.1)$$

The above update formula is not well-defined unless we further assume that $\nabla^2 f(\mathbf{x}_k)$ is positive definite. In that case, the unique minimizer of the minimization problem (5.1) is the unique stationary point:

$$\nabla f(\mathbf{x}_k) + \nabla^2 f(\mathbf{x}_k)(\mathbf{x}_{k+1} - \mathbf{x}_k) = 0,$$

which is the same as

$$\mathbf{x}_{k+1} = \mathbf{x}_k - (\nabla^2 f(\mathbf{x}_k))^{-1} \nabla f(\mathbf{x}_k). \quad (5.2)$$

The vector $-(\nabla^2 f(\mathbf{x}_k))^{-1} \nabla f(\mathbf{x}_k)$ is called the *Newton direction*, and the algorithm induced by the update formula (5.2) is called *the pure Newton's method*. Note that when $\nabla^2 f(\mathbf{x}_k)$ is positive definite for any k, pure Newton's method is essentially a scaled gradient method, and Newton's directions are descent directions.

> **Pure Newton's Method**
>
> **Input:** $\varepsilon > 0$ - tolerance parameter.
>
> **Initialization:** Pick $\mathbf{x}_0 \in \mathbb{R}^n$ arbitrarily.
> **General step:** For any $k = 0, 1, 2, \ldots$ execute the following steps:
>
> (a) Compute the Newton direction $\mathbf{d}_k$, which is the solution to the linear system $\nabla^2 f(\mathbf{x}_k)\mathbf{d}_k = -\nabla f(\mathbf{x}_k)$.
>
> (b) Set $\mathbf{x}_{k+1} = \mathbf{x}_k + \mathbf{d}_k$.
>
> (c) If $\|\nabla f(\mathbf{x}_{k+1})\| \leq \varepsilon$, then STOP, and $\mathbf{x}_{k+1}$ is the output.

At the very least, Newton's method requires that $\nabla^2 f(\mathbf{x})$ is positive definite for every $\mathbf{x} \in \mathbb{R}^n$, which in particular implies that there exists a unique optimal solution $\mathbf{x}^*$. However, this is not enough to guarantee convergence, as the following example illustrates.

Example 5.1. Consider the function $f(x) = \sqrt{1+x^2}$ defined over the real line. The minimizer of f over $\mathbb{R}$ is of course $x = 0$. The first and second derivatives of f are

$$f'(x) = \frac{x}{\sqrt{1+x^2}}, \qquad f''(x) = \frac{1}{(1+x^2)^{3/2}}.$$

Therefore, (pure) Newton's method has the form

$$x_{k+1} = x_k - \frac{f'(x_k)}{f''(x_k)} = x_k - x_k(1+x_k^2) = -x_k^3.$$

We therefore see that for $|x_0| \geq 1$ the method diverges and that for $|x_0| < 1$ the method converges very rapidly to the correct solution $x^* = 0$. ∎

Despite the fact that a lot of assumptions are required to be made in order to guarantee the convergence of the method, Newton's method does have one very attractive feature: under certain assumptions one can prove local *quadratic* rate of convergence, which means that near the optimal solution the errors $e_k = \|\mathbf{x}_k - \mathbf{x}^*\|$ (where $\mathbf{x}^*$ is the unique optimal solution) satisfy the inequality $e_{k+1} \leq M e_k^2$ for some positive $M > 0$. This property essentially means that the number of accuracy digits is doubled at each iteration.

Theorem 5.2 (quadratic local convergence of Newton's method). *Let f be a twice continuously differentiable function defined over $\mathbb{R}^n$. Assume that*

- *there exists $\dot{m} > 0$ for which $\nabla^2 f(\mathbf{x}) \succeq m\mathbf{I}$ for any $\mathbf{x} \in \mathbb{R}^n$,*
- *there exists $L > 0$ for which $\|\nabla^2 f(\mathbf{x}) - \nabla^2 f(\mathbf{y})\| \leq L\|\mathbf{x} - \mathbf{y}\|$ for any $\mathbf{x}, \mathbf{y} \in \mathbb{R}^n$.*

Let $\{\mathbf{x}_k\}_{k \geq 0}$ be the sequence generated by Newton's method, and let $\mathbf{x}^*$ be the unique minimizer of f over $\mathbb{R}^n$. Then for any $k = 0, 1, \ldots$ the inequality

$$\|\mathbf{x}_{k+1} - \mathbf{x}^*\| \leq \frac{L}{2m}\|\mathbf{x}_k - \mathbf{x}^*\|^2 \tag{5.3}$$

holds. In addition, if $\|\mathbf{x}_0 - \mathbf{x}^*\| \leq \frac{m}{L}$, then

$$\|\mathbf{x}_k - \mathbf{x}^*\| \leq \frac{2m}{L}\left(\frac{1}{2}\right)^{2^k}, \quad k = 0, 1, 2, \ldots. \tag{5.4}$$

5.1. Pure Newton's Method

Proof. Let k be a nonnegative integer. Then

$$
\begin{aligned}
\mathbf{x}_{k+1} - \mathbf{x}^* &= \mathbf{x}_k - (\nabla^2 f(\mathbf{x}_k))^{-1} \nabla f(\mathbf{x}_k) - \mathbf{x}^* \\
&\stackrel{\nabla f(\mathbf{x}^*)=0}{=} \mathbf{x}_k - \mathbf{x}^* + (\nabla^2 f(\mathbf{x}_k))^{-1} (\nabla f(\mathbf{x}^*) - \nabla f(\mathbf{x}_k)) \\
&= \mathbf{x}_k - \mathbf{x}^* + (\nabla^2 f(\mathbf{x}_k))^{-1} \int_0^1 [\nabla^2 f(\mathbf{x}_k + t(\mathbf{x}^* - \mathbf{x}_k))](\mathbf{x}^* - \mathbf{x}_k) dt \\
&= (\nabla^2 f(\mathbf{x}_k))^{-1} \int_0^1 [\nabla^2 f(\mathbf{x}_k + t(\mathbf{x}^* - \mathbf{x}_k)) - \nabla^2 f(\mathbf{x}_k)](\mathbf{x}^* - \mathbf{x}_k) dt.
\end{aligned}
$$

Since $\nabla^2 f(\mathbf{x}_k) \succeq m\mathbf{I}$, it follows that $\|(\nabla^2 f(\mathbf{x}_k))^{-1}\| \leq \frac{1}{m}$. Hence,

$$
\begin{aligned}
\|\mathbf{x}_{k+1} - \mathbf{x}^*\| &\leq \|(\nabla^2 f(\mathbf{x}_k))^{-1}\| \left\| \int_0^1 [\nabla^2 f(\mathbf{x}_k + t(\mathbf{x}^* - \mathbf{x}_k)) - \nabla^2 f(\mathbf{x}_k)](\mathbf{x}^* - \mathbf{x}_k) dt \right\| \\
&\leq \|(\nabla^2 f(\mathbf{x}_k))^{-1}\| \int_0^1 \left\| [\nabla^2 f(\mathbf{x}_k + t(\mathbf{x}^* - \mathbf{x}_k)) - \nabla^2 f(\mathbf{x}_k)](\mathbf{x}^* - \mathbf{x}_k) \right\| dt \\
&\leq \|(\nabla^2 f(\mathbf{x}_k))^{-1}\| \int_0^1 \left\| \nabla^2 f(\mathbf{x}_k + t(\mathbf{x}^* - \mathbf{x}_k)) - \nabla^2 f(\mathbf{x}_k) \right\| \cdot \|\mathbf{x}^* - \mathbf{x}_k\| dt \\
&\leq \frac{L}{m} \int_0^1 t \|\mathbf{x}_k - \mathbf{x}^*\|^2 dt = \frac{L}{2m} \|\mathbf{x}_k - \mathbf{x}^*\|^2.
\end{aligned}
$$

We will prove inequality (5.4) by induction on k. Note that for $k = 0$, we assumed that

$$\|\mathbf{x}_0 - \mathbf{x}^*\| \leq \frac{m}{L},$$

so in particular

$$\|\mathbf{x}_0 - \mathbf{x}^*\| \leq \frac{2m}{L} \left(\frac{1}{2}\right)^{2^0},$$

establishing the basis of the induction. Assume that (5.4) holds for an integer k, that is, $\|\mathbf{x}_k - \mathbf{x}^*\| \leq \frac{2m}{L} (\frac{1}{2})^{2^k}$; we will show it holds for $k+1$. Indeed, by (5.3) we have

$$\|\mathbf{x}_{k+1} - \mathbf{x}^*\| \leq \frac{L}{2m} \|\mathbf{x}_k - \mathbf{x}^*\|^2 \leq \frac{L}{2m} \left(\frac{2m}{L} \left(\frac{1}{2}\right)^{2^k}\right)^2 = \frac{2m}{L} \left(\frac{1}{2}\right)^{2^{k+1}},$$

proving the desired result. □

A very naive implementation of Newton's method in MATLAB is given below.

```
function x=pure_newton(f,g,h,x0,epsilon)
% Pure Newton's method
%
% INPUT
% ==============
% f .......... objective function
% g .......... gradient of the objective function
% h .......... Hessian of the objective function
% x0.......... initial point
% epsilon ..... tolerance parameter
% OUTPUT
% ==============
```

```
% x - solution obtained by Newton's method (up to some tolerance)

if (nargin<5)
    epsilon=1e-5;
end
x=x0;
gval=g(x);
hval=h(x);
iter=0;
while ((norm(gval)>epsilon)&&(iter<10000))
    iter=iter+1;
    x=x-hval\gval;
    fprintf('iter= %2d f(x)=%10.10f\n',iter,f(x))
    gval=g(x);
    hval=h(x);
end
if (iter==10000)
    fprintf('did not converge')
end
```

Note that the above implementation does not check the positive definiteness of the Hessian, and it essentially assumes implicitly that this property holds. As already mentioned, Newton's method requires quite a lot of assumptions in order to guarantee convergence, and hence the described implementation includes a divergence criteria (10000 iterations).

Example 5.3. Consider the minimization problem

$$\min_{x,y} 100x^4 + 0.01y^4,$$

whose optimal solution is obviously $(x,y) = (0,0)$. This is a rather poorly scaled problem, and indeed the gradient method with initial vector $\mathbf{x}_0 = (1,1)^T$ and parameters $(s,\alpha,\beta,\varepsilon) = (1,0.5,0.5,10^{-6})$ converges after the huge amount of 14612 iterations:

```
>> f=@(x)100*x(1)^4+0.01*x(2)^4;
>> g=@(x)[400*x(1)^3;0.04*x(2)^3];
>> [x,fun_val]=gradient_method_backtracking(f,g,[1;1],1,0.5,0.5,1e-6)
iter_number =    1 norm_grad = 90.513620 fun_val = 13.799181
iter_number =    2 norm_grad = 32.381098 fun_val = 3.511932
iter_number =    3 norm_grad = 11.472585 fun_val = 0.887929
       :               :                      :
iter_number = 14611 norm_grad = 0.000001 fun_val = 0.000000
iter_number = 14612 norm_grad = 0.000001 fun_val = 0.000000
```

Invoking pure Newton's method, we obtain convergence after only 17 iterations:

```
>> h=@(x)[1200*x(1)^2,0;0,0.12*x(2)^2];
>> pure_newton(f,g,h,[1;1],1e-6)
iter=  1 f(x)=19.7550617284
iter=  2 f(x)=3.9022344155
iter=  3 f(x)=0.7708117364
        :           :
iter= 15 f(x)=0.0000000027
iter= 16 f(x)=0.0000000005
iter= 17 f(x)=0.0000000001
```

Note that the basic assumptions required for the convergence of Newton's method as described in Theorem 5.2 are not satisfied. The Hessian is always positive semidefinite, but it is not always positive definite and does not satisfy a Lipschitz property. ∎

The previous example exhibited convergence even when the basic underlying assumptions of Theorem 5.2 are not satisfied. However, in general, convergence is unfortunately not guaranteed in the absence of these very restrictive assumptions.

Example 5.4. Consider the minimization problem

$$\min_{x_1, x_2} \sqrt{x_1^2 + 1} + \sqrt{x_2^2 + 1},$$

whose optimal solution is $\mathbf{x} = \mathbf{0}$. The Hessian of the function is

$$\nabla^2 f(\mathbf{x}) = \begin{pmatrix} \frac{1}{(x_1^2+1)^{3/2}} & 0 \\ 0 & \frac{1}{(x_2^2+1)^{3/2}} \end{pmatrix} \succ 0.$$

Note that despite the fact that the Hessian is positive definite, there does not exist an $m > 0$ for which $\nabla^2 f(\mathbf{x}) \succeq m\mathbf{I}$. This violation of the basic assumptions can be seen practically. Indeed, if we employ Newton's method with initial vector $\mathbf{x}_0 = (1,1)^T$ and tolerance parameter $\varepsilon = 10^{-8}$ we obtain convergence after 37 iterations:

```
>> f=@(x)sqrt(1+x(1)^2)+sqrt(1+x(2)^2);
>> g=@(x)[x(1)/sqrt(x(1)^2+1);x(2)/sqrt(x(2)^2+1)];
>> h=@(x)diag([1/(x(1)^2+1)^1.5,1/(x(2)^2+1)^1.5]);
>> pure_newton(f,g,h,[1;1],1e-8)
iter=   1  f(x)=2.8284271247
iter=   2  f(x)=2.8284271247
    :          :
iter=  30  f(x)=2.8105247315
iter=  31  f(x)=2.7757389625
iter=  32  f(x)=2.6791717153
iter=  33  f(x)=2.4507092918
iter=  34  f(x)=2.1223796622
iter=  35  f(x)=2.0020052756
iter=  36  f(x)=2.0000000081
iter=  37  f(x)=2.0000000000
```

Note that in the first 30 iterations the method is almost stuck. On the other hand, the gradient method with backtracking and parameters $(s, \alpha, \beta) = (1, 0.5, 0.5)$ converges after only 7 iterations:

```
>> [x,fun_val]=gradient_method_backtracking(f,g,[1;1],1,0.5,0.5,1e-8);
iter_number =   1  norm_grad = 0.397514  fun_val = 2.084022
iter_number =   2  norm_grad = 0.016699  fun_val = 2.000139
iter_number =   3  norm_grad = 0.000001  fun_val = 2.000000
iter_number =   4  norm_grad = 0.000001  fun_val = 2.000000
iter_number =   5  norm_grad = 0.000000  fun_val = 2.000000
iter_number =   6  norm_grad = 0.000000  fun_val = 2.000000
iter_number =   7  norm_grad = 0.000000  fun_val = 2.000000
```

If we start from the more distant point $(10, 10)^T$. The situation is much more severe. The gradient method with backtracking converges after 13 iterations:

```
>> [x,fun_val]=gradient_method_backtracking(f,g,[10;10],1,0.5,0.5,1e-8);
iter_number =    1 norm_grad = 1.405573 fun_val = 18.120635
iter_number =    2 norm_grad = 1.403323 fun_val = 16.146490
     :           :           :                  :
iter_number =   12 norm_grad = 0.000049 fun_val =  2.000000
iter_number =   13 norm_grad = 0.000000 fun_val =  2.000000
```

Newton's method, on the other hand, diverges:

```
>> pure_newton(f,g,h,[10;10],1e-8);
iter=  1  f(x)=2000.0009999997
iter=  2  f(x)=1999999999.9999990000
iter=  3  f(x)=1999999999999997300000000000.0000000
iter=  4  f(x)=1999999999999992300000000000000000000....
iter=  5  f(x)=         Inf
```

∎

As can be seen in the last example, pure Newton's method does not guarantee descent of the generated sequence of function values even when the Hessian is positive definite. This drawback can be rectified by introducing a stepsize chosen by a certain line search procedure, leading to the so-called *damped Newton's method*.

5.2 ▪ Damped Newton's Method

Below we describe the damped Newton's method with a backtracking stepsize strategy. Of course, other stepsize strategies may be used.

Damped Newton's Method

Input: $\alpha, \beta \in (0,1)$ - parameters for the backtracking procedure.
$\varepsilon > 0$ - tolerance parameter.

Initialization: Pick $\mathbf{x}_0 \in \mathbb{R}^n$ arbitrarily.
General step: For any $k = 0, 1, 2, \ldots$ execute the following steps:

(a) Compute the Newton direction $\mathbf{d}_k$, which is the solution to the linear system $\nabla^2 f(\mathbf{x}_k) \mathbf{d}_k = -\nabla f(\mathbf{x}_k)$.

(b) Set $t_k = 1$. While
$$f(\mathbf{x}_k) - f(\mathbf{x}_k + t_k \mathbf{d}_k) < -\alpha t_k \nabla f(\mathbf{x}_k)^T \mathbf{d}_k$$
set $t_k := \beta t_k$.

(c) $\mathbf{x}_{k+1} = \mathbf{x}_k + t_k \mathbf{d}_k$.

(c) If $\|\nabla f(\mathbf{x}_{k+1})\| \leq \varepsilon$, then STOP, and $\mathbf{x}_{k+1}$ is the output.

A MATLAB implementation of the method is given below.

```
function x=newton_backtracking(f,g,h,x0,alpha,beta,epsilon)
% Newton's method with backtracking
%
% INPUT
%=======================================
% f ........ objective function
% g ........ gradient of the objective function
% h ........ hessian of the objective function
% x0 ....... initial point
% alpha .... tolerance parameter for the stepsize selection strategy
% beta ..... the proportion in which the stepsize is multiplied
%            at each backtracking step (0<beta<1)
% epsilon .. tolerance parameter for stopping rule
% OUTPUT
%=======================================
% x ........ optimal solution (up to a tolerance)
%            of min f(x)
% fun_val ... optimal function value

x=x0;
gval=g(x);
hval=h(x);
d=hval\gval;
iter=0;
while ((norm(gval)>epsilon)&&(iter<10000))
    iter=iter+1;
    t=1;
    while(f(x-t*d)>f(x)-alpha*t*gval'*d)
       t=beta*t;
    end
    x=x-t*d;
    fprintf('iter= %2d f(x)=%10.10f\n',iter,f(x))
    gval=g(x);
    hval=h(x);
    d=hval\gval;
end

if (iter==10000)
    fprintf('did not converge\n')
end
```

Example 5.5. Continuing Example 5.4, invoking Newton's method with the same initial vector $\mathbf{x}_0 = (10,10)^T$ that caused the divergence of pure Newton's method, results in convergence after 17 iterations.

```
>> newton_backtracking(f,g,h,[10;10],0.5,0.5,1e-8);
iter=  1 f(x)=4.6688169339
iter=  2 f(x)=2.4101973721
iter=  3 f(x)=2.0336386321
     :           :
iter= 16 f(x)=2.0000000005
iter= 17 f(x)=2.0000000000
```

∎

5.3 • The Cholesky Factorization

An important issue that naturally arises when employing Newton's method is the one of validating whether the Hessian matrix is positive definite, and if it is, then another issue is how to solve the linear system $\nabla^2 f(\mathbf{x}_k)\mathbf{d} = -\nabla f(\mathbf{x}_k)$. These two issues are resolved by using the Cholesky factorization, which we briefly recall in this section.

Given an $n \times n$ positive definite matrix $\mathbf{A}$, a *Cholesky factorization* is a factorization of the form

$$\mathbf{A} = \mathbf{L}\mathbf{L}^T,$$

where $\mathbf{L}$ is a lower triangular $n \times n$ matrix whose diagonal is positive. Given a Cholesky factorization, the task of solving a linear system of equations of the form $\mathbf{A}\mathbf{x} = \mathbf{b}$ can be easily done by the following two steps:

A. Find the solution $\mathbf{u}$ of $\mathbf{L}\mathbf{u} = \mathbf{b}$.

B. Find the solution $\mathbf{x}$ of $\mathbf{L}^T\mathbf{x} = \mathbf{u}$.

Since $\mathbf{L}$ is a triangular matrix with a positive diagonal, steps A and B can be carried out by backward or forward substitution, which requires only an order of n^2 arithmetic operations. The computation of the Cholesky factorization requires an order of n^3 operations.

The computation of the Cholesky factor $\mathbf{L}$ is done via a simple recursive formula. Consider the following block matrix partition of the matrices $\mathbf{A}$ and $\mathbf{L}$:

$$\mathbf{A} = \begin{pmatrix} A_{11} & \mathbf{A}_{12} \\ \mathbf{A}_{12}^T & \mathbf{A}_{22} \end{pmatrix}, \qquad \mathbf{L} = \begin{pmatrix} L_{11} & 0 \\ \mathbf{L}_{21} & \mathbf{L}_{22} \end{pmatrix},$$

where $A_{11} \in \mathbb{R}, \mathbf{A}_{12} \in \mathbb{R}^{1 \times (n-1)}, \mathbf{A}_{22} \in \mathbb{R}^{(n-1) \times (n-1)}, L_{11} \in \mathbb{R}, \mathbf{L}_{21} \in \mathbb{R}^{n-1}, \mathbf{L}_{22} \in \mathbb{R}^{(n-1) \times (n-1)}$. Since $\mathbf{A} = \mathbf{L}\mathbf{L}^T$ we have

$$\begin{pmatrix} A_{11} & \mathbf{A}_{12} \\ \mathbf{A}_{12}^T & \mathbf{A}_{22} \end{pmatrix} = \begin{pmatrix} L_{11}^2 & L_{11}\mathbf{L}_{21}^T \\ L_{11}\mathbf{L}_{21} & \mathbf{L}_{21}\mathbf{L}_{21}^T + \mathbf{L}_{22}\mathbf{L}_{22}^T \end{pmatrix}.$$

Therefore, in particular,

$$L_{11} = \sqrt{A_{11}}, \qquad \mathbf{L}_{21} = \frac{1}{\sqrt{A_{11}}}\mathbf{A}_{12}^T,$$

and we can thus also write

$$\mathbf{L}_{22}\mathbf{L}_{22}^T = \mathbf{A}_{22} - \mathbf{L}_{21}\mathbf{L}_{21}^T = \mathbf{A}_{22} - \frac{1}{A_{11}}\mathbf{A}_{12}^T\mathbf{A}_{12}.$$

We are left with the task of finding a Cholesky factorization of the $(n-1) \times (n-1)$ matrix $\mathbf{A}_{22} - \frac{1}{A_{11}}\mathbf{A}_{12}^T\mathbf{A}_{12}$. Continuing in this way, we can compute the complete Cholesky factorization of the matrix. The process is illustrated in the following simple example.

Example 5.6. Let

$$\mathbf{A} = \begin{pmatrix} 9 & 3 & 3 \\ 3 & 17 & 21 \\ 3 & 21 & 107 \end{pmatrix}.$$

We will denote the Cholesky factor by

$$\mathbf{L} = \begin{pmatrix} l_{11} & 0 & 0 \\ l_{21} & l_{22} & 0 \\ l_{31} & l_{32} & l_{33} \end{pmatrix}.$$

5.3. The Cholesky Factorization

Then
$$l_{11} = \sqrt{9} = 3$$

and
$$\begin{pmatrix} l_{21} \\ l_{31} \end{pmatrix} = \frac{1}{\sqrt{9}} \begin{pmatrix} 3 \\ 3 \end{pmatrix} = \begin{pmatrix} 1 \\ 1 \end{pmatrix}.$$

We now need to find the Cholesky factorization of
$$\mathbf{L}_{22}\mathbf{L}_{22}^T = \mathbf{A}_{22} - \frac{1}{A_{11}}\mathbf{A}_{12}^T\mathbf{A}_{12} = \begin{pmatrix} 17 & 21 \\ 21 & 107 \end{pmatrix} - \frac{1}{9}\begin{pmatrix} 3 \\ 3 \end{pmatrix}\begin{pmatrix} 3 & 3 \end{pmatrix} = \begin{pmatrix} 16 & 20 \\ 20 & 106 \end{pmatrix}.$$

To do so, let us write
$$\mathbf{L}_{22} = \begin{pmatrix} l_{22} & 0 \\ l_{32} & l_{33} \end{pmatrix}.$$

Consequently, $l_{22} = \sqrt{16} = 4$ and $l_{32} = \frac{1}{\sqrt{16}} \cdot 20 = 5$. We are thus left with the task of finding the Cholesky factorization of
$$106 - \frac{1}{16} \cdot (20 \cdot 20) = 81,$$

which is of course $l_{33} = \sqrt{81} = 9$. To conclude, the Cholesky factor is given by
$$\mathbf{L} = \begin{pmatrix} 3 & 0 & 0 \\ 1 & 4 & 0 \\ 1 & 5 & 9 \end{pmatrix}. \quad \blacksquare$$

The process of computing the Cholesky factorization is well-defined as long as all the diagonal elements l_{ii} that are computed during the process are positive, so that computing their square root is possible. The positiveness of these elements is equivalent to the property that the matrix to be factored is positive definite. Therefore, the Cholesky factorization process can be viewed as a criteria for positive definiteness, and it is actually the test that is used in many algorithms.

Example 5.7. Let us check whether the matrix
$$\mathbf{A} = \begin{pmatrix} 2 & 4 & 7 \\ 4 & 6 & 7 \\ 7 & 7 & 4 \end{pmatrix}$$

is positive definite. We will invoke the Cholesky factorization process. We have
$$l_{11} = \sqrt{2}, \quad \begin{pmatrix} l_{21} \\ l_{31} \end{pmatrix} = \frac{1}{\sqrt{2}} \begin{pmatrix} 4 \\ 7 \end{pmatrix}.$$

Now we need to find the Cholesky factorization of
$$\begin{pmatrix} 6 & 7 \\ 7 & 4 \end{pmatrix} - \frac{1}{2}\begin{pmatrix} 4 \\ 7 \end{pmatrix}\begin{pmatrix} 4 & 7 \end{pmatrix} = \begin{pmatrix} -2 & -7 \\ -7 & -20.5 \end{pmatrix}.$$

At this point the process fails since we are required to find the square root of -2, which is of course not a real number. The conclusion is that $\mathbf{A}$ is not positive definite. $\blacksquare$

In MATLAB, the Cholesky factorization if performed via the function `chol`. Thus, for example, the factorization of the matrix from Example 5.6 can be done by the following MATLAB commands:

```
>> A=[9,3,3;3,17,21;3,21,107];
>> L=chol(A,'lower')
L =
     3     0     0
     1     4     0
     1     5     9
```

The function `chol` can also output a second argument which is zero if the matrix is positive definite, or positive when the matrix is not positive definite, and in the latter case a Cholesky factorization cannot be computed. For the nonpositive definite matrix of Example 5.7 we obtain

```
>> A=[2,4,7;4,6,7;7,7,4];
>> [L,p]=chol(A,'lower');
>> p
p =
     2
```

As was already mentioned several times, Newton's method (pure or not) assumes that the Hessian matrix is positive definite and we are thus left with the question of how to employ Newton's method when the Hessian is not always positive definite. There are several ways to deal with this situation, but perhaps the simplest one is to construct a hybrid method that employs either a Newton step at iterations in which the Hessian is positive definite or a gradient step when the Hessian is not positive definite. The algorithm is written in detail below and also incorporates a backtracking procedure.

Hybrid Gradient-Newton Method

Input: $\alpha, \beta \in (0,1)$ - parameters for the backtracking procedure.
$\varepsilon > 0$ - tolerance parameter.

Initialization: Pick $\mathbf{x}_0 \in \mathbb{R}^n$ arbitrarily.
General step: For any $k = 0, 1, 2, \ldots$ execute the following steps:

(a) If $\nabla^2 f(\mathbf{x}_k) \succ 0$, then take $\mathbf{d}_k$ as the Newton direction $\mathbf{d}_k$, which is the solution to the linear system $\nabla^2 f(\mathbf{x}_k)\mathbf{d}_k = -\nabla f(\mathbf{x}_k)$. Otherwise, set $\mathbf{d}_k = -\nabla f(\mathbf{x}_k)$.

(b) Set $t_k = 1$. While
$$f(\mathbf{x}_k) - f(\mathbf{x}_k + t_k \mathbf{d}_k) < -\alpha t_k \nabla f(\mathbf{x}_k)^T \mathbf{d}_k.$$
set $t_k := \beta t_k$

(c) $\mathbf{x}_{k+1} = \mathbf{x}_k + t_k \mathbf{d}_k$.

(c) If $\|\nabla f(\mathbf{x}_{k+1})\| \leq \varepsilon$, then STOP, and $\mathbf{x}_{k+1}$ is the output.

5.3. The Cholesky Factorization

Following is a MATLAB implementation of the method that also incorporates the Cholesky factorization.

```
function x=newton_hybrid(f,g,h,x0,alpha,beta,epsilon)
% Hybrid Newton's method
%
% INPUT
%=========================================
% f ......... objective function
% g ......... gradient of the objective function
% h ......... hessian of the objective function
% x0......... initial point
% alpha ..... tolerance parameter for the stepsize selection strategy
% beta ...... the proportion in which the stepsize is multiplied
%             at each backtracking step (0<beta<1)
% epsilon ... tolerance parameter for stopping rule
% OUTPUT
%=========================================
% x ......... optimal solution (up to a tolerance)
%             of min f(x)
% fun_val ... optimal function value

x=x0;
gval=g(x);
hval=h(x);
[L,p]=chol(hval,'lower');
if (p==0)
    d=L'\(L\gval);
else
    d=gval;
end
iter=0;
while ((norm(gval)>epsilon)&&(iter<10000))
    iter=iter+1;
    t=1;
    while(f(x-t*d)>f(x)-alpha*t*gval'*d)
        t=beta*t;
    end
    x=x-t*d;
    fprintf('iter= %2d f(x)=%10.10f\n',iter,f(x))
    gval=g(x);
    hval=h(x);
    [L,p]=chol(hval,'lower');
    if (p==0)
        d=L'\(L\gval);
    else
        d=gval;
    end
end

if (iter==10000)
    fprintf('did not converge\n')
end
```

Example 5.8 (Rosenbrock function). Recall that the Rosenbrock function introduced in Example 4.13 is given by $f(x_1, x_2) = 100(x_2 - x_1^2)^2 + (1 - x_1)^2$ and is severely ill-conditioned near the minimizer $(1, 1)$ (which is the unique stationary point). In Example 4.13 we employed the gradient method with a backtracking stepsize selection strategy, and it took approximately 6900 iterations to converge from the starting point $(2, 5)^T$ to a point satisfying $\|\nabla f(\mathbf{x})\| \leq 10^{-5}$. Employing hybrid Newton's method with the same starting point and stopping criteria results in a fast convergence after only 18 iterations:

```
>> f=@(x)100*(x(2)-x(1)^2)^2+(1-x(1))^2;
>> g=@(x)[-400*(x(2)-x(1)^2)*x(1)-2*(1-x(1));200*(x(2)-x(1)^2)];
>> h=@(x)[-400*x(2)+1200*x(1)^2+2,-400*x(1);-400*x(1),200];
>> x=newton_hybrid(f,g,h,[2;5],0.5,0.5,1e-5);
iter=  1  f(x)=3.2210220151
iter=  2  f(x)=1.4965858368
       :              :
iter= 16  f(x)=0.0000000000
iter= 17  f(x)=0.0000000000
```

The contour plots of the Rosenbrock function along with the 17 iterates are illustrated in Figure 5.1. ∎

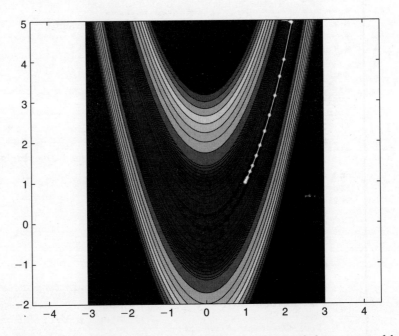

Figure 5.1. *Contour lines of the Rosenbrock function along with the* 17 *iterates of the hybrid Newton's method.*

Exercises

5.1. Find without MATLAB the Cholesky factorization of the matrix

$$\mathbf{A} = \begin{pmatrix} 1 & 2 & 4 & 7 \\ 2 & 13 & 23 & 38 \\ 4 & 23 & 77 & 122 \\ 7 & 38 & 122 & 294 \end{pmatrix}.$$

5.2. Consider the Freudenstein and Roth test function
$$f(\mathbf{x}) = f_1(\mathbf{x})^2 + f_2(\mathbf{x})^2, \qquad \mathbf{x} \in \mathbb{R}^2,$$
where
$$f_1(\mathbf{x}) = -13 + x_1 + ((5 - x_2)x_2 - 2)x_2,$$
$$f_2(\mathbf{x}) = -29 + x_1 + ((x_2 + 1)x_2 - 14)x_2.$$

(i) Show that the function f has three stationary points. Find them and prove that one is a global minimizer, one is a strict local minimum and the third is a saddle point.

(ii) Use MATLAB to employ the following three methods on the problem of minimizing f:

1. the gradient method with backtracking and parameters $(s, \alpha, \beta) = (1, 0.5, 0.5)$.
2. the hybrid Gradient-Newton Method with parameters $(s, \alpha, \beta) = (0.5, 0.5)$.
3. damped Gauss–Newton's method with a backtracking line search strategy with parameters $(s, \alpha, \beta) = (1, 0.5, 0.5)$.

All the algorithms should use the stopping criteria $\|\nabla f(\mathbf{x})\| \leq 10^{-5}$. Each algorithm should be employed four times on the following four starting points: $(-50, 7)^T, (20, 7)^T, (20, -18)^T, (5, -10)^T$. For each of the four starting points, compare the number of iterations and the point to which each method converged. If a method did not converge, explain why.

5.3. Let f be a twice continuously differentiable function satisfying $L\mathbf{I} \succeq \nabla^2 f(\mathbf{x}) \succeq m\mathbf{I}$ for some $L > m > 0$ and let $\mathbf{x}^*$ be the unique minimizer of f over $\mathbb{R}^n$.

(i) Show that
$$f(\mathbf{x}) - f(\mathbf{x}^*) \geq \frac{m}{2} \|\mathbf{x} - \mathbf{x}^*\|^2$$
for any $\mathbf{x} \in \mathbb{R}^n$.

(ii) Let $\{\mathbf{x}_k\}_{k \geq 0}$ be the sequence generated by damped Newton's method with constant stepsize $t_k = \frac{m}{L}$. Show that
$$f(\mathbf{x}_k) - f(\mathbf{x}_{k+1}) \geq \frac{m}{2L} \nabla f(\mathbf{x}_k)^T (\nabla^2 f(\mathbf{x}_k))^{-1} \nabla f(\mathbf{x}_k).$$

(iii) Show that $\mathbf{x}_k \to \mathbf{x}^*$ as $k \to \infty$.

Chapter 6

Convex Sets

In this chapter we begin our exploration of convex analysis, which is the mathematical theory essential for analyzing and understanding the theoretical and practical aspects of optimization.

6.1 ▪ Definition and Examples

We begin with the definition of a convex set.

Definition 6.1 (convex sets). *A set $C \subseteq \mathbb{R}^n$ is called* **convex** *if for any $\mathbf{x}, \mathbf{y} \in C$ and $\lambda \in [0,1]$, the point $\lambda \mathbf{x} + (1-\lambda)\mathbf{y}$ belongs to C.*

The above definition is equivalent to saying that for any $\mathbf{x}, \mathbf{y} \in C$, the line segment $[\mathbf{x}, \mathbf{y}]$ is also in C. Examples of convex and nonconvex sets in $\mathbb{R}^2$ are illustrated in Figure 6.1. We will now show some basic examples of convex sets.

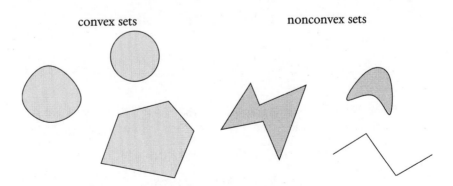

Figure 6.1. *The three left sets are convex, while the three right sets are nonconvex.*

Example 6.2 (convexity of lines). A line in $\mathbb{R}^n$ is a set of the form
$$L = \{\mathbf{z} + t\mathbf{d} : t \in \mathbb{R}\},$$
where $\mathbf{z}, \mathbf{d} \in \mathbb{R}^n$ and $\mathbf{d} \neq \mathbf{0}$. To show that L is indeed a convex set, let us take $\mathbf{x}, \mathbf{y} \in L$. Then there exist $t_1, t_2 \in \mathbb{R}$ such that $\mathbf{x} = \mathbf{z} + t_1 \mathbf{d}$ and $\mathbf{y} = \mathbf{z} + t_2 \mathbf{d}$. Therefore, for any

$\lambda \in [0, 1]$ we have

$$\lambda \mathbf{x} + (1-\lambda)\mathbf{y} = \lambda(\mathbf{z} + t_1 \mathbf{d}) + (1-\lambda)(\mathbf{z} + t_2 \mathbf{d}) = \mathbf{z} + (\lambda t_1 + (1-\lambda)t_2)\mathbf{d} \in L. \quad \blacksquare$$

Similarly we can show that for any $\mathbf{x}, \mathbf{y} \in \mathbb{R}^n$, the closed and open line segments $[\mathbf{x}, \mathbf{y}], (\mathbf{x}, \mathbf{y})$ are also convex sets. Simpler examples of convex sets are the empty set $\emptyset$ and the entire space $\mathbb{R}^n$. A *hyperplane* is a set of the form $H = \{\mathbf{x} \in \mathbb{R}^n : \mathbf{a}^T \mathbf{x} = b\}$, where $\mathbf{a} \in \mathbb{R}^n \setminus \{\mathbf{0}\}, b \in \mathbb{R}$, and the associated *half-space* is the set $H^- = \{\mathbf{x} \in \mathbb{R}^n : \mathbf{a}^T \mathbf{x} \leq b\}$. Both hyperplanes and half-spaces are convex sets.

Lemma 6.3 (convexity of hyperplanes and half-spaces). *Let $\mathbf{a} \in \mathbb{R}^n \setminus \{\mathbf{0}\}$ and $b \in \mathbb{R}$. Then the following sets are convex:*

(a) *the hyperplane $H = \{\mathbf{x} \in \mathbb{R}^n : \mathbf{a}^T \mathbf{x} = b\}$,*

(b) *the half-space $H^- = \{\mathbf{x} \in \mathbb{R}^n : \mathbf{a}^T \mathbf{x} \leq b\}$,*

(c) *the open half-space $\{\mathbf{x} \in \mathbb{R}^n : \mathbf{a}^T \mathbf{x} < b\}$.*

Proof. We will prove only the convexity of the half-space since the proof of convexity of the other two sets is almost identical. Let $\mathbf{x}, \mathbf{y} \in H^-$ and let $\lambda \in [0, 1]$. We will show that $\mathbf{z} = \lambda \mathbf{x} + (1-\lambda)\mathbf{y} \in H^-$. Indeed,

$$\mathbf{a}^T \mathbf{z} = \mathbf{a}^T [\lambda \mathbf{x} + (1-\lambda)\mathbf{y}] = \lambda(\mathbf{a}^T \mathbf{x}) + (1-\lambda)(\mathbf{a}^T \mathbf{y}) \leq \lambda b + (1-\lambda)b = b,$$

where the inequality in the above chain of equalities and inequalities follows from the fact that $\mathbf{a}^T \mathbf{x} \leq b, \mathbf{a}^T \mathbf{y} \leq b$, and $\lambda \in [0, 1]$. $\square$

Other important examples of convex sets are the closed and open balls.

Lemma 6.4 (convexity of balls). *Let $\mathbf{c} \in \mathbb{R}^n$ and $r > 0$. Let $\|\cdot\|$ be an arbitrary norm defined on $\mathbb{R}^n$. Then the open ball*

$$B(\mathbf{c}, r) = \{\mathbf{x} \in \mathbb{R}^n : \|\mathbf{x} - \mathbf{c}\| < r\}$$

and closed ball

$$B[\mathbf{c}, r] = \{\mathbf{x} \in \mathbb{R}^n : \|\mathbf{x} - \mathbf{c}\| \leq r\}$$

are convex.

Proof. We will show the convexity of the closed ball. The proof of the convexity of the open ball is almost identical. Let $\mathbf{x}, \mathbf{y} \in B[\mathbf{c}, r]$ and let $\lambda \in [0, 1]$. Then

$$\|\mathbf{x} - \mathbf{c}\| \leq r, \|\mathbf{y} - \mathbf{c}\| \leq r. \tag{6.1}$$

Let $\mathbf{z} = \lambda \mathbf{x} + (1-\lambda)\mathbf{y}$. We will show that $\mathbf{z} \in B[\mathbf{c}, r]$. Indeed,

$$\begin{aligned}
\|\mathbf{z} - \mathbf{c}\| &= \|\lambda \mathbf{x} + (1-\lambda)\mathbf{y} - \mathbf{c}\| \\
&= \|\lambda(\mathbf{x} - \mathbf{c}) + (1-\lambda)(\mathbf{y} - \mathbf{c})\| \\
&\leq \|\lambda(\mathbf{x} - \mathbf{c})\| + \|(1-\lambda)(\mathbf{y} - \mathbf{c})\| && \text{(triangle inequality)} \\
&= \lambda \|\mathbf{x} - \mathbf{c}\| + (1-\lambda)\|\mathbf{y} - \mathbf{c}\| && (0 \leq \lambda \leq 1) \\
&\leq \lambda r + (1-\lambda)r = r && \text{(equation (6.1))}.
\end{aligned}$$

Hence, $\mathbf{z} \in B[\mathbf{c}, r]$, establishing the result. $\square$

6.1. Definition and Examples

Note that the above result is true for any norm defined on $\mathbb{R}^n$. The *unit-ball* is the ball $B[\mathbf{0}, 1]$. There are different unit-balls, depending on the norm that is being used. The l_1, l_2, and l_∞ balls are illustrated in Figure 6.2. As always, unless otherwise specified, we assume in this book that the underlying norm is the l_2-norm and that the balls are with respect to the l_2-norm.

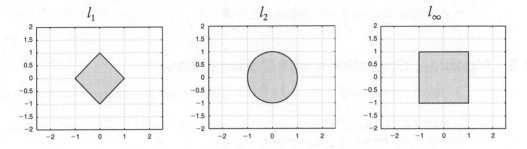

Figure 6.2. $l_1, l_2,$ and l_∞ balls in $\mathbb{R}^2$.

Another important example of convex sets are ellipsoids.

Example 6.5 (convexity of ellipsoids). An *ellipsoid* is a set of the form

$$E = \{\mathbf{x} \in \mathbb{R}^n : \mathbf{x}^T \mathbf{Q} \mathbf{x} + 2\mathbf{b}^T \mathbf{x} + c \leq 0\},$$

where $\mathbf{Q} \in \mathbb{R}^{n \times n}$ is positive semidefinite, $\mathbf{b} \in \mathbb{R}^n$, and $c \in \mathbb{R}$. Denoting

$$f(\mathbf{x}) \equiv \mathbf{x}^T \mathbf{Q} \mathbf{x} + 2\mathbf{b}^T \mathbf{x} + c,$$

the set E can be rewritten as

$$E = \{\mathbf{x} \in \mathbb{R}^n : f(\mathbf{x}) \leq 0\}.$$

To prove the convexity of E, we take $\mathbf{x}, \mathbf{y} \in E$ and $\lambda \in [0, 1]$. Then $f(\mathbf{x}) \leq 0, f(\mathbf{y}) \leq 0$, and thus the vector $\mathbf{z} = \lambda \mathbf{x} + (1 - \lambda) \mathbf{y}$ satisfies

$$\begin{aligned}\mathbf{z}^T \mathbf{Q} \mathbf{z} &= (\lambda \mathbf{x} + (1 - \lambda) \mathbf{y})^T \mathbf{Q} (\lambda \mathbf{x} + (1 - \lambda) \mathbf{y}) \\ &= \lambda^2 \mathbf{x}^T \mathbf{Q} \mathbf{x} + (1 - \lambda)^2 \mathbf{y}^T \mathbf{Q} \mathbf{y} + 2\lambda(1 - \lambda) \mathbf{x}^T \mathbf{Q} \mathbf{y}.\end{aligned} \quad (6.2)$$

Now, note that $\mathbf{x}^T \mathbf{Q} \mathbf{y} = (\mathbf{Q}^{1/2} \mathbf{x})^T (\mathbf{Q}^{1/2} \mathbf{y})$, and hence by the Cauchy-Schwarz inequality, it follows that

$$\mathbf{x}^T \mathbf{Q} \mathbf{y} \leq \|\mathbf{Q}^{1/2} \mathbf{x}\| \cdot \|\mathbf{Q}^{1/2} \mathbf{y}\| = \sqrt{\mathbf{x}^T \mathbf{Q} \mathbf{x}} \sqrt{\mathbf{y}^T \mathbf{Q} \mathbf{y}} \leq \frac{1}{2} (\mathbf{x}^T \mathbf{Q} \mathbf{x} + \mathbf{y}^T \mathbf{Q} \mathbf{y}), \quad (6.3)$$

where the last inequality follows from the fact that $\sqrt{ac} \leq \frac{1}{2}(a + c)$ for any two nonnegative scalars a, c. Plugging inequality (6.3) into (6.2), we obtain that

$$\mathbf{z}^T \mathbf{Q} \mathbf{z} \leq \lambda \mathbf{x}^T \mathbf{Q} \mathbf{x} + (1 - \lambda) \mathbf{y}^T \mathbf{Q} \mathbf{y},$$

and hence

$$\begin{aligned}f(\mathbf{z}) &= \mathbf{z}^T Q \mathbf{z} + 2\mathbf{b}^T \mathbf{z} + c \\ &\leq \lambda \mathbf{x}^T Q \mathbf{x} + (1-\lambda)\mathbf{y}^T Q \mathbf{y} + 2\lambda \mathbf{b}^T \mathbf{x} + 2(1-\lambda)\mathbf{b}^T \mathbf{y} + c \\ &= \lambda(\mathbf{x}^T Q \mathbf{x} + 2\mathbf{b}^T \mathbf{x} + c) + (1-\lambda)(\mathbf{y}^T Q \mathbf{y} + 2\mathbf{b}^T \mathbf{y} + c) \\ &= \lambda f(\mathbf{x}) + (1-\lambda) f(\mathbf{y}) \leq 0,\end{aligned}$$

establishing the desired result that $\mathbf{z} \in E$. ∎

6.2 ▪ Algebraic Operations with Convex Sets

An important property of convexity is that it is preserved under the intersection of sets.

Lemma 6.6. *Let $C_i \subseteq \mathbb{R}^n$ be a convex set for any $i \in I$, where I is an index set (possibly infinite). Then the set $\bigcap_{i \in I} C_i$ is convex.*

Proof. Suppose that $\mathbf{x}, \mathbf{y} \in \bigcap_{i \in I} C_i$ and let $\lambda \in [0,1]$. Then $\mathbf{x}, \mathbf{y} \in C_i$ for any $i \in I$, and since C_i is convex, it follows that $\lambda \mathbf{x} + (1-\lambda)\mathbf{y} \in C_i$ for any $i \in I$. Therefore, $\lambda \mathbf{x} + (1-\lambda)\mathbf{y} \in \bigcap_{i \in I} C_i$. □

Example 6.7 (convex polytopes). A direct consequence of the above result is that a set defined by a set of linear inequalities, specifically,

$$P = \{\mathbf{x} \in \mathbb{R}^n : \mathbf{A}\mathbf{x} \leq \mathbf{b}\},$$

where $\mathbf{A} \in \mathbb{R}^{m \times n}$ and $\mathbf{b} \in \mathbb{R}^m$ is convex. The convexity of P follows from the fact that it is an intersection of half-spaces:

$$P = \bigcap_{i=1}^m \{\mathbf{x} \in \mathbb{R}^n : \mathbf{A}_i \mathbf{x} \leq b_i\},$$

where $\mathbf{A}_i$ is the ith row of $\mathbf{A}$. Since half-spaces are convex (Lemma 6.3), the convexity of P follows. Sets of the form P are called *convex polytopes*. ∎

Convexity is also preserved under addition, Cartesian product, linear mappings, and inverse linear mappings. This result is now stated, and its simple proof is left as an exercise (see Exercise 6.1).

Theorem 6.8 (preservation of convexity under addition, intersection and linear mappings).

(a) *Let $C_1, C_2, \ldots, C_k \subseteq \mathbb{R}^n$ be convex sets and let $\mu_1, \mu_2, \ldots, \mu_k \in \mathbb{R}$. Then the set*

$$\mu_1 C_1 + \mu_2 C_2 + \cdots + \mu_k C_k = \left\{ \sum_{i=1}^k \mu_i \mathbf{x}_i : \mathbf{x}_i \in C_i, i = 1, 2, \ldots, k \right\}$$

is convex.

(b) *Let $C_i \subseteq \mathbb{R}^{k_i}$ be a convex set for any $i = 1, 2, \ldots, m$. Then the Cartesian product*

$$C_1 \times C_2 \times \cdots \times C_m = \{(\mathbf{x}_1, \mathbf{x}_2, \ldots, \mathbf{x}_m) : \mathbf{x}_i \in C_i, i = 1, 2, \ldots, m\}$$

is convex.

(c) Let $M \subseteq \mathbb{R}^n$ be a convex set and let $\mathbf{A} \in \mathbb{R}^{m \times n}$. Then the set

$$\mathbf{A}(M) = \{\mathbf{A}\mathbf{x} : \mathbf{x} \in M\}$$

is convex.

(d) Let $D \subseteq \mathbb{R}^m$ be a convex set, and let $\mathbf{A} \in \mathbb{R}^{m \times n}$. Then the set

$$\mathbf{A}^{-1}(D) = \{\mathbf{x} \in \mathbb{R}^n : \mathbf{A}\mathbf{x} \in D\}$$

is convex.

A direct result of part (a) of Theorem 6.8 is that if $C \subseteq \mathbb{R}^n$ is a convex set and $\mathbf{b} \in \mathbb{R}^n$, then the set

$$C + \mathbf{b} = \{\mathbf{x} + \mathbf{b} : \mathbf{x} \in C\}$$

is also convex.

6.3 • The Convex Hull

Definition 6.9 (convex combinations). *Given k vectors $\mathbf{x}_1, \mathbf{x}_2, \ldots, \mathbf{x}_k \in \mathbb{R}^n$, a **convex combination** of these k vectors is a vector of the form $\lambda_1 \mathbf{x}_1 + \lambda_2 \mathbf{x}_2 + \ldots + \lambda_k \mathbf{x}_k$, where $\lambda_1, \lambda_2, \ldots, \lambda_k$ are nonnegative numbers satisfying $\lambda_1 + \lambda_2 + \cdots + \lambda_k = 1$.*

A convex set is defined by the property that any convex combination of two points from the set is also in the set. We will now show that a convex combination of *any* number of points from a convex set is in the set.

Theorem 6.10. *Let $C \subseteq \mathbb{R}^n$ be a convex set and let $\mathbf{x}_1, \mathbf{x}_2, \ldots, \mathbf{x}_m \in C$. Then for any $\lambda \in \Delta_m$, the relation $\sum_{i=1}^m \lambda_i \mathbf{x}_i \in C$ holds.*

Proof. We will prove the result by induction on m. For $m = 1$ the result is obvious (it essentially says that $\mathbf{x}_1 \in C$ implies that $\mathbf{x}_1 \in C$...). The induction hypothesis is that for any m vectors $\mathbf{x}_1, \mathbf{x}_2, \ldots, \mathbf{x}_m \in C$ and any $\lambda \in \Delta_m$, the vector $\sum_{i=1}^m \lambda_i \mathbf{x}_i$ belongs to C. We will now prove the theorem for $m+1$ vectors. Suppose that $\mathbf{x}_1, \mathbf{x}_2, \ldots, \mathbf{x}_{m+1} \in C$ and that $\lambda \in \Delta_{m+1}$. We will show that $\mathbf{z} \equiv \sum_{i=1}^{m+1} \lambda_i \mathbf{x}_i \in C$. If $\lambda_{m+1} = 1$, then $\mathbf{z} = \mathbf{x}_{m+1} \in C$ and the result obviously follows. If $\lambda_{m+1} < 1$, then

$$\mathbf{z} = \sum_{i=1}^{m+1} \lambda_i \mathbf{x}_i$$
$$= \sum_{i=1}^{m} \lambda_i \mathbf{x}_i + \lambda_{m+1} \mathbf{x}_{m+1}$$
$$= (1 - \lambda_{m+1}) \underbrace{\sum_{i=1}^{m} \frac{\lambda_i}{1 - \lambda_{m+1}} \mathbf{x}_i}_{\mathbf{v}} + \lambda_{m+1} \mathbf{x}_{m+1}.$$

Since $\sum_{i=1}^m \frac{\lambda_i}{1-\lambda_{m+1}} = \frac{1-\lambda_{m+1}}{1-\lambda_{m+1}} = 1$, it follows that $\mathbf{v}$ (as defined in the above equation) is a convex combination of m points from C, and hence by the induction hypothesis we have that $\mathbf{v} \in C$. Thus, by the definition of a convex set, $\mathbf{z} = (1 - \lambda_{m+1})\mathbf{v} + \lambda_{m+1}\mathbf{x}_{m+1} \in C$. □

Definition 6.11 (convex hulls). *Let $S \subseteq \mathbb{R}^n$. Then the* **convex hull** *of S, denoted by* $\operatorname{conv}(S)$*, is the set comprising all the convex combinations of vectors from S:*

$$\operatorname{conv}(S) \equiv \left\{ \sum_{i=1}^{k} \lambda_i \mathbf{x}_i : \mathbf{x}_1, \mathbf{x}_2, \ldots, \mathbf{x}_k \in S, \lambda \in \Delta_k, k \in \mathbb{N} \right\}.$$

Note that in the definition of the convex hull, the number of vectors k in the convex combination representation can be any positive integer. The convex hull $\operatorname{conv}(S)$ is the "smallest" convex set containing S meaning that if another convex set T contains S, then $\operatorname{conv}(S) \subseteq T$. This property is stated and proved in the following lemma.

Lemma 6.12. *Let $S \subseteq \mathbb{R}^n$. If $S \subseteq T$ for some convex set T, then $\operatorname{conv}(S) \subseteq T$.*

Proof. Suppose that indeed $S \subseteq T$ for some convex set T. To prove that $\operatorname{conv}(S) \subseteq T$, take $\mathbf{z} \in \operatorname{conv}(S)$. Then by the definition of the convex hull, there exist $\mathbf{x}_1, \mathbf{x}_2, \ldots, \mathbf{x}_k \in S \subseteq T$ (where k is a positive integer) and $\lambda \in \Delta_k$ such that $\mathbf{z} = \sum_{i=1}^{k} \lambda_i \mathbf{x}_i$. By Theorem 6.10 and the convexity of T, it follows that any convex combination of elements from T is in T, and therefore, since $\mathbf{x}_1, \mathbf{x}_2, \ldots, \mathbf{x}_k \in T$, it follows that $\mathbf{z} \in T$, showing the desired result. □

An example of a convex hull of a nonconvex polytope is given in Figure 6.3.

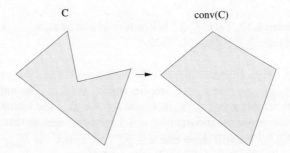

Figure 6.3. *A nonconvex set and its convex hull.*

The following well-known result, called the Carathéodory theorem, states that any element in the convex hull of a subset of a given set $S \subseteq \mathbb{R}^n$ can be represented as a convex combination of no more than $n + 1$ vectors from S.

Theorem 6.13 (Carathéodory theorem). *Let $S \subseteq \mathbb{R}^n$ and let $\mathbf{x} \in \operatorname{conv}(S)$. Then there exist $\mathbf{x}_1, \mathbf{x}_2, \ldots, \mathbf{x}_{n+1} \in S$ such that $\mathbf{x} \in \operatorname{conv}(\{\mathbf{x}_1, \mathbf{x}_2, \ldots, \mathbf{x}_{n+1}\})$; that is, there exist $\lambda \in \Delta_{n+1}$ such that*

$$\mathbf{x} = \sum_{i=1}^{n+1} \lambda_i \mathbf{x}_i.$$

6.3. The Convex Hull

Proof. Let $\mathbf{x} \in \text{conv}(S)$. By the definition of the convex hull, there exist $\mathbf{x}_1, \mathbf{x}_2, \ldots, \mathbf{x}_k \in S$ and $\lambda \in \Delta_k$ such that

$$\mathbf{x} = \sum_{i=1}^{k} \lambda_i \mathbf{x}_i.$$

We can assume that $\lambda_i > 0$ for all $i = 1, 2, \ldots, k$, since otherwise the vectors corresponding to the zero coefficients can be omitted. If $k \leq n+1$, the result is proven. Otherwise, if $k \geq n+2$, then the vectors $\mathbf{x}_2 - \mathbf{x}_1, \mathbf{x}_3 - \mathbf{x}_1, \ldots, \mathbf{x}_k - \mathbf{x}_1$, being more than n vectors in $\mathbb{R}^n$, are necessarily linearly dependent, which means that there exist $\mu_2, \mu_3, \ldots, \mu_k$ which are not all zeros such that

$$\sum_{i=2}^{k} \mu_i (\mathbf{x}_i - \mathbf{x}_1) = 0.$$

Defining $\mu_1 = -\sum_{i=2}^{k} \mu_i$, we obtain that

$$\sum_{i=1}^{k} \mu_i \mathbf{x}_i = 0,$$

where not all of the coefficients $\mu_1, \mu_2, \ldots, \mu_k$ are zeros and in addition they satisfy $\sum_{i=1}^{k} \mu_i = 0$. In particular, there exists an index i for which $\mu_i < 0$. Let $\alpha \in \mathbb{R}_+$. Then

$$\mathbf{x} = \sum_{i=1}^{k} \lambda_i \mathbf{x}_i = \sum_{i=1}^{k} \lambda_i \mathbf{x}_i + \alpha \sum_{i=1}^{k} \mu_i \mathbf{x}_i = \sum_{i=1}^{k} (\lambda_i + \alpha \mu_i) \mathbf{x}_i. \qquad (6.4)$$

We have $\sum_{i=1}^{k} (\lambda_i + \alpha \mu_i) = 1$, so the representation (6.4) is a convex combination representation if and only if

$$\lambda_i + \alpha \mu_i \geq 0 \text{ for all } i = 1, \ldots, k. \qquad (6.5)$$

Since $\lambda_i > 0$ for all i, it follows that the set of inequalities (6.5) is satisfied for all $\alpha \in [0, \varepsilon]$ where $\varepsilon = \min_{i: \mu_i < 0} \{-\frac{\lambda_i}{\mu_i}\}$. The scalar ε is well-defined since, as was already mentioned, there exists an index for which $\mu_i < 0$. If we substitute $\alpha = \varepsilon$, then (6.5) still holds, but $\lambda_j + \varepsilon \mu_j = 0$ for $j \in \text{argmin}_{i: \mu_i < 0} \{-\frac{\mu_i}{\lambda_i}\}$. This means that we have found a representation of $\mathbf{x}$ as a convex combination of $k-1$ vectors. This process can be carried on until a representation of $\mathbf{x}$ as a convex combination of no more than $n+1$ vectors is derived. $\square$

Example 6.14. For $n = 2$, consider the following four vectors:

$$\mathbf{x}_1 = \begin{pmatrix} 1 \\ 1 \end{pmatrix}, \quad \mathbf{x}_2 = \begin{pmatrix} 1 \\ 2 \end{pmatrix}, \quad \mathbf{x}_3 = \begin{pmatrix} 2 \\ 1 \end{pmatrix}, \quad \mathbf{x}_4 = \begin{pmatrix} 2 \\ 2 \end{pmatrix},$$

and let $\mathbf{x} \in \text{conv}(\{\mathbf{x}_1, \mathbf{x}_2, \mathbf{x}_3, \mathbf{x}_4\})$ be given by

$$\mathbf{x} = \frac{1}{8}\mathbf{x}_1 + \frac{1}{4}\mathbf{x}_2 + \frac{1}{2}\mathbf{x}_3 + \frac{1}{8}\mathbf{x}_4 = \begin{pmatrix} \frac{13}{8} \\ \frac{11}{8} \end{pmatrix}.$$

By the Carathéodory theorem, $\mathbf{x}$ can be expressed as a convex combination of three of the four vectors $\mathbf{x}_1, \mathbf{x}_2, \mathbf{x}_3, \mathbf{x}_4$. To find such a convex combination, let us employ the process described in the proof of the theorem. The vectors

$$\mathbf{x}_2 - \mathbf{x}_1 = \begin{pmatrix} 0 \\ 1 \end{pmatrix}, \quad \mathbf{x}_3 - \mathbf{x}_1 = \begin{pmatrix} 1 \\ 0 \end{pmatrix}, \quad \mathbf{x}_4 - \mathbf{x}_1 = \begin{pmatrix} 1 \\ 1 \end{pmatrix}$$

are linearly dependent, and the linear dependence is given by the equation

$$(\mathbf{x}_2 - \mathbf{x}_1) + (\mathbf{x}_3 - \mathbf{x}_1) - (\mathbf{x}_4 - \mathbf{x}_1) = 0.$$

We thus have the linear dependence relation

$$-\mathbf{x}_1 + \mathbf{x}_2 + \mathbf{x}_3 - \mathbf{x}_4 = 0.$$

Therefore, we can write the following for any $\alpha \geq 0$:

$$\mathbf{x} = \left(\frac{1}{8} - \alpha\right)\mathbf{x}_1 + \left(\frac{1}{4} + \alpha\right)\mathbf{x}_2 + \left(\frac{1}{2} + \alpha\right)\mathbf{x}_3 + \left(\frac{1}{8} - \alpha\right)\mathbf{x}_4.$$

The weights in the above representation add up to one, so we need only guarantee that they are nonnegative, meaning that

$$\frac{1}{8} - \alpha \geq 0, \quad \frac{1}{4} + \alpha \geq 0, \quad \frac{1}{2} + \alpha \geq 0, \quad \frac{1}{8} - \alpha \geq 0,$$

which combined with $\alpha \geq 0$ yields that $0 \leq \alpha \leq \frac{1}{8}$. Substituting $\alpha = \frac{1}{8}$, we obtain the convex combination

$$\mathbf{x} = \frac{3}{8}\mathbf{x}_2 + \frac{5}{8}\mathbf{x}_3.$$

Note that in this example two coefficients were turned into zero, so we obtained a representation with only two vectors, while the Carathéodory theorem can guarantee only a presentation by at most three vectors. ∎

6.4 ▪ Convex Cones

A set S is called a *cone* if it satisfies the following property: for any $\mathbf{x} \in S$ and $\lambda \geq 0$, the inclusion $\lambda \mathbf{x} \in S$ is satisfied. The following lemma shows that there is a very simple and elegant characterization of convex cones.

Lemma 6.15. *A set S is a convex cone if and only if the following properties hold:*

A. $\mathbf{x}, \mathbf{y} \in S \Rightarrow \mathbf{x} + \mathbf{y} \in S$.

B. $\mathbf{x} \in S, \lambda \geq 0 \Rightarrow \lambda \mathbf{x} \in S$.

Proof. (*convex cone* $\Rightarrow$ A,B). Suppose that S is a convex cone. Then property B follows from the definition of a cone. To prove property A, assume that $\mathbf{x}, \mathbf{y} \in S$. Then by the convexity of S we have that $\frac{1}{2}(\mathbf{x} + \mathbf{y}) \in S$, and hence, since S is a cone, it follows that $\mathbf{x} + \mathbf{y} = 2 \cdot \frac{1}{2}(\mathbf{x} + \mathbf{y}) \in S$.

(A,B $\Rightarrow$ *convex cone*). Now assume that S satisfies properties A and B. Then S is a cone by property B. To prove the convexity, let $\mathbf{x}, \mathbf{y} \in S$ and $\lambda \in [0,1]$. Then $\lambda\mathbf{x}, (1-\lambda)\mathbf{y} \in S$ by property B, and thus by property A, $\lambda\mathbf{x} + (1-\lambda)\mathbf{y} \in S$, establishing the convexity of S. □

The following are some well-known examples of convex cones.

6.4. Convex Cones

Example 6.16. Consider the convex polytope

$$C = \{\mathbf{x} \in \mathbb{R}^n : \mathbf{A}\mathbf{x} \leq \mathbf{0}\},$$

where $\mathbf{A} \in \mathbb{R}^{m \times n}$. The set C is clearly a convex set since it is a convex polytope (see Example 6.7). It is also a cone since

$$\mathbf{x} \in C, \lambda \geq 0 \Rightarrow \mathbf{A}\mathbf{x} \leq \mathbf{0}, \lambda \geq 0 \Rightarrow \mathbf{A}(\lambda \mathbf{x}) \leq \mathbf{0} \Rightarrow \lambda \mathbf{x} \in C.$$

Taking, for example, $m = n$ and $\mathbf{A} = -\mathbf{I}$, the set C reduces to the nonnegative orthant $\mathbb{R}^n_+$. ∎

Example 6.17 (Lorentz cone). The *Lorentz cone*, or *ice cream cone* whose boundary is described in Figure 6.4, is given by

$$L^n = \left\{ \begin{pmatrix} \mathbf{x} \\ t \end{pmatrix} \in \mathbb{R}^{n+1} : \|\mathbf{x}\| \leq t, \mathbf{x} \in \mathbb{R}^n, t \in \mathbb{R} \right\}.$$

The Lorentz cone is in fact a convex cone. To show this, let us take $\binom{\mathbf{x}}{t}, \binom{\mathbf{y}}{s} \in L^n$. Then $\|\mathbf{x}\| \leq t, \|\mathbf{y}\| \leq s$, which combined with the triangle inequality implies that

$$\|\mathbf{x} + \mathbf{y}\| \leq \|\mathbf{x}\| + \|\mathbf{y}\| \leq t + s,$$

showing that $\binom{\mathbf{x}}{t} + \binom{\mathbf{y}}{s} \in L^n$, and hence that property A holds. To show property B, take $\binom{\mathbf{x}}{t} \in L^n$ and $\lambda \geq 0$, then since $\|\mathbf{x}\| \leq t$, it readily follows that $\|\lambda \mathbf{x}\| \leq \lambda t$, so that $\lambda \binom{\mathbf{x}}{t} \in L^n$. ∎

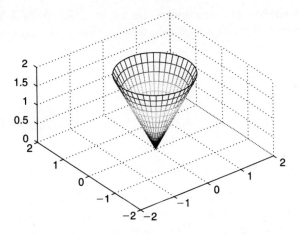

Figure 6.4. *The boundary of the ice cream cone L^2.*

Example 6.18 (nonnegative polynomials). Consider the set of all coefficients of polynomials of degree of at most $n-1$ which are nonnegative over $\mathbb{R}$:

$$K^n = \{\mathbf{x} \in \mathbb{R}^n : x_1 t^{n-1} + x_2 t^{n-2} + \cdots + x_{n-1} t + x_n \geq 0 \text{ for all } t \in \mathbb{R}\}.$$

It is easy to verify that this is a convex cone. Let us consider two special cases. When $n = 2$, then clearly

$$K^2 = \{(x_1, x_2)^T : x_1 t + x_2 \geq 0 \text{ for all } t \in \mathbb{R}\} = \{(x_1, x_2) : x_1 = 0, x_2 \geq 0\},$$

so K^2 is the nonnegative part of the x_2-axis. For $n=3$ we have

$$K^3 = \{(x_1, x_2, x_3)^T : x_1 t^2 + x_2 t + x_3 \geq 0\}.$$

A quadratic polynomial $\varphi(t) = at^2 + bt + c$ is nonnegative over $\mathbb{R}$ if and only if $a, c \geq 0$ and the discriminant $\Delta = b^2 - 4ac$ is nonpositive. We thus conclude that

$$K^3 = \{(x_1, x_2, x_3)^T : x_1, x_3 \geq 0, x_2^2 \leq 4 x_1 x_3\}. \quad \blacksquare$$

Similarly to the notion of a convex combination, we will now define the concept of a *conic combination*.

Definition 6.19 (conic combination). *Given k points $\mathbf{x}_1, \mathbf{x}_2, \ldots, \mathbf{x}_k \in \mathbb{R}^n$, a **conic combination** of these k points is a vector of the form $\lambda_1 \mathbf{x}_1 + \lambda_2 \mathbf{x}_2 + \cdots + \cdots + \lambda_k \mathbf{x}_k$, where $\lambda \in \mathbb{R}_+^k$.*

It is easy to show that any conic combination of points in a convex cone C belong to C. The proof of this elementary result is left as an exercise (Exercise 6.14).

Lemma 6.20. *Let C be a convex cone, and let $\mathbf{x}_1, \mathbf{x}_2, \ldots, \mathbf{x}_k \in C$ and $\lambda_1, \lambda_2, \ldots, \lambda_k \geq 0$. Then $\sum_{i=1}^k \lambda_i \mathbf{x}_i \in C$.*

The definition of the *conic hull* is now quite natural.

Definition 6.21 (conic hulls). *Let $S \subseteq \mathbb{R}^n$. Then the **conic hull** of S, denoted by $\mathrm{cone}(S)$, is the set comprising all the conic combinations of vectors from S:*

$$\mathrm{cone}(S) \equiv \left\{ \sum_{i=1}^k \lambda_i \mathbf{x}_i : \mathbf{x}_1, \mathbf{x}_2, \ldots, \mathbf{x}_k \in S, \lambda \in \mathbb{R}_+^k, k \in \mathbb{N} \right\}.$$

Similar to the convex hull, the conic hull of a set S is the smallest convex cone containing S. The proof of this result is left as an exercise (Exercise 6.15).

Lemma 6.22. *Let $S \subseteq \mathbb{R}^n$. If $S \subseteq T$ for some convex cone T, then $\mathrm{cone}(S) \subseteq T$.*

A natural question that arises is whether we can establish a result similar to the Carathéodory theorem on the representation of vectors in the conic hull of a set. Interestingly, we can establish an even stronger result for conic hulls: each vector in the conic hull of a set $S \subseteq \mathbb{R}^n$ can be represented as a convex combination of at most n vectors from S (recall that in Carathéodory's theorem $n+1$ vectors are required).

Theorem 6.23 (conic representation theorem). *Let $S \subseteq \mathbb{R}^n$ and let $\mathbf{x} \in \mathrm{cone}(S)$. Then there exist k linearly independent vectors $\mathbf{x}_1, \mathbf{x}_2, \ldots, \mathbf{x}_k \in S$ such that $\mathbf{x} \in \mathrm{cone}(\{\mathbf{x}_1, \mathbf{x}_2, \ldots, \mathbf{x}_k\})$; that is, there exists $\lambda \in \mathbb{R}_+^k$ such that*

$$\mathbf{x} = \sum_{i=1}^k \lambda_i \mathbf{x}_i.$$

In addition, $k \leq n$.

6.4. Convex Cones

Proof. The proof is similar to the proof of the Carathéodory theorem. Let $\mathbf{x} \in \text{cone}(S)$. By the definition of the conic hull, there exist $\mathbf{x}_1, \mathbf{x}_2, \ldots, \mathbf{x}_k \in S$ and $\lambda \in \mathbb{R}_+^k$ such that

$$\mathbf{x} = \sum_{i=1}^{k} \lambda_i \mathbf{x}_i.$$

We can assume that $\lambda_i > 0$ for all $i = 1, 2, \ldots, k$, since otherwise the vectors corresponding to the zero λ_i's can be omitted. If the vectors $\mathbf{x}_1, \mathbf{x}_2, \ldots, \mathbf{x}_k$ are linearly independent, then the result is proven. Otherwise, if the vectors are linearly dependent, then there exist $\mu_1, \mu_2, \ldots, \mu_k \in \mathbb{R}$ which are not all zeros such that

$$\sum_{i=1}^{k} \mu_i \mathbf{x}_i = 0.$$

Then for any $\alpha \in \mathbb{R}$

$$\mathbf{x} = \sum_{i=1}^{k} \lambda_i \mathbf{x}_i = \sum_{i=1}^{k} \lambda_i \mathbf{x}_i + \alpha \sum_{i=1}^{k} \mu_i \mathbf{x}_i = \sum_{i=1}^{k} (\lambda_i + \alpha \mu_i) \mathbf{x}_i. \tag{6.6}$$

The representation (6.6) is a conic representation if and only if

$$\lambda_i + \alpha \mu_i \geq 0 \text{ for all } i = 1, \ldots, k. \tag{6.7}$$

Since $\lambda_i > 0$ for all i, it follows that the set of inequalities (6.7) is satisfied for all $\alpha \in I$ where I is a closed interval with a nonempty interior. Note that one (but not both) of the endpoints of I might be infinite. If we substitute one of the finite endpoints of I, call it $\tilde{\alpha}$, into α, then we still get that (6.7) holds, but in addition $\lambda_j + \tilde{\alpha} \mu_j = 0$ for some index j. Thus we obtain a representation of $\mathbf{x}$ as a conic combination of at most $k-1$ vectors. This process can be carried on until we obtain a representation of $\mathbf{x}$ as a conic combination of linearly independent vectors. Since the vectors $\mathbf{x}_1, \mathbf{x}_2, \ldots, \mathbf{x}_k$ are linearly independent vectors in $\mathbb{R}^n$ it follows that $k \leq n$. $\square$

The latter representation theorem has an important application to convex polytopes of the form

$$P = \{\mathbf{x} \in \mathbb{R}^n : \mathbf{Ax} = \mathbf{b}, \mathbf{x} \geq 0\},$$

where $\mathbf{A} \in \mathbb{R}^{m \times n}$ and $\mathbf{b} \in \mathbb{R}^m$. We will assume without loss of generality that the rows of $\mathbf{A}$ are linearly independent. Linear systems consisting of linear equalities and nonnegativity constraints often appear as constraints in standard formulations of linear programming problems. An important property of nonempty convex polytopes of the form P is that they contain at least one vector with at most m nonzero elements (m being the number of constraints). An important notion in this respect is the one of a *basic feasible solution*.

Definition 6.24 (basic feasible solutions). *Let $P = \{\mathbf{x} \in \mathbb{R}^n : \mathbf{Ax} = \mathbf{b}, \mathbf{x} \geq 0\}$, where $\mathbf{A} \in \mathbb{R}^{m \times n}$ and $\mathbf{b} \in \mathbb{R}^m$. Suppose that the rows of $\mathbf{A}$ are linearly independent. Then $\bar{\mathbf{x}}$ is a* **basic feasible solution** *(abbreviated bfs) of P if the columns of $\mathbf{A}$ corresponding to the indices of the positive values of $\bar{\mathbf{x}}$ are linearly independent.*

Obviously, since the columns of $\mathbf{A}$ reside in $\mathbb{R}^m$, it follows that a bfs has at most m nonzero elements.

Example 6.25. Consider the linear system

$$x_1 + x_2 + x_3 = 6,$$
$$x_2 + x_3 = 3,$$
$$x_1, x_2, x_3 \geq 0.$$

An example of a bfs of the system is $(x_1, x_2, x_3) = (3, 3, 0)$. This vector is indeed a bfs since it satisfies all the constraints and the columns corresponding to the positive elements, meaning columns 1 and 2

$$\begin{pmatrix} 1 \\ 0 \end{pmatrix}, \begin{pmatrix} 1 \\ 1 \end{pmatrix},$$

are linearly independent. ∎

The existence of a bfs in P, provided that it is nonempty, follows directly from Theorem 6.23.

Theorem 6.26. *Let* $P = \{\mathbf{x} \in \mathbb{R}^n : \mathbf{A}\mathbf{x} = \mathbf{b}, \mathbf{x} \geq \mathbf{0}\}$, *where* $\mathbf{A} \in \mathbb{R}^{m \times n}$ *and* $\mathbf{b} \in \mathbb{R}^m$. *If* $P \neq \emptyset$, *then it contains at least one bfs.*

Proof. Since $P \neq \emptyset$, it follows that $\mathbf{b} \in \text{cone}(\{\mathbf{a}_1, \mathbf{a}_2, \ldots, \mathbf{a}_n\})$, where $\mathbf{a}_i$ denotes the ith column of $\mathbf{A}$. By the conic representation theorem (Theorem 6.23), we have that $\mathbf{b}$ can be represented as a conic combination of k linearly independent vectors from $\{\mathbf{a}_1, \mathbf{a}_2, \ldots, \mathbf{a}_n\}$; that is, there exist indices $i_1 < i_2 < \cdots < i_k$ and k numbers $y_{i_1}, y_{i_2}, \ldots, y_{i_k} \geq 0$ such that $\mathbf{b} = \sum_{j=1}^{k} y_{i_j} \mathbf{a}_{i_j}$ and $\mathbf{a}_{i_1}, \mathbf{a}_{i_2}, \ldots, \mathbf{a}_{i_k}$ are linearly independent. Denote $\bar{\mathbf{x}} = \sum_{j=1}^{k} y_{i_j} \mathbf{e}_{i_j}$. Then obviously $\bar{\mathbf{x}} \geq \mathbf{0}$ and in addition

$$\mathbf{A}\bar{\mathbf{x}} = \sum_{j=1}^{k} y_{i_j} \mathbf{A}\mathbf{e}_{i_j} = \sum_{j=1}^{k} y_{i_j} \mathbf{a}_{i_j} = \mathbf{b}.$$

Therefore, $\bar{\mathbf{x}}$ is contained in P and satisfies that the columns of $\mathbf{A}$ corresponding to the indices of the positive components of $\bar{\mathbf{x}}$ are linearly independent, meaning that P contains a bfs. □

6.5 • Topological Properties of Convex Sets

We begin by proving that the closure of a convex set is a convex set.

Theorem 6.27 (convexity preservation under closure). *Let* $C \subseteq \mathbb{R}^n$ *be a convex set. Then* $\text{cl}(C)$ *is a convex set.*

Proof. Let $\mathbf{x}, \mathbf{y} \in \text{cl}(C)$ and let $\lambda \in [0, 1]$. Then by the definition of the closure set, it follows that there exist sequences $\{\mathbf{x}_k\}_{k \geq 0} \subseteq C$ and $\{\mathbf{y}_k\}_{k \geq 0} \subseteq C$ for which $\mathbf{x}_k \to \mathbf{x}$ and $\mathbf{y}_k \to \mathbf{y}$ as $k \to \infty$. By the convexity of C, it follows that $\lambda \mathbf{x}_k + (1 - \lambda)\mathbf{y}_k \in C$ for any $k \geq 0$. Since $\lambda \mathbf{x}_k + (1 - \lambda)\mathbf{y}_k \to \lambda \mathbf{x} + (1 - \lambda)\mathbf{y}$, it follows that there exists a sequence in C that converges to $\lambda \mathbf{x} + (1 - \lambda)\mathbf{y}$, implying that $\lambda \mathbf{x} + (1 - \lambda)\mathbf{y} \in \text{cl}(C)$. □

Proving that the interior of a convex set is also a convex set is more tricky, and we will require the following technical and quite useful result called the *line segment principle*.

6.5. Topological Properties of Convex Sets

Lemma 6.28 (line segment principle). *Let C be a convex set, and assume that $\text{int}(C) \neq \emptyset$. Suppose that $\mathbf{x} \in \text{int}(C)$ and $\mathbf{y} \in \text{cl}(C)$. Then $(1-\lambda)\mathbf{x} + \lambda\mathbf{y} \in \text{int}(C)$ for any $\lambda \in (0,1)$.*

Proof. Since $\mathbf{x} \in \text{int}(C)$, there exists $\varepsilon > 0$ such that $B(\mathbf{x}, \varepsilon) \subseteq C$. Let $\mathbf{z} = (1-\lambda)\mathbf{x} + \lambda\mathbf{y}$. To prove that $\mathbf{z} \in \text{int}(C)$, we will show that in fact $B(\mathbf{z}, (1-\lambda)\varepsilon) \subseteq C$. Let then $\mathbf{w}$ be a vector satisfying $\|\mathbf{w} - \mathbf{z}\| < (1-\lambda)\varepsilon$. Since $\mathbf{y} \in \text{cl}(C)$, it follows that there exists $\mathbf{w}_1 \in C$ such that
$$\|\mathbf{w}_1 - \mathbf{y}\| < \frac{(1-\lambda)\varepsilon - \|\mathbf{w} - \mathbf{z}\|}{\lambda}. \tag{6.8}$$

Set $\mathbf{w}_2 = \frac{1}{1-\lambda}(\mathbf{w} - \lambda\mathbf{w}_1)$. Then
$$\begin{aligned}
\|\mathbf{w}_2 - \mathbf{x}\| &= \left\|\frac{\mathbf{w} - \lambda\mathbf{w}_1}{1-\lambda} - \mathbf{x}\right\| \\
&= \frac{1}{1-\lambda}\|(\mathbf{w} - \mathbf{z}) + \lambda(\mathbf{y} - \mathbf{w}_1)\| \\
&\leq \frac{1}{1-\lambda}(\|\mathbf{w} - \mathbf{z}\| + \lambda\|\mathbf{w}_1 - \mathbf{y}\|) \\
&\stackrel{(6.8)}{<} \varepsilon,
\end{aligned}$$

and hence, since $B(\mathbf{x}, \varepsilon) \subseteq C$, it follows that $\mathbf{w}_2 \in C$. Finally, since $\mathbf{w} = \lambda\mathbf{w}_1 + (1-\lambda)\mathbf{w}_2$ with $\mathbf{w}_1, \mathbf{w}_2 \in C$, we have that $\mathbf{w} \in C$, and the line segment principle is thus proved. $\square$

The immediate consequence of the line segment principle is the convexity of interiors of convex sets.

Theorem 6.29 (convexity of interiors of convex sets). *Let $C \subseteq \mathbb{R}^n$ be a convex set. Then $\text{int}(C)$ is convex.*

Proof. If $\text{int}(C) = \emptyset$, then the theorem is obviously true. Otherwise, let $\mathbf{x}_1, \mathbf{x}_2 \in \text{int}(C)$, and let $\lambda \in (0,1)$. Then by the line segment principle we have that $\lambda\mathbf{x}_1 + (1-\lambda)\mathbf{x}_2 \in \text{int}(C)$, establishing the convexity of $\text{int}(C)$. $\square$

Other topological properties that are implied by the line segment property are given in the next result.

Lemma 6.30. *Let C be a convex set with a nonempty interior. Then*

(a) $\text{cl}(\text{int}(C)) = \text{cl}(C)$,

(b) $\text{int}(\text{cl}(C)) = \text{int}(C)$.

Proof. (a) Obviously, since $\text{int}(C) \subseteq C$, the inclusion $\text{cl}(\text{int}(C)) \subseteq \text{cl}(C)$ holds. To prove the opposite, let $\mathbf{x} \in \text{cl}(C)$ and let $\mathbf{y} \in \text{int}(C)$. Then by the line segment principle, $\mathbf{x}_k = \frac{1}{k}\mathbf{y} + (1 - \frac{1}{k})\mathbf{x} \in \text{int}(C)$ for any $k \geq 1$. Since $\mathbf{x}$ is the limit of the sequence $\{\mathbf{x}_k\}_{k \geq 1} \subseteq \text{int}(C)$, it follows that $\mathbf{x} \in \text{cl}(\text{int}(C))$.

(b) The inclusion $\text{int}(C) \subseteq \text{int}(\text{cl}(C))$ follows immediately from the inclusion $C \subseteq \text{cl}(C)$. To show the reverse inclusion, we take $\mathbf{x} \in \text{int}(\text{cl}(C))$ and show that $\mathbf{x} \in \text{int}(C)$. Since $\mathbf{x} \in \text{int}(\text{cl}(C))$, there exists $\varepsilon > 0$ such that $B(\mathbf{x}, \varepsilon) \subseteq \text{cl}(C)$. Let $\mathbf{y} \in \text{int}(C)$. If $\mathbf{y} = \mathbf{x}$, then the result is proved. Otherwise, define

$$\mathbf{z} = \mathbf{x} + \alpha(\mathbf{x} - \mathbf{y}),$$

where $\alpha = \frac{\varepsilon}{2\|\mathbf{x}-\mathbf{y}\|}$. Since $\|\mathbf{z}-\mathbf{x}\| = \frac{\varepsilon}{2}$, it follows that $\mathbf{z} \in \mathrm{cl}(C)$. Therefore, $(1-\lambda)\mathbf{y} + \lambda\mathbf{z} \in \mathrm{int}(C)$ for any $\lambda \in [0,1)$ and specifically for $\lambda = \lambda_\alpha = \frac{1}{1+\alpha}$. Noting that $(1-\lambda_\alpha)\mathbf{y} + \lambda_\alpha\mathbf{z} = \mathbf{x}$, we conclude that $\mathbf{x} \in \mathrm{int}(C)$. $\square$

In general, the convex hull of a closed set is not necessarily a closed set. A classical example for this fact is the closed set

$$S = \{(0,0)^T\} \cup \{(x,y)^T : xy \geq 1, x \geq 0, y \geq 0\},$$

whose convex hull is given by the set

$$\mathrm{conv}(S) = \{(0,0)^T\} \cup \mathbb{R}^2_{++},$$

which is neither closed nor open. However, closedness is preserved under the convex hull operation if the set is compact. As will be shown in the proof of the following result, this nontrivial fact follows from the Carathéodory theorem.

Proposition 6.31 (closedness of convex hulls of compact sets). *Let $S \subseteq \mathbb{R}^n$ be a compact set. Then $\mathrm{conv}(S)$ is compact.*

Proof. To prove the boundedness of $\mathrm{conv}(S)$, note that since S is compact, there exists $M > 0$ such that $\|\mathbf{x}\| \leq M$ for any $\mathbf{x} \in S$. Now, let $\mathbf{y} \in \mathrm{conv}(S)$. Then by the Carathéodory theorem it follows that there exist $\mathbf{x}_1, \mathbf{x}_2, \ldots, \mathbf{x}_{n+1} \in S$ and $\lambda \in \Delta_{n+1}$ for which $\mathbf{y} = \sum_{i=1}^{n+1} \lambda_i \mathbf{x}_i$, and therefore

$$\|\mathbf{y}\| = \left\|\sum_{i=1}^{n+1} \lambda_i \mathbf{x}_i\right\| \leq \sum_{i=1}^{n+1} \lambda_i \|\mathbf{x}_i\| \leq M \sum_{i=1}^{n+1} \lambda_i = M,$$

establishing the boundedness of $\mathrm{conv}(S)$. To prove the closedness of $\mathrm{conv}(S)$, let $\{\mathbf{y}_k\}_{k \geq 1} \subseteq \mathrm{conv}(S)$ be a sequence of vectors from $\mathrm{conv}(S)$ converging to $\mathbf{y} \in \mathbb{R}^n$. Our objective is to show that $\mathbf{y} \in \mathrm{conv}(S)$. By the Carathéodory theorem we have that for any $k \geq 1$ there exist vectors $\mathbf{x}_1^k, \mathbf{x}_2^k, \ldots, \mathbf{x}_{n+1}^k \in S$ and $\lambda^k \in \Delta_{n+1}$ such that

$$\mathbf{y}_k = \sum_{i=1}^{n+1} \lambda_i^k \mathbf{x}_i^k. \tag{6.9}$$

By the compactness of S and Δ_{n+1}, it follows that $\{(\lambda^k, \mathbf{x}_1^k, \mathbf{x}_2^k, \ldots, \mathbf{x}_{n+1}^k)\}_{k \geq 1}$ has a convergent subsequence $\{(\lambda^{k_j}, \mathbf{x}_1^{k_j}, \mathbf{x}_2^{k_j}, \ldots, \mathbf{x}_{n+1}^{k_j})\}_{j \geq 1}$ whose limit will be denoted by

$$(\lambda, \mathbf{x}_1, \mathbf{x}_2, \ldots, \mathbf{x}_{n+1})$$

with $\lambda \in \Delta_{n+1}, \mathbf{x}_1, \mathbf{x}_2, \ldots, \mathbf{x}_{n+1} \in S$. Taking the limit $j \to \infty$ in

$$\mathbf{y}_{k_j} = \sum_{i=1}^{n+1} \lambda_i^{k_j} \mathbf{x}_i^{k_j},$$

we obtain that

$$\mathbf{y} = \sum_{i=1}^{n+1} \lambda_i \mathbf{x}_i,$$

meaning that $\mathbf{y} \in \text{conv}(S)$ as required. $\square$

Another topological result which might seem simple but actually requires the conic representation theorem is the closedness of the conic hull of a finite set of points.

Lemma 6.32 (closedness of the conic hull of a finite set). *Let $\mathbf{a}_1, \mathbf{a}_2, \ldots, \mathbf{a}_k \in \mathbb{R}^n$. Then $\text{cone}(\{\mathbf{a}_1, \mathbf{a}_2, \ldots, \mathbf{a}_k\})$ is closed.*

Proof. By the conic representation theorem, each element of $\text{cone}(\{\mathbf{a}_1, \mathbf{a}_2, \ldots, \mathbf{a}_k\})$ can be represented as a conic combination of a linearly independent subset of $\{\mathbf{a}_1, \mathbf{a}_2, \ldots, \mathbf{a}_k\}$. Therefore, if $S_1, S_2, \ldots, S_N$ are all the subsets of $\{\mathbf{a}_1, \mathbf{a}_2, \ldots, \mathbf{a}_k\}$ comprising linearly independent vectors, then

$$\text{cone}(\{\mathbf{a}_1, \mathbf{a}_2, \ldots, \mathbf{a}_k\}) = \bigcup_{i=1}^{N} \text{cone}(S_i).$$

It is enough to show that $\text{cone}(S_i)$ is closed for any $i \in \{1, 2, \ldots, N\}$. Indeed, let $i \in \{1, 2, \ldots, N\}$. Then

$$S_i = \{\mathbf{b}_1, \mathbf{b}_2, \ldots, \mathbf{b}_m\}$$

for some linearly independent vectors $\mathbf{b}_1, \mathbf{b}_2, \ldots, \mathbf{b}_m$. We can write $\text{cone}(S_i)$ as

$$\text{cone}(S_i) = \{\mathbf{By} : \mathbf{y} \in \mathbb{R}_+^m\},$$

where $\mathbf{B}$ is the matrix whose columns are $\mathbf{b}_1, \mathbf{b}_2, \ldots, \mathbf{b}_m$. Suppose that $\mathbf{x}_k \in \text{cone}(S_i)$ for all $k \geq 1$ and that $\mathbf{x}_k \to \bar{\mathbf{x}}$. We need to show that $\bar{\mathbf{x}} \in \text{cone}(S_i)$. Since $\mathbf{x}_k \in \text{cone}(S_i)$, it follows that there exists $\mathbf{y}_k \in \mathbb{R}_+^m$ such that

$$\mathbf{x}_k = \mathbf{B}\mathbf{y}_k. \tag{6.10}$$

Therefore, using the fact that the columns of $\mathbf{B}$ are linearly independent, we can deduce that

$$\mathbf{y}_k = (\mathbf{B}^T \mathbf{B})^{-1} \mathbf{B}^T \mathbf{x}_k.$$

Taking the limit as $k \to \infty$ in the last equation, we obtain that $\mathbf{y}_k \to \bar{\mathbf{y}}$ where $\bar{\mathbf{y}} = (\mathbf{B}^T \mathbf{B})^{-1} \mathbf{B}^T \bar{\mathbf{x}}$, and since $\mathbf{y}_k \in \mathbb{R}_+^m$ for all k, we also have $\bar{\mathbf{y}} \in \mathbb{R}_+^m$. Thus, taking the limit in (6.10), we conclude that $\bar{\mathbf{x}} = \mathbf{B}\bar{\mathbf{y}}$ with $\bar{\mathbf{y}} \in \mathbb{R}_+^m$, and hence $\bar{\mathbf{x}} \in \text{cone}(S_i)$. $\square$

6.6 • Extreme Points

Definition 6.33 (extreme points). *Let $S \subseteq \mathbb{R}^n$ be a convex set. A point $\mathbf{x} \in S$ is called an* **extreme point** *of S if there do not exist $\mathbf{x}_1, \mathbf{x}_2 \in S(\mathbf{x}_1 \neq \mathbf{x}_2)$ and $\lambda \in (0,1)$, such that $\mathbf{x} = \lambda \mathbf{x}_1 + (1-\lambda)\mathbf{x}_2$.*

That is, an extreme point is a point in the set that cannot be represented as a nontrivial convex combination of two different points in S. The set of extreme points is denoted by $\text{ext}(S)$. The set of extreme points of a convex polytope consists of all its vertices; see, for example, Figure 6.5.

We can fully characterize the extreme points of convex polytopes of the form $P = \{\mathbf{x} \in \mathbb{R}^n : \mathbf{A}\mathbf{x} = \mathbf{b}, \mathbf{x} \geq \mathbf{0}\}$, where $\mathbf{A} \in \mathbb{R}^{m \times n}$ has linearly independent rows and $\mathbf{b} \in \mathbb{R}^m$.

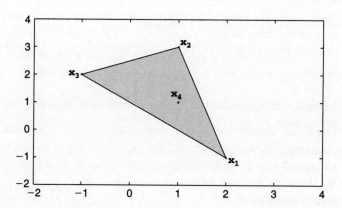

Figure 6.5. *The filled area is the convex set* $S = \text{conv}\{x_1, x_2, x_3, x_4\}$, *where* $x_1 = (2, -1)^T, x_2 = (1, 3)^T, x_3 = (-1, 2)^T, x_4 = (1, 1)^T$. *The extreme points set is* $\text{ext}(S) = \{x_1, x_2, x_3\}$.

Recall (see Section 6.4) that $\bar{x}$ is called a basic feasible solution of P if the columns of A corresponding to the indices of the positive values of $\bar{x}$ are linearly independent. In Section 6.4 it was shown that if P is not empty, then it has at least one basic feasible solution. Interestingly, the extreme points of P are *exactly* the basic feasible solutions of P, which means that the linear independence of the columns of A corresponding to the positive variables is an algebraic characterization of extreme points.

Theorem 6.34 (equivalence between extreme points and basic feasible solutions). *Let* $P = \{x \in \mathbb{R}^n : Ax = b, x \geq 0\}$, *where* $A \in \mathbb{R}^{m \times n}$ *has linearly independent rows and* $b \in \mathbb{R}^m$. *Then* $\bar{x}$ *is a basic feasible solution of* P *if and only if it is an extreme point of* P.

Proof. Suppose that $\bar{x}$ is a basic feasible solution and assume without loss of generality that its first k components are positive while the others are zeros: $\bar{x}_1 > 0, \bar{x}_2 > 0, \ldots, \bar{x}_k > 0, \bar{x}_{k+1} = \bar{x}_{k+2} = \cdots = \bar{x}_n = 0$. Since $\bar{x}$ is a basic feasible solution, the first k columns of A denoted by $a_1, a_2, \ldots, a_k$ are linearly independent. Suppose in contradiction that $\bar{x} \notin \text{ext}(P)$. Then there exist two different vectors $y, z \in P$ and $\lambda \in (0, 1)$ such that $\bar{x} = \lambda y + (1 - \lambda) z$. Combining this with the fact that $y, z \geq 0$, we can conclude that the last $n - k$ variables in y and z are zeros. We can therefore write

$$\sum_{i=1}^{k} y_i a_i = b,$$

$$\sum_{i=1}^{k} z_i a_i = b.$$

Subtracting the second inequality from the first, we obtain

$$\sum_{i=1}^{k} (y_i - z_i) a_i = 0,$$

and since $y \neq z$, we obtain that the vectors $a_1, a_2, \ldots, a_k$ are linearly dependent, which is a contradiction to the assumption that they are linearly independent. To prove the reverse direction, let us suppose that $\tilde{x} \in P$ is an extreme point and assume in contradiction that $\tilde{x}$ is not a basic feasible solution. This means that the columns corresponding to the positive components are linearly dependent. Without loss of generality, let us assume that the positive variables are exactly the first k components. The linear dependance of the first k columns means that there exists a nonzero vector $\mathbf{y} \in \mathbb{R}^k$ for which

$$\sum_{i=1}^{k} y_i \mathbf{a}_i = 0.$$

The above identity can also be written as $\mathbf{A}\tilde{\mathbf{y}} = 0$, where $\tilde{\mathbf{y}} = \binom{\mathbf{y}}{0_{n-k}}$. Since the first k components of $\tilde{\mathbf{x}}$ are positive, it follows that there exist an $\varepsilon > 0$ for which $\mathbf{x}_1 = \tilde{\mathbf{x}} + \varepsilon \tilde{\mathbf{y}} \geq 0$ and $\mathbf{x}_2 = \tilde{\mathbf{x}} - \varepsilon \tilde{\mathbf{y}} \geq 0$. In addition $\mathbf{A}\mathbf{x}_1 = \mathbf{A}\tilde{\mathbf{x}} + \varepsilon \mathbf{A}\tilde{\mathbf{y}} = \mathbf{b} + \varepsilon \cdot 0 = \mathbf{b}$, and similarly $\mathbf{A}\mathbf{x}_2 = \mathbf{b}$. We thus conclude that $\mathbf{x}_1, \mathbf{x}_2 \in P$; these two vectors are different since $\tilde{\mathbf{y}}$ is not the zeros vector, and finally, $\tilde{\mathbf{x}} = \frac{1}{2}\mathbf{x}_1 + \frac{1}{2}\mathbf{x}_2$, contradicting the assumption that $\tilde{\mathbf{x}}$ is an extreme point. $\square$

We finish this section with a very important and well-known theorem called the Krein-Milman theorem, stating that a compact convex set is the convex hull of its extreme points. We will state this theorem without a proof.

Theorem 6.35 (Krein-Milman). *Let $S \subseteq \mathbb{R}^n$ be a compact convex set. Then*

$$S = \mathrm{conv}(\mathrm{ext}(S)).$$

Exercises

6.1. Prove Theorem 6.8.

6.2. Give an example of two convex sets C_1, C_2 whose union $C_1 \cup C_2$ is not convex.

6.3. Show that the following set is not convex:

$$S = \{\mathbf{x} \in \mathbb{R}^2 : x_1^2 - x_2^2 + x_1 + x_2 \leq 4\}.$$

6.4. Prove that

$$\mathrm{conv}\{\mathbf{e}_1, \mathbf{e}_2, -\mathbf{e}_1, -\mathbf{e}_2\} = \{\mathbf{x} \in \mathbb{R}^2 : |x_1| + |x_2| \leq 1\},$$

where $\mathbf{e}_1 = (1,0)^T, \mathbf{e}_2 = (0,1)^T$.

6.5. Let K be a convex, bounded, and symmetric[2] set such that $0 \in \mathrm{int} K$. Define the Minkowski functional as

$$p(\mathbf{x}) = \inf\left\{\lambda > 0 : \frac{\mathbf{x}}{\lambda} \in K\right\}, \quad \mathbf{x} \in \mathbb{R}^n.$$

Show that $p(\cdot)$ is a norm.

6.6. Show the following properties of the convex hull:

  (i) If $S \subseteq T$, then $\mathrm{conv}(S) \subseteq \mathrm{conv}(T)$.

  (ii) For any $S \subseteq \mathbb{R}^n$, the identity $\mathrm{conv}(\mathrm{conv}(S)) = \mathrm{conv}(S)$ holds.

  (iii) For any $S_1, S_2 \subseteq \mathbb{R}^n$ the following holds:

$$\mathrm{conv}(S_1 + S_2) = \mathrm{conv}(S_1) + \mathrm{conv}(S_2).$$

---

[2] A set S is called *symmetric* if $\mathbf{x} \in S$ implies $-\mathbf{x} \in S$.

6.7. Let C be a convex set. Prove that cone(C) is a convex set.

6.8. Show that the conic hull of the set
$$S = \{(x_1, x_2) : (x_1 - 1)^2 + x_2^2 = 1\}$$
is the set
$$\{(x_1, x_2) : x_1 > 0\} \cup \{(0, 0)\}.$$

Remark: This is an example illustrating the fact that the conic hull of a closed set is not necessarily a closed set.

6.9. Let $\mathbf{a}, \mathbf{b} \in \mathbb{R}^n (\mathbf{a} \neq \mathbf{b})$. For what values of μ is the set
$$S_\mu = \{\mathbf{x} \in \mathbb{R}^n : \|\mathbf{x} - \mathbf{a}\| \leq \mu \|\mathbf{x} - \mathbf{b}\|\}$$
convex?

6.10. Let $C \subseteq \mathbb{R}^n$ be a nonempty convex set. For each $\mathbf{x} \in C$ define the *normal cone* of C at $\mathbf{x}$ by
$$N_C(\mathbf{x}) = \{\mathbf{w} \in \mathbb{R}^n : \langle \mathbf{w}, \mathbf{y} - \mathbf{x} \rangle \leq 0 \text{ for all } \mathbf{y} \in C\},$$
and define $N_C(\mathbf{x}) = \emptyset$ when $\mathbf{x} \notin C$. Show that $N_C(\mathbf{x})$ is a closed convex cone.

6.11. Let $C \subseteq \mathbb{R}^n$ be cone. The **dual cone** is defined by
$$C^* = \{\mathbf{y} \in \mathbb{R}^n : \langle \mathbf{y}, \mathbf{x} \rangle \geq 0 \text{ for all } \mathbf{x} \in C\}.$$

(i) Prove that C^* is a closed convex cone (even if C is nonconvex).

(ii) Prove that if C_1, C_2 are cones satisfying $C_1 \subseteq C_2$, then $C_2^* \subseteq C_1^*$.

(iii) Show that $(L^n)^* = L^n$.

(iv) Show that the dual cone of $K = \{\binom{\mathbf{x}}{t} : \|\mathbf{x}\|_1 \leq t\}$ is
$$K^* = \left\{\binom{\mathbf{x}}{t} : \|\mathbf{x}\|_\infty \leq t\right\}.$$

6.12. A cone K is called *pointed* if it contains no lines, meaning that $\mathbf{x}, -\mathbf{x} \in K \Rightarrow \mathbf{x} = \mathbf{0}$. Show that if K has a nonempty interior, then K^* is pointed.

6.13. Consider the optimization problem
$$(P_\mathbf{a}) \quad \min\{\mathbf{a}^T \mathbf{x} : \mathbf{x} \in S\},$$
where $S \subseteq \mathbb{R}^n$. Let $\mathbf{x}^* \in S$ and let $K \subseteq \mathbb{R}^n$ be the set of all vectors $\mathbf{a}$ for which $\mathbf{x}^*$ is an optimal solution of $(P_\mathbf{a})$. Show that K is a convex cone.

6.14. Prove Lemma 6.20.

6.15. Prove Lemma 6.22.

6.16. Find all the basic feasible solutions of the system
$$-4x_2 + x_3 = 6,$$
$$2x_1 - 2x_2 - x_4 = 1,$$
$$x_1, x_2, x_3, x_4 \geq 0.$$

6.17. Let S be a convex set. Prove that $\mathbf{x} \in S$ is an extreme point of S if and only if $S \setminus \{\mathbf{x}\}$ is convex.

6.18. Let $S = \{\mathbf{x} \in \mathbb{R}^n : \|\mathbf{x}\|_\infty \leq 1\}$. Show that
$$\text{ext}(S) = \{\mathbf{x} \in \mathbb{R}^n : x_i^2 = 1, i = 1, 2, \ldots, n\}.$$

6.19. Let $S = \{\mathbf{x} \in \mathbb{R}^n : \|\mathbf{x}\|_2 \leq 1\}$. Show that
$$\text{ext}(S) = \{\mathbf{x} \in \mathbb{R}^n : \|\mathbf{x}\|_2 = 1\}.$$

6.20. Let $X_i \subseteq \mathbb{R}^{n_i}, i = 1, 2, \ldots, k$. Prove that
$$\text{ext}(X_1 \times X_2 \times \cdots \times X_k) = \text{ext}(X_1) \times \text{ext}(X_2) \times \cdots \times \text{ext}(X_k).$$

Chapter 7
Convex Functions

7.1 • Definition and Examples

In the last chapter we introduced the notion of a convex set. This chapter is devoted to the concept of convex functions, which is fundamental in the theory of optimization.

Definition 7.1 (convex functions). *A function $f : C \to \mathbb{R}$ defined on a convex set $C \subseteq \mathbb{R}^n$ is called* **convex** *(or* **convex over** *C) if*

$$f(\lambda \mathbf{x} + (1-\lambda)\mathbf{y}) \leq \lambda f(\mathbf{x}) + (1-\lambda)f(\mathbf{y}) \text{ for any } \mathbf{x}, \mathbf{y} \in C, \lambda \in [0,1]. \tag{7.1}$$

The fundamental inequality (7.1) is illustrated in Figure 7.1.

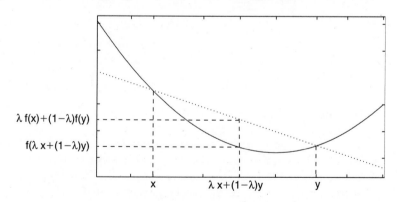

Figure 7.1. *Illustration of the inequality $f(\lambda \mathbf{x} + (1-\lambda)\mathbf{y}) \leq \lambda f(\mathbf{x}) + (1-\lambda)f(\mathbf{y})$.*

In case when no domain is specified, then we naturally assume that f is defined over the entire space $\mathbb{R}^n$. If we do not allow equality in (7.1) when $\mathbf{x} \neq \mathbf{y}$ and $\lambda \in (0,1)$, the function is called *strictly convex*.

Definition 7.2 (strictly convex functions). *A function $f : C \to \mathbb{R}$ defined on a convex set $C \subseteq \mathbb{R}^n$ is called* **strictly convex** *if*

$$f(\lambda \mathbf{x} + (1-\lambda)\mathbf{y}) < \lambda f(\mathbf{x}) + (1-\lambda)f(\mathbf{y}) \text{ for any } \mathbf{x} \neq \mathbf{y} \in C, \lambda \in (0,1).$$

Another important concept is *concavity*. A function is called *concave* if $-f$ is convex. Similarly, f is called *strictly concave* if $-f$ is strictly convex. We can of course write a more direct definition of concavity based on the definition of convexity. A function f is concave over a convex set $C \subseteq \mathbb{R}^n$ if and only if for any $\mathbf{x}, \mathbf{y} \in C$ and $\lambda \in [0,1]$ we have

$$f(\lambda \mathbf{x} + (1-\lambda)\mathbf{y}) \geq \lambda f(\mathbf{x}) + (1-\lambda)f(\mathbf{y}).$$

Equipped only with the definition of convexity, we can give some elementary examples of convex functions. We begin by showing the convexity of affine functions, which are functions of the form $f(\mathbf{x}) = \mathbf{a}^T \mathbf{x} + b$, where $\mathbf{a} \in \mathbb{R}^n$ and $b \in \mathbb{R}$. (If $b = 0$, then f is also called *linear*.)

Example 7.3 (convexity of affine functions). Let $f(\mathbf{x}) = \mathbf{a}^T \mathbf{x} + b$, where $\mathbf{a} \in \mathbb{R}^n$ and $b \in \mathbb{R}$. To show that f is convex, take $\mathbf{x}, \mathbf{y} \in \mathbb{R}^n$ and $\lambda \in [0,1]$. Then

$$\begin{aligned}
f(\lambda \mathbf{x} + (1-\lambda)\mathbf{y}) &= \mathbf{a}^T(\lambda \mathbf{x} + (1-\lambda)\mathbf{y}) + b \\
&= \lambda(\mathbf{a}^T \mathbf{x}) + (1-\lambda)(\mathbf{a}^T \mathbf{y}) + \lambda b + (1-\lambda)b \\
&= \lambda(\mathbf{a}^T \mathbf{x} + b) + (1-\lambda)(\mathbf{a}^T \mathbf{y} + b) \\
&= \lambda f(\mathbf{x}) + (1-\lambda)f(\mathbf{y}),
\end{aligned}$$

and thus in particular $f(\lambda \mathbf{x} + (1-\lambda)\mathbf{y}) \leq \lambda f(\mathbf{x}) + (1-\lambda)f(\mathbf{y})$, and convexity follows. Of course, if f is an affine function, then so is $-f$, which implies that affine functions (and in fact, as shown in Exercise 7.3, only affine functions) are both convex and concave. ∎

Example 7.4 (convexity of norms). Let $\|\cdot\|$ be a norm on $\mathbb{R}^n$. We will show that the norm function $f(\mathbf{x}) = \|\mathbf{x}\|$ is convex. Indeed, let $\mathbf{x}, \mathbf{y} \in \mathbb{R}^n$ and $\lambda \in [0,1]$. Then by the triangle inequality we have

$$\begin{aligned}
f(\lambda \mathbf{x} + (1-\lambda)\mathbf{y}) &= \|\lambda \mathbf{x} + (1-\lambda)\mathbf{y}\| \\
&\leq \|\lambda \mathbf{x}\| + \|(1-\lambda)\mathbf{y}\| \\
&= \lambda\|\mathbf{x}\| + (1-\lambda)\|\mathbf{y}\| \\
&= \lambda f(\mathbf{x}) + (1-\lambda)f(\mathbf{y}),
\end{aligned}$$

establishing the convexity of f. ∎

The basic property characterizing a convex function is that the function value of a convex combination of *two* points $\mathbf{x}$ and $\mathbf{y}$ is smaller than or equal to the corresponding convex combination of the function values $f(\mathbf{x})$ and $f(\mathbf{y})$. An interesting result is that convexity implies that this property can be generalized to convex combinations of *any* number of vectors. This is the so-called Jensen's inequality.

Theorem 7.5 (Jensen's inequality). *Let $f : C \to \mathbb{R}$ be a convex function defined on the convex set $C \subseteq \mathbb{R}^n$. Then for any $\mathbf{x}_1, \mathbf{x}_2, \ldots, \mathbf{x}_k \in C$ and $\lambda \in \Delta_k$, the following inequality holds:*

$$f\left(\sum_{i=1}^k \lambda_i \mathbf{x}_i\right) \leq \sum_{i=1}^k \lambda_i f(\mathbf{x}_i). \tag{7.2}$$

Proof. We will prove the inequality (7.2) by induction on k. For $k = 1$ the result is obvious (it amounts to $f(\mathbf{x}_1) \leq f(\mathbf{x}_1)$ for any $\mathbf{x}_1 \in C$). The induction hypothesis is that for any k

vectors $x_1, x_2, \ldots, x_k \in C$ and any $\lambda \in \Delta_k$, the inequality (7.2) holds. We will now prove the theorem for $k+1$ vectors. Suppose that $x_1, x_2, \ldots, x_{k+1} \in C$ and that $\lambda \in \Delta_{k+1}$. We will show that $f(z) \leq \sum_{i=1}^{k+1} \lambda_i f(x_i)$, where $z = \sum_{i=1}^{k+1} \lambda_i x_i$. If $\lambda_{k+1} = 1$, then $z = x_{k+1}$ and (7.2) is obvious. If $\lambda_{k+1} < 1$, then

$$f(z) = f\left(\sum_{i=1}^{k+1} \lambda_i x_i\right)$$

$$= f\left(\sum_{i=1}^{k} \lambda_i x_i + \lambda_{k+1} x_{k+1}\right) \tag{7.3}$$

$$= f\left((1-\lambda_{k+1}) \underbrace{\sum_{i=1}^{k} \frac{\lambda_i}{1-\lambda_{k+1}} x_i}_{v} + \lambda_{k+1} x_{k+1}\right) \tag{7.4}$$

$$\leq (1-\lambda_{k+1}) f(v) + \lambda_{k+1} f(x_{k+1}). \tag{7.5}$$

Since $\sum_{i=1}^{k} \frac{\lambda_i}{1-\lambda_{k+1}} = \frac{1-\lambda_{k+1}}{1-\lambda_{k+1}} = 1$, it follows that v is a convex combination of k points from C, and hence by the induction hypothesis we have that $f(v) \leq \sum_{i=1}^{k} \frac{\lambda_i}{1-\lambda_{k+1}} f(x_i)$, which combined with (7.5) yields

$$f(z) \leq \sum_{i=1}^{k+1} \lambda_i f(x_i). \quad \square$$

7.2 ▪ First Order Characterizations of Convex Functions

Convex functions are not necessarily differentiable, but in case they are, we can replace the Jensen's inequality definition with other characterizations which utilize the gradient of the function. An important characterizing inequality is *the gradient inequality*, which essentially states that the tangent hyperplanes of convex functions are always underestimates of the function.

Theorem 7.6 (the gradient inequality). *Let $f : C \to \mathbb{R}$ be a continuously differentiable function defined on a convex set $C \subseteq \mathbb{R}^n$. Then f is convex over C if and only if*

$$f(x) + \nabla f(x)^T (y-x) \leq f(y) \text{ for any } x, y \in C. \tag{7.6}$$

Proof. Suppose first that f is convex. Let $x, y \in C$ and $\lambda \in (0,1]$. If $x = y$, then (7.6) trivially holds. We will therefore assume that $x \neq y$. Then

$$f(\lambda y + (1-\lambda) x) \leq \lambda f(y) + (1-\lambda) f(x),$$

and hence

$$\frac{f(x + \lambda(y-x)) - f(x)}{\lambda} \leq f(y) - f(x).$$

Taking $\lambda \to 0^+$, the left-hand side converges to the directional derivative of f at x in the direction $y - x$, so that

$$f'(x; y-x) \leq f(y) - f(x).$$

Since f is continuously differentiable, it follows that $f'(\mathbf{x};\mathbf{y}-\mathbf{x}) = \nabla f(\mathbf{x})^T(\mathbf{y}-\mathbf{x})$, and hence (7.6) follows.

To prove the reverse direction, assume the gradient inequality holds. Let $\mathbf{z},\mathbf{w} \in C$, and let $\lambda \in (0,1)$. We will show that $f(\lambda \mathbf{z} + (1-\lambda)\mathbf{w}) \leq \lambda f(\mathbf{z}) + (1-\lambda)f(\mathbf{w})$. Let $\mathbf{u} = \lambda \mathbf{z} + (1-\lambda)\mathbf{w} \in C$. Then

$$\mathbf{z} - \mathbf{u} = \frac{\mathbf{u} - (1-\lambda)\mathbf{w}}{\lambda} - \mathbf{u} = -\frac{1-\lambda}{\lambda}(\mathbf{w} - \mathbf{u}).$$

Invoking the gradient inequality on the pairs $\mathbf{z},\mathbf{u}$ and $\mathbf{w},\mathbf{u}$, we obtain

$$f(\mathbf{u}) + \nabla f(\mathbf{u})^T(\mathbf{z} - \mathbf{u}) \leq f(\mathbf{z}),$$
$$f(\mathbf{u}) - \frac{\lambda}{1-\lambda}\nabla f(\mathbf{u})^T(\mathbf{z} - \mathbf{u}) \leq f(\mathbf{w}).$$

Multiplying the first inequality by $\frac{\lambda}{1-\lambda}$ and adding it to the second one, we obtain

$$\frac{1}{1-\lambda}f(\mathbf{u}) \leq \frac{\lambda}{1-\lambda}f(\mathbf{z}) + f(\mathbf{w}),$$

which after multiplication by $1-\lambda$ amounts to the desired inequality:

$$f(\mathbf{u}) \leq \lambda f(\mathbf{z}) + (1-\lambda)f(\mathbf{w}). \quad \Box$$

A modification of the above proof will show that a function is *strictly convex* if and only if the gradient inequality is satisfied with strict inequality for any $\mathbf{x} \neq \mathbf{y}$.

Theorem 7.7 (the gradient inequality for strictly convex function). *Let $f : C \rightarrow \mathbb{R}$ be a continuously differentiable function defined on a convex set $C \subseteq \mathbb{R}^n$. Then f is strictly convex over C if and only if*

$$f(\mathbf{x}) + \nabla f(\mathbf{x})^T(\mathbf{y} - \mathbf{x}) < f(\mathbf{y}) \textit{ for any } \mathbf{x},\mathbf{y} \in C \textit{ satisfying } \mathbf{x} \neq \mathbf{y}.$$

Geometrically, the gradient inequality essentially states that for convex functions, the tangent hyperplane is below the surface of the function. A two-dimensional illustration is given in Figure 7.2.

A direct result of the gradient inequality is that the first order optimality condition $\nabla f(\mathbf{x}^*) = \mathbf{0}$ is sufficient for global optimality.

Proposition 7.8 (sufficiency of stationarity under convexity). *Let f be a continuously differentiable function which is convex over a convex set $C \subseteq \mathbb{R}^n$. Suppose that $\nabla f(\mathbf{x}^*) = \mathbf{0}$ for some $\mathbf{x}^* \in C$. Then $\mathbf{x}^*$ is a global minimizer of f over C.*

Proof. Let $\mathbf{z} \in C$. Plugging $\mathbf{x} = \mathbf{x}^*$ and $\mathbf{y} = \mathbf{z}$ in the gradient inequality (7.6), we obtain that

$$f(\mathbf{z}) \geq f(\mathbf{x}^*) + \nabla f(\mathbf{x}^*)^T(\mathbf{z} - \mathbf{x}^*),$$

which by the fact that $\nabla f(\mathbf{x}^*) = \mathbf{0}$ implies that $f(\mathbf{z}) \geq f(\mathbf{x}^*)$, thus establishing that $\mathbf{x}^*$ is a global minimizer of f over C. $\quad \Box$

7.2. First Order Characterizations of Convex Functions

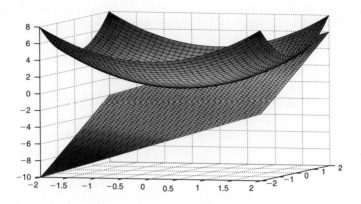

Figure 7.2. *The function $f(x,y) = x^2 + y^2$ and its tangent hyperplane at $(1,1)$, which is a lower bound of the function's surface.*

We note that Proposition 7.8 establishes only the sufficiency of the stationarity condition $\nabla f(\mathbf{x}^*) = 0$ for guaranteeing that $\mathbf{x}^*$ is a global optimal solution. When C is not the entire space, this condition is not necessary. However, when $C = \mathbb{R}^n$, then by Theorem 2.6, this is also a necessary condition, and we can thus write the following statement.

Theorem 7.9 (necessity and sufficiency of stationarity). *Let $f : \mathbb{R}^n \to \mathbb{R}$ be a continuously differentiable convex function. Then $\nabla f(\mathbf{x}^*) = 0$ if and only if $\mathbf{x}^*$ is a global minimum point of f over $\mathbb{R}^n$.*

Using the gradient inequality we can now establish the conditions under which a quadratic function is convex/strictly convex.

Theorem 7.10 (convexity and strict convexity of quadratic functions with positive semidefinite matrices). *Let $f : \mathbb{R}^n \to \mathbb{R}$ be the quadratic function given by $f(\mathbf{x}) = \mathbf{x}^T \mathbf{A} \mathbf{x} + 2\mathbf{b}^T \mathbf{x} + c$, where $\mathbf{A} \in \mathbb{R}^{n \times n}$ is symmetric, $\mathbf{b} \in \mathbb{R}^n$, and $c \in \mathbb{R}$. Then f is (strictly) convex if and only if $\mathbf{A} \succeq 0$ ($\mathbf{A} \succ 0$).*

Proof. By Theorem 7.6 the convexity of f is equivalent to the validity of the gradient inequality:
$$f(\mathbf{y}) \geq f(\mathbf{x}) + \nabla f(\mathbf{x})^T (\mathbf{y} - \mathbf{x}) \text{ for any } \mathbf{x}, \mathbf{y} \in \mathbb{R}^n,$$
which can be written explicitly as
$$\mathbf{y}^T \mathbf{A} \mathbf{y} + 2\mathbf{b}^T \mathbf{y} + c \geq \mathbf{x}^T \mathbf{A} \mathbf{x} + 2\mathbf{b}^T \mathbf{x} + c + 2(\mathbf{A}\mathbf{x} + \mathbf{b})^T (\mathbf{y} - \mathbf{x}) \text{ for any } \mathbf{x}, \mathbf{y} \in \mathbb{R}^n.$$

After some rearrangement of terms, we can rewrite the latter inequality as
$$(\mathbf{y} - \mathbf{x})^T \mathbf{A} (\mathbf{y} - \mathbf{x}) \geq 0 \text{ for any } \mathbf{x}, \mathbf{y} \in \mathbb{R}^n. \tag{7.7}$$

Making the transformation $\mathbf{d} = \mathbf{y} - \mathbf{x}$, we conclude that inequality (7.7) is equivalent to the inequality $\mathbf{d}^T \mathbf{A} \mathbf{d} \geq 0$ for any $\mathbf{d} \in \mathbb{R}^n$, which is the same as saying that $\mathbf{A} \succeq 0$. To prove the strict convexity variant, note that strict convexity of f is the same as
$$f(\mathbf{y}) > f(\mathbf{x}) + \nabla f(\mathbf{x})^T (\mathbf{y} - \mathbf{x}) \text{ for any } \mathbf{x}, \mathbf{y} \in \mathbb{R}^n \text{ such that } \mathbf{x} \neq \mathbf{y}.$$

The same arguments as above imply that this is equivalent to

$$\mathbf{d}^T \mathbf{A} \mathbf{d} > 0 \text{ for any } 0 \neq \mathbf{d} \in \mathbb{R}^n,$$

which is the same as $\mathbf{A} \succ \mathbf{0}$. □

Examples of convex and nonconvex quadratic functions are illustrated in Figure 7.3.

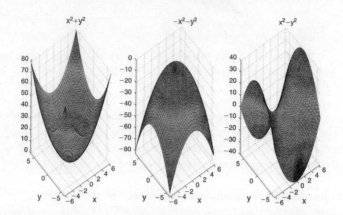

Figure 7.3. *The left quadratic function is convex ($f(x,y) = x^2 + y^2$), while the middle ($-x^2 - y^2$) and right ($x^2 - y^2$) functions are nonconvex.*

Another type of a first order characterization of convexity is the monotonicity property of the gradient. In the one-dimensional case, this means that the derivative is nondecreasing, but another definition of monotonicity is required in the n-dimensional case.

Theorem 7.11 (monotonicity of the gradient). *Suppose that f is a continuously differentiable function over a convex set $C \subseteq \mathbb{R}^n$. Then f is convex over C if and only if*

$$(\nabla f(\mathbf{x}) - \nabla f(\mathbf{y}))^T (\mathbf{x} - \mathbf{y}) \geq 0 \text{ for any } \mathbf{x}, \mathbf{y} \in C. \tag{7.8}$$

Proof. Assume first that f is convex over C. Then by the gradient inequality we have for any $\mathbf{x}, \mathbf{y} \in C$

$$f(\mathbf{x}) \geq f(\mathbf{y}) + \nabla f(\mathbf{y})^T (\mathbf{x} - \mathbf{y}),$$
$$f(\mathbf{y}) \geq f(\mathbf{x}) + \nabla f(\mathbf{x})^T (\mathbf{y} - \mathbf{x}).$$

By summing the two inequalities, the inequality (7.8) follows. To prove the opposite direction, suppose that (7.8) holds and let $\mathbf{x}, \mathbf{y} \in C$. Let g be the one-dimensional function defined by

$$g(t) = f(\mathbf{x} + t(\mathbf{y} - \mathbf{x})), \quad t \in [0,1].$$

By the fundamental theorem of calculus we have

$$f(\mathbf{y}) = g(1) = g(0) + \int_0^1 g'(t)dt$$

$$= f(\mathbf{x}) + \int_0^1 (\mathbf{y}-\mathbf{x})^T \nabla f(\mathbf{x}+t(\mathbf{y}-\mathbf{x}))dt$$

$$= f(\mathbf{x}) + \nabla f(\mathbf{x})^T(\mathbf{y}-\mathbf{x}) + \int_0^1 (\mathbf{y}-\mathbf{x})^T (\nabla f(\mathbf{x}+t(\mathbf{y}-\mathbf{x})) - \nabla f(\mathbf{x}))dt$$

$$\geq f(\mathbf{x}) + \nabla f(\mathbf{x})^T(\mathbf{y}-\mathbf{x}),$$

where the last inequality follows from the fact that for any $t > 0$ we have by the monotonicity of ∇f that

$$(\mathbf{y}-\mathbf{x})^T(\nabla f(\mathbf{x}+t(\mathbf{y}-\mathbf{x})) - \nabla f(\mathbf{x})) = \frac{1}{t}(\nabla f(\mathbf{x}+t(\mathbf{y}-\mathbf{x})) - \nabla f(\mathbf{x}))^T(\mathbf{x}+t(\mathbf{y}-\mathbf{x}) - \mathbf{x}) \geq 0.$$

□

7.3 ▪ Second Order Characterization of Convex Functions

For twice continuously differentiable functions, convexity can be characterized by the positive semidefiniteness of the Hessian matrix.

Theorem 7.12 (second order characterization of convexity). *Let f be a twice continuously differentiable function over an open convex set $C \subseteq \mathbb{R}^n$. Then f is convex over C if and only if $\nabla^2 f(\mathbf{x}) \succeq 0$ for any $\mathbf{x} \in C$.*

Proof. Suppose that $\nabla^2 f(\mathbf{x}) \succeq 0$ for all $\mathbf{x} \in C$. We will prove the gradient inequality, which by Theorem 7.6 is enough in order to establish convexity. Let $\mathbf{x}, \mathbf{y} \in C$. Then by the linear approximation theorem (Theorem 1.24) we have that there exists $\mathbf{z} \in [\mathbf{x}, \mathbf{y}]$ (and hence $\mathbf{z} \in C$) for which

$$f(\mathbf{y}) = f(\mathbf{x}) + \nabla f(\mathbf{x})^T(\mathbf{y}-\mathbf{x}) + \frac{1}{2}(\mathbf{y}-\mathbf{x})^T \nabla^2 f(\mathbf{z})(\mathbf{y}-\mathbf{x}). \tag{7.9}$$

Since $\nabla^2 f(\mathbf{z}) \succeq 0$, it follows that $(\mathbf{y}-\mathbf{x})^T \nabla^2 f(\mathbf{z})(\mathbf{y}-\mathbf{x}) \geq 0$, and hence by (7.9), the inequality $f(\mathbf{y}) \geq f(\mathbf{x}) + \nabla f(\mathbf{x})^T(\mathbf{y}-\mathbf{x})$ holds.

To prove the opposite direction, assume that f is convex over C. Let $\mathbf{x} \in C$ and let $\mathbf{y} \in \mathbb{R}^n$. Since C is open, it follows that $\mathbf{x} + \lambda \mathbf{y} \in C$ for $0 < \lambda < \varepsilon$, where ε is a small enough positive number. Invoking the gradient inequality we have

$$f(\mathbf{x}+\lambda\mathbf{y}) \geq f(\mathbf{x}) + \lambda \nabla f(\mathbf{x})^T \mathbf{y}. \tag{7.10}$$

In addition, by the quadratic approximation theorem (Theorem 1.25) we have that

$$f(\mathbf{x}+\lambda\mathbf{y}) = f(\mathbf{x}) + \lambda \nabla f(\mathbf{x})^T \mathbf{y} + \frac{\lambda^2}{2} \mathbf{y}^T \nabla^2 f(\mathbf{x})\mathbf{y} + o(\lambda^2 \|\mathbf{y}\|^2),$$

which combined with (7.10) yields the inequality

$$\frac{\lambda^2}{2} \mathbf{y}^T \nabla^2 f(\mathbf{x})\mathbf{y} + o(\lambda^2 \|\mathbf{y}\|^2) \geq 0$$

for any $\lambda \in (0,\varepsilon)$. Dividing the latter inequality by λ^2 we have

$$\frac{1}{2}\mathbf{y}^T\nabla^2 f(\mathbf{x})\mathbf{y} + \frac{o(\lambda^2\|\mathbf{y}\|^2)}{\lambda^2} \geq 0.$$

Finally, taking $\lambda \to 0^+$, we conclude that

$$\mathbf{y}^T\nabla^2 f(\mathbf{x})\mathbf{y} \geq 0$$

for any $\mathbf{y} \in \mathbb{R}^n$, implying that $\nabla^2 f(\mathbf{x}) \succeq 0$ for any $\mathbf{x} \in C$. $\square$

We also present the corresponding result for strictly convex functions stating that if the Hessian is positive definite, then the function is strictly convex. The proof of this result is similar to the one given in Theorem 7.12 and is hence left as an exercise (see Exercise 7.6).

Theorem 7.13 (sufficient second order condition for strict convexity). *Let f be a twice continuously differentiable function over a convex set $C \subseteq \mathbb{R}^n$, and suppose that $\nabla^2 f(\mathbf{x}) \succ 0$ for any $\mathbf{x} \in C$. Then f is strictly convex over C.*

Note that the positive definiteness of the Hessian is only a sufficient condition for strict convexity and is not necessary. Indeed, the function $f(x) = x^4$ is strictly convex, but its second order derivative $f''(x) = 12x^2$ is equal to zero for $x = 0$. The Hessian test immediately establishes the strict convexity of the one-dimensional functions x^2, e^x, e^{-x} and also of $-\ln(x), x\ln(x)$ over $\mathbb{R}_{++}$. A much more complicated example is that of the so-called *log-sum-exp* function, whose convexity can be shown by the Hessian test.

Example 7.14 (convexity of the log-sum-exp function). Consider the function

$$f(\mathbf{x}) = \ln(e^{x_1} + e^{x_2} + \cdots + e^{x_n}),$$

called the *log-sum-exp* function and defined over the entire space $\mathbb{R}^n$. We will prove its convexity using the Hessian test. The partial derivatives of f are given by

$$\frac{\partial f}{\partial x_i}(\mathbf{x}) = \frac{e^{x_i}}{\sum_{k=1}^n e^{x_k}}, \quad i = 1, 2, \ldots, n,$$

and therefore

$$\frac{\partial^2 f}{\partial x_i \partial x_j}(\mathbf{x}) = \begin{cases} -\frac{e^{x_i} e^{x_j}}{\left(\sum_{k=1}^n e^{x_k}\right)^2}, & i \neq j, \\ -\frac{e^{x_i} e^{x_i}}{\left(\sum_{k=1}^n e^{x_k}\right)^2} + \frac{e^{x_i}}{\sum_{k=1}^n e^{x_k}}, & i = j. \end{cases}$$

We can thus write the Hessian matrix as

$$\nabla^2 f(\mathbf{x}) = \text{diag}(\mathbf{w}) - \mathbf{w}\mathbf{w}^T,$$

where $w_i = \frac{e^{x_i}}{\sum_{j=1}^n e^{x_j}}$. In particular, $\mathbf{w} \in \Delta_n$. To prove the positive semidefiniteness of $\nabla^2 f(\mathbf{x})$, take $\mathbf{0} \neq \mathbf{v} \in \mathbb{R}^n$ and consider the expression

$$\mathbf{v}^T \nabla^2 f(\mathbf{x}) \mathbf{v} = \sum_{i=1}^n w_i v_i^2 - (\mathbf{v}^T\mathbf{w})^2.$$

The latter expression is nonnegative since employing the Cauchy–Schwarz inequality on the vectors $\mathbf{s}, \mathbf{t}$ defined by

$$s_i = \sqrt{w_i} v_i, \quad t_i = \sqrt{w_i}, \quad i = 1, 2, \ldots, n,$$

yields

$$(\mathbf{v}^T \mathbf{w})^2 = (\mathbf{s}^T \mathbf{t})^2 \leq \|\mathbf{s}\|^2 \|\mathbf{t}\|^2 = \left(\sum_{i=1}^n w_i v_i^2\right)\left(\sum_{i=1}^n w_i\right) \stackrel{\mathbf{w} \in \Delta_n}{=} \sum_{i=1}^n w_i v_i^2,$$

establishing the inequality $\mathbf{v}^T \nabla^2 f(\mathbf{x}) \mathbf{v} \geq 0$. Since the latter inequality is valid for any $\mathbf{v} \in \mathbb{R}^n$, it follows that $\nabla^2 f(\mathbf{x})$ is indeed positive semidefinite. ∎

Example 7.15 (quadratic-over-linear). Let

$$f(x_1, x_2) = \frac{x_1^2}{x_2},$$

defined over $\mathbb{R} \times \mathbb{R}_{++} = \{(x_1, x_2) : x_2 > 0\}$. The Hessian of f is given by

$$\nabla^2 f(x_1, x_2) = 2 \begin{pmatrix} \frac{1}{x_2} & -\frac{x_1}{x_2^2} \\ -\frac{x_1}{x_2^2} & \frac{x_1^2}{x_2^3} \end{pmatrix}.$$

By Proposition 2.20, since the Hessian is a 2×2 matrix, to prove that it is positive semidefinite, it is enough to show that the trace and determinant are nonnegative, and indeed,

$$\text{Tr}[\nabla^2 f(x_1, x_2)] = 2\left[\frac{1}{x_2} + \frac{x_1^2}{x_2^3}\right] > 0,$$

$$\det[\nabla^2 f(x_1, x_2)] = 4\left[\frac{1}{x_2} \cdot \frac{x_1^2}{x_2^3} - \left(\frac{x_1}{x_2^2}\right)^2\right] = 0,$$

establishing the positive semidefiniteness of $\nabla^2 f(x_1, x_2)$ and hence the convexity of f. ∎

7.4 ▪ Operations Preserving Convexity

There are several important operations that preserve the convexity property. First, the sum of convex functions is a convex function and a multiplication of a convex function by a nonnegative number results with a convex function.

Theorem 7.16 (preservation of convexity under summation and multiplication by nonnegative scalars).

(a) *Let f be a convex function defined over a convex set $C \subseteq \mathbb{R}^n$ and let $\alpha \geq 0$. Then αf is a convex function over C.*

(b) *Let $f_1, f_2, \ldots, f_p$ be convex functions over a convex set $C \subseteq \mathbb{R}^n$. Then the sum function $f_1 + f_2 + \cdots + f_p$ is convex over C.*

Proof. (a) Denote $g(\mathbf{x}) \equiv \alpha f(\mathbf{x})$. We will prove the convexity of g by definition. Let $\mathbf{x}, \mathbf{y} \in C$ and $\lambda \in [0,1]$. Then

$$\begin{aligned}
g(\lambda \mathbf{x} + (1-\lambda)\mathbf{y}) &= \alpha f(\lambda \mathbf{x} + (1-\lambda)\mathbf{y}) && \text{(definition of } g\text{)}\\
&\leq \alpha \lambda f(\mathbf{x}) + \alpha(1-\lambda) f(\mathbf{y}) && \text{(convexity of } f\text{)}\\
&= \lambda g(\mathbf{x}) + (1-\lambda) g(\mathbf{y}) && \text{(definition of } g\text{)}.
\end{aligned}$$

(b) Let $\mathbf{x}, \mathbf{y} \in C$ and $\lambda \in [0,1]$. For each $i = 1, 2, \ldots, p$, since f_i is convex, we have

$$f_i(\lambda \mathbf{x} + (1-\lambda)\mathbf{y}) \leq \lambda f_i(\mathbf{x}) + (1-\lambda) f_i(\mathbf{y}).$$

Summing the latter inequality over $i = 1, 2, \ldots, k$ yields the inequality

$$g(\lambda \mathbf{x} + (1-\lambda)\mathbf{y}) \leq \lambda g(\mathbf{x}) + (1-\lambda) g(\mathbf{y})$$

for all $\mathbf{x}, \mathbf{y} \in C$ and $\lambda \in [0,1]$, where $g = f_1 + f_2 + \cdots + f_p$. We have thus established that the sum function is convex. $\square$

Another important operation preserving convexity is affine change of variables.

Theorem 7.17 (preservation of convexity under affine change of variables). *Let $f : C \to \mathbb{R}$ be a convex function defined on a convex set $C \subseteq \mathbb{R}^n$. Let $\mathbf{A} \in \mathbb{R}^{n \times m}$ and $\mathbf{b} \in \mathbb{R}^n$. Then the function g defined by*

$$g(\mathbf{y}) = f(\mathbf{A}\mathbf{y} + \mathbf{b})$$

is convex over the convex set $D = \{\mathbf{y} \in \mathbb{R}^m : \mathbf{A}\mathbf{y} + \mathbf{b} \in C\}$.

Proof. First of all, note that D is indeed a convex set since it can be represented as an inverse linear mapping of a translation of C (see Theorem 6.8):

$$D = \mathbf{A}^{-1}(C - \mathbf{b}).$$

Let $\mathbf{y}_1, \mathbf{y}_2 \in D$. Define

$$\begin{aligned}
\mathbf{x}_1 &= \mathbf{A}\mathbf{y}_1 + \mathbf{b}, & (7.11)\\
\mathbf{x}_2 &= \mathbf{A}\mathbf{y}_2 + \mathbf{b}, & (7.12)
\end{aligned}$$

which by the definition of D satisfy $\mathbf{x}_1, \mathbf{x}_2 \in C$. Let $\lambda \in [0,1]$. By the convexity of f we have

$$f(\lambda \mathbf{x}_1 + (1-\lambda)\mathbf{x}_2) \leq \lambda f(\mathbf{x}_1) + (1-\lambda) f(\mathbf{x}_2).$$

Plugging the expressions (7.11) and (7.12) of $\mathbf{x}_1$ and $\mathbf{x}_2$ into the latter inequality, we obtain that

$$f(\mathbf{A}(\lambda \mathbf{y}_1 + (1-\lambda)\mathbf{y}_2) + \mathbf{b}) \leq \lambda f(\mathbf{A}\mathbf{y}_1 + \mathbf{b}) + (1-\lambda) f(\mathbf{A}\mathbf{y}_2 + \mathbf{b}),$$

which is the same as

$$g(\lambda \mathbf{y}_1 + (1-\lambda)\mathbf{y}_2) \leq \lambda g(\mathbf{y}_1) + (1-\lambda) g(\mathbf{y}_2),$$

thus establishing the convexity of g. $\square$

7.4. Operations Preserving Convexity

Example 7.18 (generalized quadratic-over-linear). Let $A \in \mathbb{R}^{m\times n}, b \in \mathbb{R}^m, c \in \mathbb{R}^n$, and $d \in \mathbb{R}$. We assume that $c \neq 0$. We will show that the quadratic-over-linear function

$$g(x) = \frac{\|Ax+b\|^2}{c^T x + d}$$

is convex over $D = \{x \in \mathbb{R}^n : c^T x + d > 0\}$. We begin by proving the convexity of the function

$$h(y, t) = \frac{\|y\|^2}{t}$$

over the convex set $C \equiv \{\binom{y}{t} \in \mathbb{R}^{m+1} : y \in \mathbb{R}^m, t > 0\}$. For that, note that $h = \sum_{i=1}^m h_i$ where

$$h_i(y, t) = \frac{y_i^2}{t}.$$

By the convexity of the quadratic-over-linear function $\varphi(x, z) = \frac{x^2}{z}$ over $\{(x,z) : x \in \mathbb{R}, z > 0\}$ (see Example 7.15), it follows that h_i is convex for any i (specifically, h_i is generated from φ by the linear transformation $x = y_i, z = t$). Hence, h is convex over C. The function f can be represented as

$$f(x) = h(Ax + b, c^T x + d).$$

Consequently, since f is the function h which has gone through an affine change of variables, it is convex over the domain $\{x \in \mathbb{R}^n : c^T x + d > 0\}$. ∎

Example 7.19. Consider the function

$$f(x_1, x_2) = x_1^2 + 2x_1 x_2 + 3x_2^2 + 2x_1 - 3x_2 + e^{x_1}.$$

To prove the convexity of f, note that $f = f_1 + f_2$, where

$$f_1(x_1, x_2) = x_1^2 + 2x_1 x_2 + 3x_2^2 + 2x_1 - 3x_2,$$
$$f_2(x_1, x_2) = e^{x_1}.$$

The function f_1 is convex since it is a quadratic function with an associated matrix $A = \begin{pmatrix} 1 & 1 \\ 1 & 3 \end{pmatrix}$ which is positive semidefinite since $\text{Tr}(A) = 4 > 0, \det(A) = 2 > 0$. The function f_2 is convex since it is generated from the one-dimensional convex function $\varphi(t) = e^t$ by the linear transformation $t = x_1$. ∎

Example 7.20. The function $f(x_1, x_2, x_3) = e^{x_1 - x_2 + x_3} + e^{2x_2} + x_1$ is convex over $\mathbb{R}^3$ as a sum of three convex functions: the function $e^{x_1 - x_2 + x_3}$, which is convex since it is constructed by making the linear change of variables $t = x_1 - x_2 + x_3$ in the one-dimensional function $\varphi(t) = e^t$. For the same reason, e^{2x_2} is convex. Finally, the function x_1, being linear, is convex. ∎

Example 7.21. The function $f(x_1, x_2) = -\ln(x_1 x_2)$ is convex over $\mathbb{R}^2_{++}$ since it can be written as

$$f(x_1, x_2) = -\ln(x_1) - \ln(x_2),$$

and the convexity of $-\ln(x_1)$ and $-\ln(x_2)$ follows from the convexity of $\varphi(t) = -\ln(t)$ over $\mathbb{R}_{++}$. ∎

In general, convexity is not preserved under composition of convex functions. For example, let $g(t) = t^2$ and $h(t) = t^2 - 4$. Then g and h are convex. However, their composition

$$s(t) = g(h(t)) = (t^2 - 4)^2$$

is not convex, as illustrated in Figure 7.4. (This can also be seen by the fact that $s''(t) = 12t^2 - 16$ and hence $s''(t) < 0$ for all $|t| < \sqrt{\frac{4}{3}}$.) The next result shows that convexity is preserved in the case of a composition of a *nondecreasing* convex function with a convex function.

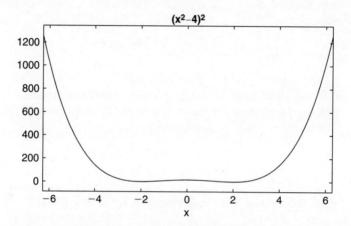

Figure 7.4. *The nonconvex function $(t^2 - 4)^2$.*

Theorem 7.22 (preservation of convexity under composition with a nondecreasing convex function). *Let $f : C \to \mathbb{R}$ be a convex function over the convex set $C \subseteq \mathbb{R}^n$. Let $g : I \to \mathbb{R}$ be a one-dimensional nondecreasing convex function over the interval $I \subseteq \mathbb{R}$. Assume that the image of C under f is contained in I: $f(C) \subseteq I$. Then the composition of g with f defined by*

$$h(\mathbf{x}) \equiv g(f(\mathbf{x})), \quad \mathbf{x} \in C,$$

is a convex function over C.

Proof. Let $\mathbf{x}, \mathbf{y} \in C$ and let $\lambda \in [0, 1]$. Then

$$\begin{aligned}
h(\lambda \mathbf{x} + (1-\lambda)\mathbf{y}) &= g(f(\lambda \mathbf{x} + (1-\lambda)\mathbf{y})) && \text{(definition of } h\text{)} \\
&\leq g(\lambda f(\mathbf{x}) + (1-\lambda) f(\mathbf{y})) && \text{(convexity of } f \text{ and monotonicity of } g\text{)} \\
&\leq \lambda g(f(\mathbf{x})) + (1-\lambda) g(f(\mathbf{y})) && \text{(convexity of } g\text{)} \\
&= \lambda h(\mathbf{x}) + (1-\lambda) h(\mathbf{y}) && \text{(definition of } h\text{),}
\end{aligned}$$

thus establishing the convexity of h. $\square$

Example 7.23. The function $h(\mathbf{x}) = e^{\|\mathbf{x}\|^2}$ is convex since it can be represented as $h(\mathbf{x}) = g(f(\mathbf{x}))$, where $g(t) = e^t$ is a nondecreasing convex function and $f(\mathbf{x}) = \|\mathbf{x}\|^2$ is a convex function. ∎

7.4. Operations Preserving Convexity

Example 7.24. The function $h(\mathbf{x}) = (\|\mathbf{x}\|^2+1)^2$ is a convex function over $\mathbb{R}^n$ since it can be represented as $h(\mathbf{x}) = g(f(\mathbf{x}))$, where $g(t) = t^2$ and $f(\mathbf{x}) = \|\mathbf{x}\|^2+1$. Both f and g are convex, but note that g is not a nondecreasing function. However, the image of $\mathbb{R}^n$ under f is the interval $[1,\infty)$ on which the function g is nondecreasing. Consequently, the composition $h(\mathbf{x}) = g(f(\mathbf{x}))$ is convex. ∎

Another important operation that preserves convexity is the pointwise maximum of convex functions.

Theorem 7.25 (pointwise maximum of convex functions). *Let $f_1,\ldots,f_p : C \to \mathbb{R}$ be p convex functions over the convex set $C \subseteq \mathbb{R}^n$. Then the maximum function*

$$f(\mathbf{x}) \equiv \max_{i=1,2,\ldots,p} f_i(\mathbf{x})$$

is a convex function over C.

Proof. Let $\mathbf{x}, \mathbf{y} \in C$ and let $\lambda \in [0,1]$. Then

$$\begin{aligned}
f(\lambda \mathbf{x} + (1-\lambda)\mathbf{y}) &= \max_{i=1,2,\ldots,p} f_i(\lambda \mathbf{x} + (1-\lambda)\mathbf{y}) && \text{(definition of } f\text{)} \\
&\leq \max_{i=1,2,\ldots,p} \{\lambda f_i(\mathbf{x}) + (1-\lambda) f_i(\mathbf{y})\} && \text{(convexity of } f_i\text{)} \\
&\leq \lambda \max_{i=1,2,\ldots,p} f_i(\mathbf{x}) + (1-\lambda) \max_{i=1,2,\ldots,p} f_i(\mathbf{y}) && (*) \\
&= \lambda f(\mathbf{x}) + (1-\lambda) f(\mathbf{y}) && \text{(definition of } f\text{).}
\end{aligned}$$

The inequality (*) follows from the fact that for any two sequences $\{a_i\}_{i=1}^p, \{b_i\}_{i=1}^p$ one has

$$\max_{i=1,2,\ldots,p} (a_i + b_i) \leq \max_{i=1,2,\ldots,p} a_i + \max_{i=1,2,\ldots,p} b_i. \quad \square$$

Example 7.26 (convexity of the maximum function). Let

$$f(\mathbf{x}) = \max\{x_1, x_2, \ldots, x_n\}.$$

Then since f is the maximum of n linear functions, which are in particular convex, it follows by Theorem 7.25 that it is convex. ∎

Example 7.27 (convexity of the sum of the k largest values). Given a vector $\mathbf{x} = (x_1, x_2, \ldots, x_n)^T$. Let $x_{[i]}$ denote the ith largest value in $\mathbf{x}$. In particular, $x_{[1]} = \max\{x_1, x_2, \ldots, x_n\}$ and $x_{[n]} = \min\{x_1, x_2, \ldots, x_n\}$. As stated in the previous example, the function $h(\mathbf{x}) = x_{[1]}$ is convex. However, in general the function $h(\mathbf{x}) = x_{[i]}$ is not convex. On the other hand, the function

$$h_k(\mathbf{x}) = x_{[1]} + x_{[2]} + \cdots + x_{[k]},$$

that is, the function producing the sum of the k largest components, is in fact convex. To see this, note that h_k can be rewritten as

$$h_k(\mathbf{x}) = \max\{x_{i_1} + x_{i_2} + \cdots + x_{i_k} : i_1, i_2, \ldots, i_k \in \{1, 2, \ldots, n\} \text{ are different}\},$$

so that h_k, as a maximum of linear (and hence convex) functions, is a convex function. ∎

Another operation preserving convexity is partial minimization.

Theorem 7.28. *Let $f : C \times D \to \mathbb{R}$ be a convex function defined over the set $C \times D$ where $C \subseteq \mathbb{R}^m$ and $D \subseteq \mathbb{R}^n$ are convex sets. Let*

$$g(\mathbf{x}) = \min_{\mathbf{y} \in D} f(\mathbf{x}, \mathbf{y}), \quad \mathbf{x} \in C,$$

where we assume that the minimal value in the above definition is real. Then g is convex over C.

Proof. Let $\mathbf{x}_1, \mathbf{x}_2 \in C$ and $\lambda \in [0, 1]$. Take $\varepsilon > 0$. Then there exist $\mathbf{y}_1, \mathbf{y}_2 \in D$ such that

$$f(\mathbf{x}_1, \mathbf{y}_1) \leq g(\mathbf{x}_1) + \varepsilon, \tag{7.13}$$
$$f(\mathbf{x}_2, \mathbf{y}_2) \leq g(\mathbf{x}_2) + \varepsilon. \tag{7.14}$$

By the convexity of f we have

$$\begin{aligned} f(\lambda \mathbf{x}_1 + (1-\lambda)\mathbf{x}_2, \lambda \mathbf{y}_1 + (1-\lambda)\mathbf{y}_2) &\leq \lambda f(\mathbf{x}_1, \mathbf{y}_1) + (1-\lambda)f(\mathbf{x}_2, \mathbf{y}_2) \\ &\overset{(7.13),(7.14)}{\leq} \lambda(g(\mathbf{x}_1) + \varepsilon) + (1-\lambda)(g(\mathbf{x}_2) + \varepsilon) \\ &= \lambda g(\mathbf{x}_1) + (1-\lambda)g(\mathbf{x}_2) + \varepsilon. \end{aligned}$$

By the definition of g we can conclude that

$$g(\lambda \mathbf{x}_1 + (1-\lambda)\mathbf{x}_2) \leq \lambda g(\mathbf{x}_1) + (1-\lambda)g(\mathbf{x}_2) + \varepsilon.$$

Since the above inequality holds for any $\varepsilon > 0$, it follows that $g(\lambda \mathbf{x}_1 + (1-\lambda)\mathbf{x}_2) \leq \lambda g(\mathbf{x}_1) + (1-\lambda)g(\mathbf{x}_2)$, and the convexity of g is established. $\square$

Note that in the above theorem, we only assumed that the minimal value is real, but we did not assume that it is attained.

Example 7.29 (convexity of the distance function). Let $C \subseteq \mathbb{R}^n$ be a convex set. The *distance* function defined by

$$d(\mathbf{x}, C) = \min\{\|\mathbf{x} - \mathbf{y}\| : \mathbf{y} \in C\}$$

is convex since the function $f(\mathbf{x}, \mathbf{y}) = \|\mathbf{x} - \mathbf{y}\|$ is convex over $\mathbb{R}^n \times C$, and thus by Theorem 7.28 it follows that $d(\cdot, C)$ is convex. ∎

7.5 ▪ Level Sets of Convex Functions

We begin with the definition of a *level set*.

Definition 7.30 (level sets). *Let $f : S \to \mathbb{R}$ be a function defined over a set $S \subseteq \mathbb{R}^n$. Then the **level set** of f with level α is given by*

$$\mathrm{Lev}(f, \alpha) = \{\mathbf{x} \in S : f(\mathbf{x}) \leq \alpha\}.$$

A fundamental property of convex functions is that their level sets are necessarily convex.

Theorem 7.31 (convexity of level sets of convex functions). *Let $f : C \to \mathbb{R}$ be a convex function defined over a convex set $C \subseteq \mathbb{R}^n$. Then for any $\alpha \in \mathbb{R}$ the level set $\mathrm{Lev}(f, \alpha)$ is convex.*

7.5. Level Sets of Convex Functions

Proof. Let $\mathbf{x}, \mathbf{y} \in \text{Lev}(f, \alpha)$ and $\lambda \in [0,1]$. Then $f(\mathbf{x}), f(\mathbf{y}) \leq \alpha$. By the convexity of C we have that $\lambda \mathbf{x} + (1-\lambda)\mathbf{y} \in C$, which combined with the convexity of f yields

$$f(\lambda \mathbf{x} + (1-\lambda)\mathbf{y}) \leq \lambda f(\mathbf{x}) + (1-\lambda)f(\mathbf{y}) \leq \lambda \alpha + (1-\lambda)\alpha = \alpha,$$

establishing the fact that $\lambda \mathbf{x} + (1-\lambda)\mathbf{y} \in \text{Lev}(f, \alpha)$ and subsequently the convexity of $\text{Lev}(f, \alpha)$. $\square$

Example 7.32. Consider the following subset of $\mathbb{R}^n$:

$$D = \left\{ \mathbf{x} : (\mathbf{x}^T \mathbf{Q} \mathbf{x} + 1)^2 + \ln\left(\sum_{i=1}^{n} e^{x_i}\right) \leq 3 \right\},$$

where $\mathbf{Q} \succeq 0$ is an $n \times n$ matrix. The set D is convex as a level set of a convex function. Specifically, $D = \text{Lev}(f, 3)$, where

$$f(\mathbf{x}) = (\mathbf{x}^T \mathbf{Q} \mathbf{x} + 1)^2 + \ln\left(\sum_{i=1}^{n} e^{x_i}\right).$$

The function f is indeed convex as the sum of two convex functions: the log-sum-exp function, which was shown to be convex in Example 7.14, and the function $g(\mathbf{x}) = (\mathbf{x}^T \mathbf{Q} \mathbf{x} + 1)^2$, which is convex as a composition of the nondecreasing convex function $\varphi(t) = (t+1)^2$ defined on $\mathbb{R}_+$ with the convex quadratic function $\mathbf{x}^T \mathbf{Q} \mathbf{x}$. ∎

All convex functions have convex level sets, but the reverse claim is not true. That is, there do exist nonconvex functions whose level sets are all convex. Functions satisfying the property that all their level sets are convex are called *quasi-convex* functions.

Definition 7.33 (quasi-convex functions). *A function $f : C \to \mathbb{R}$ defined over the convex set $C \subseteq \mathbb{R}^n$ is called* **quasi-convex** *if for any $\alpha \in \mathbb{R}$ the set $\text{Lev}(f, \alpha)$ is convex.*

The following example demonstrates the fact that quasi-convex functions may be nonconvex.

Example 7.34. The one-dimensional function $f(x) = \sqrt{|x|}$ is obviously not convex (see Figure 7.5), but its level sets are convex: for any $\alpha < 0$ we have that $\text{Lev}(f, \alpha) = \emptyset$, and for any $\alpha \geq 0$ the corresponding level set is convex:

$$\text{Lev}(f, \alpha) = \{x : \sqrt{|x|} \leq \alpha\} = \{x : |x| \leq \alpha^2\} = [-\alpha^2, \alpha^2].$$

We deduce that the nonconvex function f is quasi-convex. ∎

Example 7.35 (linear-over-linear). Consider the function

$$f(\mathbf{x}) = \frac{\mathbf{a}^T \mathbf{x} + b}{\mathbf{c}^T \mathbf{x} + d},$$

where $\mathbf{a}, \mathbf{c} \in \mathbb{R}^n$ and $b, d \in \mathbb{R}$. To avoid trivial cases, we assume that $\mathbf{c} \neq 0$ and that the function is defined over the open half-space

$$C = \{\mathbf{x} \in \mathbb{R}^n : \mathbf{c}^T \mathbf{x} + d > 0\}.$$

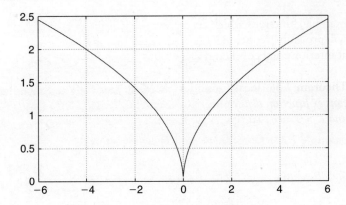

Figure 7.5. *The quasi-convex function* $\sqrt{|x|}$.

In general, f is not a convex function, but it is not difficult to show that it is quasi-convex. Indeed, let $\alpha \in \mathbb{R}$. Then the corresponding level set is given by

$$\text{Lev}(f, \alpha) = \{\mathbf{x} \in C : f(\mathbf{x}) \leq \alpha\} = \{\mathbf{x} \in \mathbb{R}^n : \mathbf{c}^T \mathbf{x} + d > 0, (\mathbf{a} - \alpha \mathbf{c})^T \mathbf{x} + (b - \alpha d) \leq 0\},$$

which is convex due to the fact that it is an intersection of two half-spaces (which are in particular convex sets) when $\mathbf{a} \neq \alpha \mathbf{c}$, and when $\mathbf{a} = \alpha \mathbf{c}$ it is either a half-space (if $b - \alpha d \leq 0$) or the empty set (if $b - \alpha d > 0$). ∎

7.6 • Continuity and Differentiability of Convex Functions

Convex functions are not necessarily continuous when defined on nonopen sets. Let us consider, for example, the function

$$f(x) = \begin{cases} 1, & x = 0, \\ x^2, & 0 < x \leq 1, \end{cases}$$

defined over the interval $[0, 1]$. It is easy to see that this is a convex function, and obviously it is not a continuous function (as also illustrated in Figure 7.6). The main result is that convex functions are always continuous at interior points of their domain. Thus, for

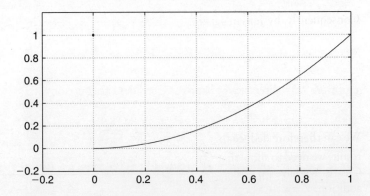

Figure 7.6. *A noncontinuous convex function over the interval* $[0, 1]$.

7.6. Continuity and Differentiability of Convex Functions

example, functions which are convex over the entire space $\mathbb{R}^n$ are always continuous. We will prove an even stronger result: convex functions are always local Lipschitz continuous at interior points of their domain.

Theorem 7.36 (local Lipschitz continuity of convex functions). *Let $f : C \to \mathbb{R}$ be a convex function defined over a convex set $C \subseteq \mathbb{R}^n$. Let $\mathbf{x}_0 \in \text{int}(C)$. Then there exist $\varepsilon > 0$ and $L > 0$ such that $B[\mathbf{x}_0, \varepsilon] \subseteq C$ and*

$$|f(\mathbf{x}) - f(\mathbf{x}_0)| \leq L \|\mathbf{x} - \mathbf{x}_0\| \tag{7.15}$$

for all $\mathbf{x} \in B[\mathbf{x}_0, \varepsilon]$.

Proof. Since $\mathbf{x}_0 \in \text{int}(C)$, it follows that there exists $\varepsilon > 0$ such that

$$B_\infty[\mathbf{x}_0, \varepsilon] \equiv \{\mathbf{x} \in \mathbb{R}^n : \|\mathbf{x} - \mathbf{x}_0\|_\infty \leq \varepsilon\} \subseteq C.$$

Next we show that f is upper bounded over $B_\infty[\mathbf{x}_0, \varepsilon]$. Let $\mathbf{v}_1, \mathbf{v}_2, \ldots, \mathbf{v}_{2^n}$ be the 2^n extreme points of $B_\infty[\mathbf{x}_0, \varepsilon]$; these are the vectors $\mathbf{v}_i = \mathbf{x}_0 + \varepsilon \mathbf{w}_i$, where $\mathbf{w}_1, \ldots, \mathbf{w}_{2^n}$ are the vectors in $\{-1, 1\}^n$. Then obviously, for any $\mathbf{x} \in B_\infty[\mathbf{x}_0, \varepsilon]$ there exists $\lambda \in \Delta_{2^n}$ such that $\mathbf{x} = \sum_{i=1}^{2^n} \lambda_i \mathbf{v}_i$, and hence, by Jensen's inequality,

$$f(\mathbf{x}) = f\left(\sum_{i=1}^{2^n} \lambda_i \mathbf{v}_i\right) \leq \sum_{i=1}^{2^n} \lambda_i f(\mathbf{v}_i) \leq M,$$

where $M = \max_{i=1,2,\ldots,2^n} f(\mathbf{v}_i)$. Since $\|\mathbf{x}\|_\infty \leq \|\mathbf{x}\|_2$ for any $\mathbb{R}^n$ it holds that

$$B_2[\mathbf{x}_0, \varepsilon] = B[\mathbf{x}_0, \varepsilon] = \{\mathbf{x} \in \mathbb{R}^n : \|\mathbf{x} - \mathbf{x}_0\|_2 \leq \varepsilon\} \subseteq B_\infty[\mathbf{x}_0, \varepsilon].$$

We therefore conclude that $f(\mathbf{x}) \leq M$ for any $\mathbf{x} \in B[\mathbf{x}_0, \varepsilon]$. Let $\mathbf{x} \in B[\mathbf{x}_0, \varepsilon]$ be such that $\mathbf{x} \neq \mathbf{x}_0$. (The result (7.15) is obvious when $\mathbf{x} = \mathbf{x}_0$.) Define

$$\mathbf{z} = \mathbf{x}_0 + \frac{1}{\alpha}(\mathbf{x} - \mathbf{x}_0),$$

where $\alpha = \frac{1}{\varepsilon}\|\mathbf{x} - \mathbf{x}_0\|$. Then obviously $\alpha \leq 1$ and $\mathbf{z} \in B[\mathbf{x}_0, \varepsilon]$, and in particular $f(\mathbf{z}) \leq M$. In addition,

$$\mathbf{x} = \alpha \mathbf{z} + (1 - \alpha) \mathbf{x}_0.$$

Consequently, by Jensen's inequality we have

$$\begin{aligned} f(\mathbf{x}) &\leq \alpha f(\mathbf{z}) + (1 - \alpha) f(\mathbf{x}_0) \\ &\leq f(\mathbf{x}_0) + \alpha (M - f(\mathbf{x}_0)) \\ &= f(\mathbf{x}_0) + \frac{M - f(\mathbf{x}_0)}{\varepsilon} \|\mathbf{x} - \mathbf{x}_0\|. \end{aligned}$$

We can therefore deduce that $f(\mathbf{x}) - f(\mathbf{x}_0) \leq L\|\mathbf{x} - \mathbf{x}_0\|$, where $L = \frac{M - f(\mathbf{x}_0)}{\varepsilon}$. To prove the result, we need to show that $f(\mathbf{x}) - f(\mathbf{x}_0) \geq -L\|\mathbf{x} - \mathbf{x}_0\|$. For that, define $\mathbf{u} = \mathbf{x}_0 + \frac{1}{\alpha}(\mathbf{x}_0 - \mathbf{x})$. Clearly we have $\|\mathbf{u} - \mathbf{x}_0\| = \varepsilon$ and hence $\mathbf{u} \in B[\mathbf{x}_0, \varepsilon]$ and in particular $f(\mathbf{u}) \leq M$. In addition, $\mathbf{x} = \mathbf{x}_0 + \alpha(\mathbf{x}_0 - \mathbf{u})$. Therefore,

$$f(\mathbf{x}) = f(\mathbf{x}_0 + \alpha(\mathbf{x}_0 - \mathbf{u})) \geq f(\mathbf{x}_0) + \alpha(f(\mathbf{x}_0) - f(\mathbf{u})). \tag{7.16}$$

The latter inequality is valid since

$$\mathbf{x}_0 = \frac{1}{1+\alpha}(\mathbf{x}_0 + \alpha(\mathbf{x}_0 - \mathbf{u})) + \frac{\alpha}{1+\alpha}\mathbf{u},$$

and hence, by Jensen's inequality

$$f(\mathbf{x}_0) \leq \frac{1}{1+\alpha} f(\mathbf{x}_0 + \alpha(\mathbf{x}_0 - \mathbf{u})) + \frac{\alpha}{1+\alpha} f(\mathbf{u}),$$

which is the same as the inequality in (7.16) (after some rearrangement of terms). Now, continuing (7.16),

$$\begin{aligned} f(\mathbf{x}) &\geq f(\mathbf{x}_0) + \alpha(f(\mathbf{x}_0) - f(\mathbf{u})) \\ &\geq f(\mathbf{x}_0) - \alpha(M - f(\mathbf{x}_0)) \\ &= f(\mathbf{x}_0) - \frac{M - f(\mathbf{x}_0)}{\varepsilon} \|\mathbf{x} - \mathbf{x}_0\| \\ &= f(\mathbf{x}_0) - L\|\mathbf{x} - \mathbf{x}_0\|, \end{aligned}$$

and the desired result is established. $\square$

Convex functions are not necessarily differentiable, but on the other hand, as will be shown in the next result, all the directional derivatives at interior points exist.

Theorem 7.37 (existence of directional derivatives for convex functions). *Let $f : C \to \mathbb{R}$ be a convex function defined over the convex set $C \subseteq \mathbb{R}^n$. Let $\mathbf{x} \in \mathrm{int}(C)$. Then for any $\mathbf{d} \neq \mathbf{0}$, the directional derivative $f'(\mathbf{x}; \mathbf{d})$ exists.*

Proof. Let $\mathbf{x} \in \mathrm{int}(C)$ and $\mathbf{d} \neq \mathbf{0}$. Then the directional derivative (if exists) is the limit

$$\lim_{t \to 0^+} \frac{g(t) - g(0)}{t}, \tag{7.17}$$

where $g(t) = f(\mathbf{x} + t\mathbf{d})$. Defining $h(t) \equiv \frac{g(t) - g(0)}{t}$, the limit (7.17) can be equivalently written as

$$\lim_{t \to 0^+} h(t).$$

Note that g, as well as h, is defined for small enough values of t by the fact that $\mathbf{x} \in \mathrm{int}(C)$. In fact, we will take an $\varepsilon > 0$ for which $\mathbf{x} + t\mathbf{d}, \mathbf{x} - t\mathbf{d} \in C$ for all $t \in [0, \varepsilon]$. Now, let $0 < t_1 < t_2 \leq \varepsilon$. Then

$$\mathbf{x} + t_1 \mathbf{d} = \left(1 - \frac{t_1}{t_2}\right) \mathbf{x} + \frac{t_1}{t_2}(\mathbf{x} + t_2 \mathbf{d}),$$

and thus, by the convexity of f we have

$$f(\mathbf{x} + t_1 \mathbf{d}) \leq \left(1 - \frac{t_1}{t_2}\right) f(\mathbf{x}) + \frac{t_1}{t_2} f(\mathbf{x} + t_2 \mathbf{d}).$$

The latter inequality can be rewritten (after some rearrangement of terms) as

$$\frac{f(\mathbf{x} + t_1 \mathbf{d}) - f(\mathbf{x})}{t_1} \leq \frac{f(\mathbf{x} + t_2 \mathbf{d}) - f(\mathbf{x})}{t_2},$$

which is the same as $h(t_1) \leq h(t_2)$. We thus conclude that the function h is monotone nondecreasing over $\mathbb{R}_{++}$. All that is left is to prove that h is bounded below over $(0, \varepsilon]$. Indeed, taking $0 < t \leq \varepsilon$, note that

$$\mathbf{x} = \frac{\varepsilon}{\varepsilon + t}(\mathbf{x} + t\mathbf{d}) + \frac{t}{\varepsilon + t}(\mathbf{x} - \varepsilon\mathbf{d}).$$

Hence, by the convexity of f we have

$$f(\mathbf{x}) \leq \frac{\varepsilon}{\varepsilon + t} f(\mathbf{x} + t\mathbf{d}) + \frac{t}{\varepsilon + t} f(\mathbf{x} - \varepsilon\mathbf{d}),$$

which after some rearrangement of terms can be seen to be equivalent to the inequality

$$h(t) = \frac{f(\mathbf{x} + t\mathbf{d}) - f(\mathbf{x})}{t} \geq \frac{f(\mathbf{x}) - f(\mathbf{x} - \varepsilon\mathbf{d})}{\varepsilon},$$

showing that h is bounded below over $(0, \varepsilon]$. Since h is nondecreasing and bounded below over $(0, \varepsilon]$ it follows that the limit $\lim_{t \to 0+} h(t)$ exists, meaning that the directional derivative $f'(\mathbf{x}; \mathbf{d})$ exists. $\square$

7.7 ▪ Extended Real-Valued Functions

Until now we have discussed functions that are *real-valued*, meaning that they take their values in $\mathbb{R} = (-\infty, \infty)$. It is also quite natural to consider functions that are defined over the entire space $\mathbb{R}^n$ that take values in $\mathbb{R} \cup \{\infty\} = (-\infty, \infty]$. Such a function is called an *extended real-valued function*. One very important example of an extended real-valued function is the *indicator function*, which is defined as follows: given a set $S \subseteq \mathbb{R}^n$, the indicator function $\delta_S : \mathbb{R}^n \to \mathbb{R} \cup \{\infty\}$ is given by

$$\delta_S(\mathbf{x}) = \begin{cases} 0 & \text{if } \mathbf{x} \in S, \\ \infty & \text{if } \mathbf{x} \notin S. \end{cases}$$

The *effective domain* of an extended real-valued function is the set of vectors for which the function takes a real value:

$$\text{dom}(f) = \{\mathbf{x} \in \mathbb{R}^n : f(\mathbf{x}) < \infty\}.$$

An extended real-valued function $f : \mathbb{R}^n \to \mathbb{R} \cup \{\infty\}$ is called *proper* if it is not always equal to ∞, meaning that there exists $\mathbf{x}_0 \in \mathbb{R}^n$ such that $f(\mathbf{x}_0) < \infty$. Similarly to the definition for real-valued functions, an extended real-valued function is convex if for any $\mathbf{x}, \mathbf{y} \in \mathbb{R}^n$ and $\lambda \in [0, 1]$ the following inequality holds:

$$f(\lambda \mathbf{x} + (1 - \lambda)\mathbf{y}) \leq \lambda f(\mathbf{x}) + (1 - \lambda) f(\mathbf{y}),$$

where we use the usual arithmetic with ∞:

$$a + \infty = \infty \text{ for any } a \in \mathbb{R},$$
$$a \cdot \infty = \infty \text{ for any } a \in \mathbb{R}_{++}.$$

In addition, we have the much less obvious rule that $0 \cdot \infty = 0$. The above definition of convexity of extended real-valued functions is equivalent to saying that $\text{dom}(f)$ is a convex set and that the restriction of f to its effective domain dom(f); that is, the function $g : \text{dom}(f) \to \mathbb{R}$ defined by $g(\mathbf{x}) = f(\mathbf{x})$ for any $\mathbf{x} \in \text{dom}(f)$ is a convex real-valued function over dom(f). As an example, the indicator function $\delta_C(\cdot)$ of a set $C \subseteq \mathbb{R}^n$ is convex if and only if C is a convex set.

An important set associated with extended real-valued functions is its *epigraph*. Suppose that $f : \mathbb{R}^n \to \mathbb{R} \cup \{\infty\}$. Then the epigraph set epi(f) $\subseteq \mathbb{R}^{n+1}$ is defined by

$$\text{epi}(f) = \left\{ \begin{pmatrix} \mathbf{x} \\ t \end{pmatrix} : f(\mathbf{x}) \leq t \right\}.$$

An example of an epigraph can be seen in Figure 7.7. It is not difficult to show that an extended real-valued (or a real-valued) function f is convex if and only if its epigraph set epi(f) is convex (see Exercise 7.29). An important property of convex extended real-valued functions that convexity is preserved under the maximum operation. As was already mentioned, we do not use the "sup" notation in this book and we always refer to the maximum of a function or the maximum over a given index set.

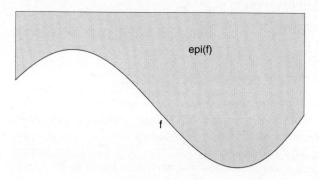

Figure 7.7. *The epigraph of a one-dimensional function.*

Theorem 7.38 (preservation of convexity under maximum). *Let $f_i : \mathbb{R}^n \to \mathbb{R} \cup \{\infty\}$ be an extended real-valued convex function for any $i \in I$ (I being an arbitrary index set). Then the function $f(\mathbf{x}) = \max_{i \in I} f_i(\mathbf{x})$ is an extended real-valued convex function.*

Proof. The result follows from the fact that epi(f) = $\bigcap_{i \in I} \text{epi}(f_i)$. The convexity of f_i for any $i \in I$ implies the convexity of epi(f_i) for any $i \in I$. Consequently, epi(f), as an intersection of convex sets, is convex, and hence the convexity of f is established. □

The differences between Theorems 7.38 and 7.25 are that the functions in Theorem 7.38 are not necessarily real-valued and that the index set I can be infinite.

Example 7.39 (support functions). Let $S \subseteq \mathbb{R}^n$. The *support function of S* is the function

$$\sigma_S(\mathbf{x}) = \max_{\mathbf{y} \in S} \mathbf{x}^T \mathbf{y}.$$

Since for each $\mathbf{y} \in S$, the function $f_\mathbf{y}(\mathbf{x}) \equiv \mathbf{y}^T \mathbf{x}$ is a convex function over $\mathbb{R}^n$ (being linear), it follows by Theorem 7.38 that σ_S is an extended real-valued convex function. ∎

As an example of a support function, let us consider the unit (Euclidean) ball $S = B[0,1] = \{y \in \mathbb{R}^n : \|y\| \leq 1\}$. Let $x \in \mathbb{R}^n$. We will show that

$$\sigma_S(x) = \|x\|. \tag{7.18}$$

Obviously, if $x = 0$, then $\sigma_S(x) = 0$ and hence (7.18) holds for $x = 0$. If $x \neq 0$, then for any $y \in S$ and $x \in \mathbb{R}^n$, we have by the Cauchy–Schwarz inequality that

$$x^T y \leq \|x\| \cdot \|y\| \leq \|x\|.$$

On the other hand, taking $\tilde{y} = \frac{1}{\|x\|} x \in S$, we have

$$x^T \tilde{y} = \|x\|,$$

and the desired formula (7.18) follows.

Example 7.40. Consider the function

$$f(t) = d^T A(t)^{-1} d,$$

where $d \in \mathbb{R}^n$ and $A(t) = \sum_{i=1}^m t_i A_i$ with $A_1, \ldots, A_m$ being $n \times n$ positive definite matrices. We will show that this function is convex over $\mathbb{R}_{++}^m$. Indeed, for any $x \in \mathbb{R}^n$, the function

$$g_x(t) = \begin{cases} 2d^T x - x^T A(t) x, & t \in \mathbb{R}_{++}^m, \\ \infty & \text{else} \end{cases}$$

is convex over $\mathbb{R}^m$ since it is an affine function over its convex domain. The corresponding max function is

$$\max_{x \in \mathbb{R}^n} \{2d^T x - x^T A(t) x\} = d^T A(t)^{-1} d$$

for $t \in \mathbb{R}_{++}^m$ and ∞ elsewhere. Therefore, the extended real-valued function

$$\tilde{f}(t) = \begin{cases} d^T A(t)^{-1} d, & t \in \mathbb{R}_{++}^m, \\ \infty & \text{else} \end{cases}$$

is convex over $\mathbb{R}^m$, which is the same as saying that f is convex over $\mathbb{R}_{++}^m$. ∎

7.8 • Maxima of Convex Functions

In the next chapter we will learn that problems consisting of minimizing a convex function over a convex set are in some sense "easy," but in this section we explore some important properties of the much more difficult problem of maximizing a convex function over a convex feasible set. First, we show that the maximum of a nonconstant convex function defined on a convex set C cannot be attained at an interior point of the set C.

Theorem 7.41. *Let $f : C \to \mathbb{R}$ be a convex function which is not constant over the convex set C. Then f does not attain a maximum at a point in* $\text{int}(C)$.

Proof. Assume in contradiction that $x^* \in \text{int}(C)$ is a global maximizer of f over C. Since the function is not constant, there exists $y \in C$ such that $f(y) < f(x^*)$. Since $x^* \in \text{int}(C)$, there exists $\varepsilon > 0$ such that $\mathbf{z} = \mathbf{x}^* + \varepsilon(\mathbf{x}^* - \mathbf{y}) \in C$. Since $\mathbf{x}^* = \frac{\varepsilon}{\varepsilon+1}\mathbf{y} + \frac{1}{\varepsilon+1}\mathbf{z}$, it follows by the convexity of f that

$$f(\mathbf{x}^*) \leq \frac{\varepsilon}{\varepsilon+1} f(\mathbf{y}) + \frac{1}{\varepsilon+1} f(\mathbf{z}),$$

and hence $f(\mathbf{z}) \geq \varepsilon(f(\mathbf{x}^*) - f(\mathbf{y})) + f(\mathbf{x}^*) > f(\mathbf{x}^*)$, which is a contradiction to the optimality of $\mathbf{x}^*$. □

When the underlying set is also compact, then the next result shows that there exists at least one maximizer that is an extreme point of the set.

Theorem 7.42. *Let $f : C \to \mathbb{R}$ be a convex and continuous function over the convex and compact set $C \subseteq \mathbb{R}^n$. Then there exists at least one maximizer of f over C that is an extreme point of C.*

Proof. Let $\mathbf{x}^*$ be a maximizer of f over C (whose existence is guaranteed by the Weierstrass theorem, Theorem 2.30). If $\mathbf{x}^*$ is an extreme point of C, then the result is established. Otherwise, if $\mathbf{x}^*$ is not an extreme point, then by the Krein–Milman theorem (Theorem 6.35), $C = \text{conv}(\text{ext}(C))$, which means that there exist $\mathbf{x}_1, \mathbf{x}_2, \ldots, \mathbf{x}_k \in \text{ext}(C)$ and $\lambda \in \Delta_k$ such that

$$\mathbf{x}^* = \sum_{i=1}^{k} \lambda_i \mathbf{x}_i,$$

and $\lambda_i > 0$ for all $i = 1, 2, \ldots, k$. Hence, by the convexity of f we have

$$f(\mathbf{x}^*) \leq \sum_{i=1}^{k} \lambda_i f(\mathbf{x}_i),$$

or equivalently

$$\sum_{i=1}^{k} \lambda_i (f(\mathbf{x}_i) - f(\mathbf{x}^*)) \geq 0. \tag{7.19}$$

Since $\mathbf{x}^*$ is a maximizer of f over C, we have $f(\mathbf{x}_i) \leq f(\mathbf{x}^*)$ for all $i = 1, 2, \ldots, k$. This means that inequality (7.19) states that a sum of nonpositive numbers is nonnegative, implying that each of the terms is zero, that is, $f(\mathbf{x}_i) = f(\mathbf{x}^*)$. Consequently, the extreme points $\mathbf{x}_1, \mathbf{x}_2, \ldots, \mathbf{x}_k$ are all maximizers of f over C. □

Example 7.43. Consider the problem

$$\max\{\mathbf{x}^T \mathbf{Q} \mathbf{x} : \|\mathbf{x}\|_\infty \leq 1\},$$

where $\mathbf{Q} \succeq 0$. Since the objective function is convex, and the feasible set is convex and compact, it follows that there exists a maximizer at an extreme point of the feasible set. The set of extreme points of the feasible set is $\{-1, 1\}^n$, and hence we conclude that there exists a maximizer that satisfies that each of its components is equal to 1 or -1. ∎

Example 7.44 (computation of $\|\mathbf{A}\|_{1,1}$)**.** Let $\mathbf{A} \in \mathbb{R}^{m \times n}$. Recall that (see Example 1.8)

$$\|\mathbf{A}\|_{1,1} = \max\{\|\mathbf{A}\mathbf{x}\|_1 : \|\mathbf{x}\|_1 \leq 1\}.$$

Since the optimization problem consists of maximizing a convex function (composition of a norm function with a linear function) over a compact convex set, there exists a maximizer which is an extreme point of the l_1 ball. Note that there are exactly $2n$ extreme points to the l_1 ball: $\mathbf{e}_1, -\mathbf{e}_1, \mathbf{e}_2, -\mathbf{e}_2, \ldots, \mathbf{e}_n, -\mathbf{e}_n$. In addition,

$$\|\mathbf{A}\mathbf{e}_j\|_1 = \|\mathbf{A}(-\mathbf{e}_j)\|_1 = \sum_{i=1}^{m} |A_{i,j}|,$$

and thus

$$\|\mathbf{A}\|_{1,1} = \max_{j=1,2,\ldots,n} \|\mathbf{A}\mathbf{e}_j\|_1 = \max_{j=1,2,\ldots,n} \sum_{i=1}^{m} |A_{i,j}|.$$

This is exactly the maximum absolute column sum norm introduced in Example 1.8. ∎

7.9 ▪ Convexity and Inequalities

Convexity is a powerful tool for proving inequalities. For example, the arithmetic geometric mean (AGM) inequality follows directly from the convexity of the scalar function $-\ln(x)$ over $\mathbb{R}_{++}$.

Proposition 7.45 (AGM inequality). *For any $x_1, x_2, \ldots, x_n \geq 0$ the following inequality holds:*

$$\frac{1}{n}\sum_{i=1}^{n} x_i \geq \sqrt[n]{\prod_{i=1}^{n} x_i}. \tag{7.20}$$

More generally, for any $\lambda \in \Delta_n$ one has

$$\sum_{i=1}^{n} \lambda_i x_i \geq \prod_{i=1}^{n} x_i^{\lambda_i}. \tag{7.21}$$

Proof. Employing Jensen's inequality on the convex function $f(x) = -\ln(x)$, we have that for any $x_1, x_2, \ldots, x_n > 0$ and $\lambda \in \Delta_n$

$$f\left(\sum_{i=1}^{n} \lambda_i x_i\right) \leq \sum_{i=1}^{n} \lambda_i f(x_i),$$

and hence

$$-\ln\left(\sum_{i=1}^{n} \lambda_i x_i\right) \leq -\sum_{i=1}^{n} \lambda_i \ln(x_i)$$

or

$$\ln\left(\sum_{i=1}^{n} \lambda_i x_i\right) \geq \sum_{i=1}^{n} \lambda_i \ln(x_i).$$

Taking the exponential function of both sides we have

$$\sum_{i=1}^{n} \lambda_i x_i \geq e^{\sum_{i=1}^{n} \lambda_i \ln(x_i)},$$

which is the same as the generalized AGM inequality (7.21). Plugging in $\lambda_i = \frac{1}{n}$ for all i yields the special case (7.20). We have proven the AGM inequalities only for the case when $x_1, x_2, \ldots, x_n$ are all positive. However, they are trivially satisfied if there exists an i for which $x_i = 0$, and hence the inequalities are valid for any $x_1, x_2, \ldots, x_n \geq 0$. □

A direct result of the generalized AGM inequality is *Young's inequality*.

Lemma 7.46 (Young's inequality). *For any $s, t \geq 0$ and $p, q > 1$ satisfying $\frac{1}{p} + \frac{1}{q} = 1$ it holds that*

$$st \leq \frac{s^p}{p} + \frac{t^q}{q}. \tag{7.22}$$

Proof. By the generalized AGM inequality we have for any $x, y \geq 0$

$$x^{\frac{1}{p}} y^{\frac{1}{q}} \leq \frac{x}{p} + \frac{y}{q}.$$

Setting $x = s^p, y = t^q$ in the latter inequality, the result follows. □

We can now prove several important inequalities. The first one is Hölder's inequality, which is a generalization of the Cauchy–Schwarz inequality.

Lemma 7.47 (Hölder's inequality). *For any $\mathbf{x}, \mathbf{y} \in \mathbb{R}^n$ and $p, q \geq 1$ satisfying $\frac{1}{p} + \frac{1}{q} = 1$ it holds that*

$$|\mathbf{x}^T \mathbf{y}| \leq \|\mathbf{x}\|_p \|\mathbf{y}\|_q.$$

Proof. First, if $\mathbf{x} = \mathbf{0}$ or $\mathbf{y} = \mathbf{0}$, then the inequality is trivial. Suppose then that $\mathbf{x} \neq \mathbf{0}$ and $\mathbf{y} \neq \mathbf{0}$. For any $i \in \{1, 2, \ldots, n\}$, setting $s = \frac{|x_i|}{\|\mathbf{x}\|_p}$ and $t = \frac{|y_i|}{\|\mathbf{x}\|_q}$ in (7.22) yields the inequality

$$\frac{|x_i y_i|}{\|\mathbf{x}\|_p \|\mathbf{y}\|_q} \leq \frac{1}{p} \frac{|x_i|^p}{\|\mathbf{x}\|_p^p} + \frac{1}{q} \frac{|y_i|^q}{\|\mathbf{y}\|_q^q}.$$

Summing the above inequality over $i = 1, 2, \ldots, n$ we obtain

$$\frac{\sum_{i=1}^n |x_i y_i|}{\|\mathbf{x}\|_p \|\mathbf{y}\|_p} \leq \frac{1}{p} \frac{\sum_{i=1}^n |x_i|^p}{\|\mathbf{x}\|_p^p} + \frac{1}{q} \frac{\sum_{i=1}^n |y_i|^q}{\|\mathbf{y}\|_q^q} = \frac{1}{p} + \frac{1}{q} = 1.$$

Hence, by the triangle inequality we have

$$|\mathbf{x}^T \mathbf{y}| \leq \sum_{i=1}^n |x_i y_i| \leq \|\mathbf{x}\|_p \|\mathbf{y}\|_q. \quad \square$$

Of course, for $p = q = 2$ Hölder's inequality is just the Cauchy–Schwarz inequality. Another inequality that can be deduced as a result of convexity is Minkowski's inequality, stating that the p-norm (for $p \geq 1$) satisfies the triangle inequality.

Lemma 7.48 (Minkowski's inequality). *Let $p \geq 1$. Then for any $\mathbf{x}, \mathbf{y} \in \mathbb{R}^n$ the inequality*

$$\|\mathbf{x} + \mathbf{y}\|_p \leq \|\mathbf{x}\|_p + \|\mathbf{y}\|_p$$

holds.

Proof. For $p=1$, the inequality follows by summing up the inequalities $|x_i+y_i| \leq |x_i|+|y_i|$. Suppose then that $p>1$. We can assume that $\mathbf{x} \neq \mathbf{0}, \mathbf{y} \neq \mathbf{0}$, and $\mathbf{x}+\mathbf{y} \neq \mathbf{0}$. Otherwise, the inequality is trivial. The function $\varphi(t) = t^p$ is convex over $\mathbb{R}_+$ since $\varphi''(t) = p(p-1)t^{p-2} > 0$ for $t > 0$. Therefore, by the definition of convexity we have that for any $\lambda_1, \lambda_2 \geq 0$ with $\lambda_1 + \lambda_2 = 1$ one has

$$(\lambda_1 t + \lambda_2 s)^p \leq \lambda_1 t^p + \lambda_2 s^p.$$

Let $i \in \{1,2,\ldots,n\}$. Plugging $\lambda_1 = \frac{\|\mathbf{x}\|_p}{\|\mathbf{x}\|_p + \|\mathbf{y}\|_p}, \lambda_2 = \frac{\|\mathbf{y}\|_p}{\|\mathbf{x}\|_p + \|\mathbf{y}\|_p}, t = \frac{|x_i|}{\|\mathbf{x}\|_p}$, and $s = \frac{|y_i|}{\|\mathbf{y}\|_p}$ in the above inequality yields

$$\frac{1}{(\|\mathbf{x}\|_p + \|\mathbf{y}\|_p)^p}(|x_i|+|y_i|)^p \leq \frac{\|\mathbf{x}\|_p}{\|\mathbf{x}\|_p + \|\mathbf{y}\|_p} \frac{|x_i|^p}{\|\mathbf{x}\|_p^p} + \frac{\|\mathbf{y}\|_p}{\|\mathbf{x}\|_p + \|\mathbf{y}\|_p} \frac{|y_i|^p}{\|\mathbf{y}\|_p^p}.$$

Summing the above inequality over $i = 1,2,\ldots,n$, we obtain that

$$\frac{1}{(\|\mathbf{x}\|_p + \|\mathbf{y}\|_p)^p} \sum_{i=1}^n (|x_i|+|y_i|)^p \leq \frac{\|\mathbf{x}\|_p}{\|\mathbf{x}\|_p + \|\mathbf{y}\|_p} + \frac{\|\mathbf{y}\|_p}{\|\mathbf{x}\|_p + \|\mathbf{y}\|_p} = 1,$$

and hence

$$\sum_{i=1}^n (|x_i|+|y_i|)^p \leq (\|\mathbf{x}\|_p + \|\mathbf{y}\|_p)^p.$$

Finally,

$$\|\mathbf{x}+\mathbf{y}\|_p = \sqrt[p]{\sum_{i=1}^n |x_i+y_i|^p} \leq \sqrt[p]{\sum_{i=1}^n (|x_i|+|y_i|)^p} \leq \|\mathbf{x}\|_p + \|\mathbf{y}\|_p. \quad \Box$$

Exercises

7.1. For each of the following sets determine whether they are convex or not (explaining your choice).
   (i) $C_1 = \{\mathbf{x} \in \mathbb{R}^n : \|\mathbf{x}\|^2 = 1\}$.
   (ii) $C_2 = \{\mathbf{x} \in \mathbb{R}^n : \max_{i=1,2,\ldots,n} x_i \leq 1\}$.
   (iii) $C_3 = \{\mathbf{x} \in \mathbb{R}^n : \min_{i=1,2,\ldots,n} x_i \leq 1\}$.
   (iv) $C_4 = \{\mathbf{x} \in \mathbb{R}^n_{++} : \prod_{i=1}^n x_i \geq 1\}$.

7.2. Show that the set

$$M = \{\mathbf{x} \in \mathbb{R}^n : \mathbf{x}^T \mathbf{Q} \mathbf{x} \leq (\mathbf{a}^T \mathbf{x})^2, \mathbf{a}^T \mathbf{x} \geq 0\},$$

where $\mathbf{Q}$ is an $n \times n$ positive definite matrix and $\mathbf{a} \in \mathbb{R}^n$ is a convex cone.

7.3. Let $f: \mathbb{R}^n \to \mathbb{R}$ be a convex as well as concave function. Show that f is an affine function; that is, there exist $\mathbf{a} \in \mathbb{R}^n$ and $b \in \mathbb{R}$ such that $f(\mathbf{x}) = \mathbf{a}^T \mathbf{x} + b$ for any $\mathbf{x} \in \mathbb{R}^n$.

7.4. Let $f: \mathbb{R}^n \to \mathbb{R}$ be a continuously differentiable convex function. Show that for any $\varepsilon > 0$, the function
$$g_\varepsilon(\mathbf{x}) = f(\mathbf{x}) + \varepsilon \|\mathbf{x}\|^2$$
is coercive.

7.5. Let $f: \mathbb{R}^n \to \mathbb{R}$. Prove that f is convex if and only if for any $\mathbf{x} \in \mathbb{R}^n$ and $\mathbf{d} \neq \mathbf{0}$, the one-dimensional function $g_{\mathbf{x},\mathbf{d}}(t) = f(\mathbf{x} + t\mathbf{d})$ is convex.

7.6. Prove Theorem 7.13.

7.7. Let $C \subseteq \mathbb{R}^n$ be a convex set. Let f be a convex function over C, and let g be a strictly convex function over C. Show that the sum function $f + g$ is strictly convex over C.

7.8. (i) Let f be a convex function defined on a convex set C. Suppose that f is *not* strictly convex on C. Prove that there exist $\mathbf{x}, \mathbf{y} \in \mathbb{R}^n (\mathbf{x} \neq \mathbf{y})$ such that f is affine over the segment $[\mathbf{x}, \mathbf{y}]$.

(ii) Prove that the function $f(x) = x^4$ is strictly convex on $\mathbb{R}$ and that $g(x) = x^p$ for $p > 1$ is strictly convex over $\mathbb{R}_+$.

7.9. Show that the log-sum-exp function $f(\mathbf{x}) = \ln\left(\sum_{i=1}^n e^{x_i}\right)$ is *not* strictly convex over $\mathbb{R}^n$.

7.10. Show that the following functions are convex over the specified domain C:

(i) $f(x_1, x_2, x_3) = -\sqrt{x_1 x_2} + 2x_1^2 + 2x_2^2 + 3x_3^2 - 2x_1 x_2 - 2x_2 x_3$ over $\mathbb{R}^3_{++}$.

(ii) $f(\mathbf{x}) = \|\mathbf{x}\|^4$ over $\mathbb{R}^n$.

(iii) $f(\mathbf{x}) = \sum_{i=1}^n x_i \ln(x_i) - \left(\sum_{i=1}^n x_i\right) \ln\left(\sum_{i=1}^n x_i\right)$ over $\mathbb{R}^n_{++}$.

(iv) $f(\mathbf{x}) = \sqrt{\mathbf{x}^T \mathbf{Q} \mathbf{x} + 1}$ over $\mathbb{R}^n$, where $\mathbf{Q} \succeq \mathbf{0}$ is an $n \times n$ matrix.

(v) $f(x_1, x_2, x_3) = \max\{\sqrt{x_1^2 + x_2^2 + 20x_3^2 - x_1 x_2 - 4x_2 x_3 + 1}, (x_1^2 + x_2^2 + x_1 + x_2 + 2)^2\}$ over $\mathbb{R}^3$.

(vi) $f(x_1, x_2) = (2x_1^2 + 3x_2^2)\left(\frac{1}{2}x_1^2 + \frac{1}{3}x_2^2\right)$.

7.11. Let $\mathbf{A} \in \mathbb{R}^{m \times n}$, and let $f: \mathbb{R}^n \to \mathbb{R}$ be defined by
$$f(\mathbf{x}) = \ln\left(\sum_{i=1}^m e^{\mathbf{A}_i \mathbf{x}}\right),$$
where $\mathbf{A}_i$ is the ith row of $\mathbf{A}$. Prove that f is convex over $\mathbb{R}^n$.

7.12. Prove that the following set is a convex subset of $\mathbb{R}^{n+2}$:
$$C = \left\{ \begin{pmatrix} \mathbf{x} \\ y \\ z \end{pmatrix} : \|\mathbf{x}\|^2 \leq yz, \mathbf{x} \in \mathbb{R}^n, y, z \in \mathbb{R}_+ \right\}.$$

7.13. Show that the function $f(x_1, x_2, x_3) = -e^{(-x_1 + x_2 - 2x_3)^2}$ is not convex over $\mathbb{R}^n$.

7.14. Prove that the geometric mean function $f(\mathbf{x}) = \sqrt[n]{\prod_{i=1}^n x_i}$ is concave over $\mathbb{R}^n_{++}$. Is it strictly concave over $\mathbb{R}^n_{++}$?

7.15. (i) Let $f : \mathbb{R}^n \to \mathbb{R}$ be a convex function which is nondecreasing with respect to each of its variables separately; that is, for any $i \in \{1, 2, \ldots, n\}$ and fixed $x_1, x_2, \ldots, x_{i-1}, x_{i+1}, \ldots, x_n$, the one-dimensional function

$$g_i(y) = f(x_1, x_2, \ldots, x_{i-1}, y, x_{i+1}, \ldots, x_n)$$

is nondecreasing with respect to y. Let $h_1, h_2, \ldots, h_n : \mathbb{R}^p \to \mathbb{R}$ be convex functions. Prove that the composite function

$$r(z_1, z_2, \ldots, z_p) = f(h_1(z_1, z_2, \ldots, z_p), \ldots, h_n(z_1, z_2, \ldots, z_p))$$

is convex.

(ii) Prove that the function $f(x_1, x_2) = \ln(e^{x_1^2 + x_2^2} + e^{\sqrt{x_1^2 + 1}})$ is convex over $\mathbb{R}^2$.

7.16. Let f be a convex function over $\mathbb{R}^n$ and let $\mathbf{x}, \mathbf{y} \in \mathbb{R}^n$ and $\alpha > 0$. Define $\mathbf{z} = \mathbf{x} + \frac{1}{\alpha}(\mathbf{x} - \mathbf{y})$. Prove that

$$f(\mathbf{y}) \geq f(\mathbf{x}) + \alpha(f(\mathbf{x}) - f(\mathbf{z})).$$

7.17. Let f be a convex function over a convex set $C \subseteq \mathbb{R}^n$. Let $\mathbf{x}_1, \mathbf{x}_3 \in C$ and let $\mathbf{x}_2 \in [\mathbf{x}_1, \mathbf{x}_3]$. Prove that if $\mathbf{x}_1, \mathbf{x}_2, \mathbf{x}_3$ are different from each other, then

$$\frac{f(\mathbf{x}_3) - f(\mathbf{x}_2)}{\|\mathbf{x}_3 - \mathbf{x}_2\|} \geq \frac{f(\mathbf{x}_2) - f(\mathbf{x}_1)}{\|\mathbf{x}_2 - \mathbf{x}_1\|}.$$

7.18. Let $\phi : \mathbb{R}_{++} \to \mathbb{R}$ be a convex function. Then the function $f : \mathbb{R}_{++}^2 \to \mathbb{R}$ defined by

$$f(x, y) = y\phi\left(\frac{x}{y}\right), \quad x, y > 0,$$

is convex over $\mathbb{R}_{++}^2$.

7.19. Prove that the function $f(x, y) = -x^p y^{1-p}$ $(0 < p < 1)$ is convex over $\mathbb{R}_{++}^2$.

7.20. Let $f : C \to \mathbb{R}$ be a function defined over the convex set $C \subseteq \mathbb{R}^n$. Prove that f is quasi-convex if and only if

$$f(\lambda \mathbf{x} + (1 - \lambda)\mathbf{y}) \leq \max\{f(\mathbf{x}), f(\mathbf{y})\}, \quad \text{for any } \mathbf{x}, \mathbf{y} \in C, \lambda \in [0, 1].$$

7.21. Let $f(\mathbf{x}) = \frac{g(\mathbf{x})}{h(\mathbf{x})}$, where g is a convex function defined over a convex set $C \subseteq \mathbb{R}^n$ and $h(\mathbf{x}) = \mathbf{a}^T \mathbf{x} + b$ for some $\mathbf{a} \in \mathbb{R}^n$ and $b \in \mathbb{R}$. Assume that $h(\mathbf{x}) > 0$ for all $\mathbf{x} \in C$. Show that f is quasi-convex over C.

7.22. Show an example of two quasi-convex functions whose sum is *not* a quasi-convex function.

7.23. Let $f(\mathbf{x}) = \mathbf{x}^T \mathbf{A} \mathbf{x} + 2\mathbf{b}^T \mathbf{x} + c$, where $\mathbf{A}$ is an $n \times n$ symmetric matrix, $\mathbf{b} \in \mathbb{R}^n$, and $c \in \mathbb{R}$. Show that f is *quasi-convex* if and only if $\mathbf{A} \succeq 0$.

7.24. A function $f : C \to \mathbb{R}$ is called *log-concave* over the convex set $C \subseteq \mathbb{R}^n$ if $f(\mathbf{x}) > 0$ for any $\mathbf{x} \in C$ and $\ln(f)$ is a concave function over C.

(i) Show that the function $f(\mathbf{x}) = \frac{1}{\sum_{i=1}^n \frac{1}{x_i}}$ is log-concave over $\mathbb{R}_{++}^n$.

(ii) Let f be a twice continuously differentiable function over $\mathbb{R}$ with $f(x) > 0$ for all $x \in \mathbb{R}$. Show that f is log-concave if and only if $f''(x)f(x) \leq (f'(x))^2$ for all $x \in \mathbb{R}$.

7.25. Prove that if f and g are convex, twice differentiable, nondecreasing, and positive on $\mathbb{R}$, then the product fg is convex over $\mathbb{R}$. Show by an example that the positivity assumption is necessary to establish the convexity.

7.26. Let C be a convex subset of $\mathbb{R}^n$. A function f is called *strongly convex* over C if there exists $\sigma > 0$ such that the function $f(\mathbf{x}) - \frac{\sigma}{2}\|\mathbf{x}\|^2$ is convex over C. The parameter σ is called *the strong convexity parameter*. In the following questions C is a given convex subset of $\mathbb{R}^n$.

(i) Prove that f is strongly convex over C with parameter σ if and only if
$$f(\lambda \mathbf{x} + (1-\lambda)\mathbf{y}) \leq \lambda f(\mathbf{x}) + (1-\lambda)f(\mathbf{y}) - \frac{\sigma}{2}\lambda(1-\lambda)\|\mathbf{x}-\mathbf{y}\|^2$$
for any $\mathbf{x},\mathbf{y} \in C$ and $\lambda \in [0,1]$.

(ii) Prove that a strongly convex function over C is also strictly convex over C.

(iii) Suppose that f is continuously differentiable over C. Prove that f is strongly convex over C with parameter σ if and only if
$$f(\mathbf{y}) \geq f(\mathbf{x}) + \nabla f(\mathbf{x})^T(\mathbf{y}-\mathbf{x}) + \frac{\sigma}{2}\|\mathbf{x}-\mathbf{y}\|^2$$
for any $\mathbf{x},\mathbf{y} \in C$.

(iv) Suppose that f is continuously differentiable over C. Prove that f is strongly convex over C with parameter σ if and only if
$$(\nabla f(\mathbf{x}) - \nabla f(\mathbf{y}))^T(\mathbf{x}-\mathbf{y}) \geq \sigma \|\mathbf{x}-\mathbf{y}\|^2$$
for any $\mathbf{x},\mathbf{y} \in C$.

(v) Suppose that f is twice continuously differentiable over C. Show that f is strongly convex over C with parameter σ if and only if $\nabla^2 f(\mathbf{x}) \succeq \sigma \mathbf{I}$ for any $\mathbf{x} \in C$.

7.27. (i) Show that the function $f(\mathbf{x}) = \sqrt{1+\|\mathbf{x}\|^2}$ is strictly convex over $\mathbb{R}^n$ but is not *strongly convex* over $\mathbb{R}^n$.

(ii) Show that the quadratic function $f(\mathbf{x}) = \mathbf{x}^T \mathbf{A}\mathbf{x} + 2\mathbf{b}^T\mathbf{x} + c$ with $\mathbf{A} = \mathbf{A}^T \in \mathbb{R}^{n \times n}, \mathbf{b} \in \mathbb{R}^n, c \in \mathbb{R}$ is strongly convex if and only if $\mathbf{A} \succ \mathbf{0}$, and in that case the strong convexity parameter is $2\lambda_{\min}(\mathbf{A})$.

7.28. Let $f \in C_L^{1,1}(\mathbb{R}^n)$ be a convex function. For a fixed $\mathbf{x} \in \mathbb{R}^n$ define the function
$$g_{\mathbf{x}}(\mathbf{y}) = f(\mathbf{y}) - \nabla f(\mathbf{x})^T \mathbf{y}.$$

(i) Prove that $\mathbf{x}$ is a minimizer of $g_{\mathbf{x}}$ over $\mathbb{R}^n$.

(ii) Show that for any $\mathbf{x},\mathbf{y} \in \mathbb{R}^n$
$$g_{\mathbf{x}}(\mathbf{x}) \leq g_{\mathbf{x}}(\mathbf{y}) - \frac{1}{2L}\|\nabla g_{\mathbf{x}}(\mathbf{y})\|^2.$$

(iii) Show that for any $\mathbf{x},\mathbf{y} \in \mathbb{R}^n$
$$f(\mathbf{x}) \leq f(\mathbf{y}) + \nabla f(\mathbf{x})^T(\mathbf{y}-\mathbf{x}) - \frac{1}{2L}\|\nabla f(\mathbf{x}) - \nabla f(\mathbf{y})\|^2.$$

(iv) Prove that for any $\mathbf{x}, \mathbf{y} \in \mathbb{R}^n$

$$\langle \nabla f(\mathbf{x}) - \nabla f(\mathbf{y}), \mathbf{x} - \mathbf{y} \rangle \geq \frac{1}{L} \|\nabla f(\mathbf{x}) - \nabla f(\mathbf{y})\|^2 \text{ for any } \mathbf{x}, \mathbf{y} \in \mathbb{R}^n.$$

7.29. Let $f : \mathbb{R}^n \to \mathbb{R} \cup \{\infty\}$ be an extended real-valued function. Show that f is convex if and only if $\text{epi}(f)$ is convex.

7.30. Show that the support function of the set $S = \{\mathbf{x} \in \mathbb{R}^n : \mathbf{x}^T \mathbf{Q} \mathbf{x} \leq 1\}$, where $\mathbf{Q} \succ 0$, is $\sigma_S(\mathbf{y}) = \sqrt{\mathbf{y}^T \mathbf{Q}^{-1} \mathbf{y}}$.

7.31. Let $S = \{\mathbf{x} \in \mathbb{R}^n : \mathbf{a}^T \mathbf{x} \leq b\}$, where $0 \neq \mathbf{a} \in \mathbb{R}^n$ and $b \in \mathbb{R}$. Find the support function σ_S.

7.32. Let $p > 1$. Show that the support function of $S = \{\mathbf{x} \in \mathbb{R}^n : \|\mathbf{x}\|_p \leq 1\}$ is $\sigma_S(\mathbf{y}) = \|\mathbf{y}\|_q$, where q is defined by the relation $\frac{1}{p} + \frac{1}{q} = 1$.

7.33. Let $f_0, f_1, \ldots, f_m$ be convex functions over $\mathbb{R}^n$ and consider the perturbation function

$$F(\mathbf{b}) = \min_{\mathbf{x}} \{f_0(\mathbf{x}) : f_i(\mathbf{x}) \leq b_i, i = 1, 2, \ldots, m\}.$$

Assume that for any $\mathbf{b} \in \mathbb{R}^m$ the minimization problem in the above definition of $F(\mathbf{b})$ has an optimal solution. Show that F is convex over $\mathbb{R}^m$.

7.34. Let $C \subseteq \mathbb{R}^n$ be a convex set and let $\phi_1, \ldots, \phi_m$ be convex functions over C. Let U be the following subset of $\mathbb{R}^m$:

$$U = \{\mathbf{y} \in \mathbb{R}^m : \phi_1(\mathbf{x}) \leq y_1, \ldots, \phi_m(\mathbf{x}) \leq y_m \text{ for some } \mathbf{x} \in C\}.$$

Show that U is a convex set.

7.35. (i) Show that the extreme points of the unit simplex Δ_n are the unit-vectors $\mathbf{e}_1, \mathbf{e}_2, \ldots, \mathbf{e}_n$.

(ii) Find the optimal solution of the problem

$$\begin{array}{ll} \max & 57x_1^2 + 65x_2^2 + 17x_3^2 + 96x_1x_2 - 32x_1x_3 + 8x_2x_3 + 27x_1 - 84x_2 + 20x_3 \\ \text{s.t.} & x_1 + x_2 + x_3 = 1 \\ & x_1, x_2, x_3 \geq 0. \end{array}$$

7.36. Prove that for any $x_1, x_2, \ldots, x_n \in \mathbb{R}_+$ the following inequality holds:

$$\frac{\sum_{i=1}^n x_i}{n} \leq \sqrt{\frac{\sum_{i=1}^n x_i^2}{n}}.$$

7.37. Prove that for any $x_1, x_2, \ldots, x_n \in \mathbb{R}_{++}$ the following inequality holds:

$$\frac{\sum_{i=1}^n x_i^2}{\sum_{i=1}^n x_i} \leq \sqrt{\frac{\sum_{i=1}^n x_i^3}{\sum_{i=1}^n x_i}}.$$

7.38. Let $x_1, x_2, \ldots, x_n > 0$ satisfy $\sum_{i=1}^n x_i = 1$. Prove that

$$\sum_{i=1}^n \frac{x_i}{\sqrt{1 - x_i}} \geq \sqrt{\frac{n}{n-1}}.$$

7.39. Prove that for any $a, b, c > 0$ the following inequality holds:
$$\frac{9}{a+b+c} \leq 2\left(\frac{1}{a+b} + \frac{1}{b+c} + \frac{1}{c+a}\right).$$

7.40. (i) Prove that the function $f(x) = \frac{1}{1+e^x}$ is strictly convex over $[0, \infty)$.

(ii) Prove that for any $a_1, a_2, \ldots, a_n \geq 1$ the inequality
$$\sum_{i=1}^{n} \frac{1}{1+a_i} \geq \frac{n}{1+\sqrt[n]{a_1 a_2 \cdots a_n}}$$
holds.

Chapter 8
Convex Optimization

8.1 • Definition

A *convex optimization* problem (or just a *convex problem*) is a problem consisting of minimizing a convex function over a convex set. More explicitly, a convex problem is of the form

$$\begin{aligned} \min \quad & f(\mathbf{x}) \\ \text{s.t.} \quad & \mathbf{x} \in C, \end{aligned} \qquad (8.1)$$

where C is a convex set and f is a convex function over C. Problem (8.1) is in a sense implicit, and we will often consider more explicit formulations of convex problems such as convex optimization problems in *functional form*, which are convex problems of the form

$$\begin{aligned} \min \quad & f(\mathbf{x}) \\ \text{s.t.} \quad & g_i(\mathbf{x}) \leq 0, \quad i = 1, 2, \ldots, m, \\ & h_j(\mathbf{x}) = 0, \quad j = 1, 2, \ldots, p, \end{aligned} \qquad (8.2)$$

where $f, g_1, \ldots, g_m : \mathbb{R}^n \to \mathbb{R}$ are convex functions and $h_1, h_2, \ldots, h_p : \mathbb{R}^n \to \mathbb{R}$ are affine functions. Note that the above problem does fit into the general form (8.1) of convex problems. Indeed, the objective function is convex and the feasible set is a convex set since it can be written as

$$C = \left(\bigcap_{i=1}^{m} \text{Lev}(g_i, 0) \right) \cap \left(\bigcap_{j=1}^{p} \{\mathbf{x} : h_j(\mathbf{x}) = 0\} \right),$$

which implies that C is a convex set as an intersection of level sets of convex functions, which are necessarily convex sets, and hyperplanes, which are also convex sets.

The following result shows a very important property of convex problems: all local minimum points are also *global* minimum points.

Theorem 8.1 (local=global in convex optimization). *Let $f : C \to \mathbb{R}$ be a convex function defined on the convex set C. Let $\mathbf{x}^* \in C$ be a local minimum of f over C. Then $\mathbf{x}^*$ is a global minimum of f over C.*

Proof. Since $\mathbf{x}^*$ is a local minimum of f over C, it follows that there exists $r > 0$ such that $f(\mathbf{x}) \geq f(\mathbf{x}^*)$ for any $\mathbf{x} \in C$ satisfying $\mathbf{x} \in B[\mathbf{x}^*, r]$. Now let $\mathbf{y} \in C$ satisfy $\mathbf{y} \neq \mathbf{x}^*$. Our objective is to show that $f(\mathbf{y}) \geq f(\mathbf{x}^*)$. Let $\lambda \in (0, 1]$ be such that $\mathbf{x}^* + \lambda(\mathbf{y} - \mathbf{x}^*) \in B[\mathbf{x}^*, r]$.

An example of such λ is $\lambda = \frac{r}{\|y-x^*\|}$. Since $x^* + \lambda(y-x^*) \in B[x^*, r] \cap C$, it follows that $f(x^*) \leq f(x^* + \lambda(y-x^*))$, and hence by Jensen's inequality

$$f(x^*) \leq f(x^* + \lambda(y-x^*)) \leq (1-\lambda)f(x^*) + \lambda f(y).$$

Thus, $\lambda f(x^*) \leq \lambda f(y)$, and hence the desired inequality $f(x^*) \leq f(y)$ follows. $\square$

A slight modification of the above result shows that any local minimum of a *strictly* convex function over a convex set is a *strict* global minimum of the function over the set.

Theorem 8.2. *Let $f : C \to \mathbb{R}$ be a strictly convex function defined on the convex set C. Let $x^* \in C$ be a local minimum of f over C. Then x^* is a strict global minimum of f over C.*

The optimal set of the convex problem (8.1) is the set of all its minimizers, that is, $\operatorname{argmin}\{f(x) : x \in C\}$. This definition of an *optimal set* is also valid for general problems. An important property of convex problems is that their optimal sets are also convex.

Theorem 8.3 (convexity of the optimal set in convex optimization). *Let $f : C \to \mathbb{R}$ be a convex function defined over the convex set $C \subseteq \mathbb{R}^n$. Then the set of optimal solutions of the problem*

$$\min\{f(x) : x \in C\}, \tag{8.3}$$

which we denote by X^*, is convex. If, in addition, f is strictly convex over C, then there exists at most one optimal solution of the problem (8.3).

Proof. If $X^* = \emptyset$, the result follows trivially. Suppose that $X^* \neq \emptyset$ and denote the optimal value by f^*. Let $x, y \in X^*$ and $\lambda \in [0, 1]$. Then by Jensen's inequality $f(\lambda x + (1-\lambda)y) \leq \lambda f^* + (1-\lambda)f^* = f^*$, and hence $\lambda x + (1-\lambda)y$ is also optimal, i.e., belongs to X^*, establishing the convexity of X^*. Suppose now that f is strictly convex and X^* is nonempty; to show that X^* is a singleton, suppose in contradiction that there exist $x, y \in X^*$ such that $x \neq y$. Then $\frac{1}{2}x + \frac{1}{2}y \in C$, and by the strict convexity of f we have

$$f\left(\frac{1}{2}x + \frac{1}{2}y\right) < \frac{1}{2}f(x) + \frac{1}{2}f(y) = \frac{1}{2}f^* + \frac{1}{2}f^* = f^*,$$

which is a contradiction to the fact that f^* is the optimal value. $\square$

Convex optimization problems consist of minimizing convex functions over convex sets, but we will also refer to problems consisting of maximizing concave functions over convex sets as *convex problems*. (Indeed, they can be recast as minimization problems of convex functions by multiplying the objective function by minus one.)

Example 8.4. The problem

$$\begin{aligned} \min \quad & -2x_1 + x_2 \\ \text{s.t.} \quad & x_1^2 + x_2^2 \leq 3 \end{aligned}$$

is convex since the objective function is linear, and thus convex, and the single inequality constraint corresponds to the convex function $f(x_1, x_2) = x_1^2 + x_2^2 - 3$, which is a convex quadratic function. On the other hand, the problem

$$\begin{aligned} \min \quad & x_1^2 - x_2 \\ \text{s.t.} \quad & x_1^2 + x_2^2 = 3 \end{aligned}$$

is nonconvex. The objective function is convex, but the constraint is a nonlinear equality constraint and therefore nonconvex. Note that the feasible set is the boundary of the ball with center $(0,0)$ and radius $\sqrt{3}$. ∎

8.2 • Examples

8.2.1 • Linear Programming

A linear programming (LP) problem is an optimization problem consisting of minimizing a linear objective function subject to linear equalities and inequalities:

$$
\text{(LP)} \quad \begin{aligned} \min \quad & \mathbf{c}^T \mathbf{x} \\ \text{s.t.} \quad & \mathbf{A}\mathbf{x} \leq \mathbf{b}, \\ & \mathbf{B}\mathbf{x} = \mathbf{g}. \end{aligned}
$$

Here $\mathbf{A} \in \mathbb{R}^{m \times n}, \mathbf{b} \in \mathbb{R}^m, \mathbf{B} \in \mathbb{R}^{p \times n}, \mathbf{g} \in \mathbb{R}^p, \mathbf{c} \in \mathbb{R}^n$. This is of course a convex optimization problem since affine functions are convex. An interesting observation concerning LP problems is based on the fact that linear functions are both convex and concave. Consider the following LP problem:

$$
\begin{aligned} \max \quad & \mathbf{c}^T \mathbf{x} \\ \text{s.t.} \quad & \mathbf{A}\mathbf{x} = \mathbf{b}, \\ & \mathbf{x} \geq \mathbf{0}. \end{aligned}
$$

In the literature the latter formulation is many times called the "standard formulation." The above problem is on one hand a convex optimization problem as a maximization of a concave function over a convex set, but on the other hand, it is also a problem of maximizing a convex function over a convex set. We can therefore deduce by Theorem 7.42 that if the feasible set is nonempty and compact, then there exists at least one optimal solution which is an extreme point of the feasible set. By Theorem 6.34, this means that there exists at least one optimal solution which is a basic feasible solution. A more general result dropping the compactness assumption is called the "fundamental theorem of linear programming," and it states that if the problem has an optimal solution, then it necessarily has an optimal basic feasible solution.

Although the class of LP problems seems to be quite restrictive due to the linearity of all the involved functions, it encompasses a huge amount of applications and has a great impact on many fields in applied mathematics. Following is an example of a scheduling problem that can be recast as an LP problem.

Example 8.5. For a new position in a company, we need to schedule job interviews for n candidates numbered $1, 2, \ldots, n$ in this order (candidate i is scheduled to be the ith interview). Assume that the starting time of candidate i must be in the interval $[\alpha_i, \beta_i]$, where $\alpha_i < \beta_i$. To assure that the problem is feasible we assume that $\alpha_i \leq \beta_j$ for any $j > i$. The objective is to formulate the problem of finding n starting times of interviews so that the minimal starting time difference between consecutive interviews is maximal.

Let t_i denote the starting time of interview i. The objective function is the minimal difference between consecutive starting times of interviews:

$$ f(\mathbf{t}) = \min\{t_2 - t_1, t_3 - t_2, \ldots, t_n - t_{n-1}\}, $$

and the corresponding optimization problem is

$$
\begin{aligned} \max \quad & \left[\min\{t_2 - t_1, t_3 - t_2, \ldots, t_n - t_{n-1}\}\right] \\ \text{s.t.} \quad & \alpha_i \leq t_i \leq \beta_i, \quad i = 1, 2, \ldots, n. \end{aligned}
$$

Note that we did not incorporate the constraints that $t_i \leq t_{i+1}$ for $i = 1, 2, \ldots, n-1$ since the feasibility condition will guarantee in any case that these constraints will be satisfied in an optimal solution. The problem is convex since it consists of maximizing a concave function subject to affine (and hence convex) constraints. To show that the objective function is indeed concave, note that by Theorem 7.25 the maximum of convex functions is a convex function. The corresponding result for concave functions (that can be obtained by simply looking at minus of the function) is that the minimum of concave functions is a concave function. Therefore, since the objective function is a minimum of linear (and hence concave) functions, it is a concave function. In order to formulate the problem as an LP problem, we reformulate the problem as

$$\begin{aligned}
\max_{\mathbf{t},s} \quad & s \\
\text{s.t.} \quad & \min\{t_2 - t_1, t_3 - t_2, \ldots, t_n - t_{n-1}\} = s, \\
& \alpha_i \leq t_i \leq \beta_i, \quad i = 1, 2, \ldots, n.
\end{aligned} \qquad (8.4)$$

We now claim that problem (8.4) is equivalent to the corresponding problem with an inequality constraint instead of an equality:

$$\begin{aligned}
\max_{\mathbf{t},s} \quad & s \\
\text{s.t.} \quad & \min\{t_2 - t_1, t_3 - t_2, \ldots, t_n - t_{n-1}\} \geq s, \\
& \alpha_i \leq t_i \leq \beta_i, \quad i = 1, 2, \ldots, n.
\end{aligned} \qquad (8.5)$$

By "equivalent" we mean that any optimal solution of (8.5) satisfies the inequality constraint as an equality constraint. Indeed, suppose in contradiction that there exists an optimal solution $(\mathbf{t}^*, s^*)$ of (8.5) that satisfies the inequality constraints strictly, meaning that $\min\{t_2^* - t_1^*, t_3^* - t_2^*, \ldots, t_n^* - t_{n-1}^*\} > s^*$. Then we can easily check that the solution $(\mathbf{t}^*, \tilde{s})$, where $\tilde{s} = \min\{t_2^* - t_1^*, t_3^* - t_2^*, \ldots, t_n^* - t_{n-1}^*\}$ is also feasible for (8.5) and has a larger objective function value, which is a contradiction to the optimality of $(\mathbf{t}^*, s^*)$. Finally, we can rewrite the inequality $\min\{t_2 - t_1, t_3 - t_2, \ldots, t_n - t_{n-1}\} \geq s$ as $t_{i+1} - t_i \geq s$ for any $i = 1, 2, \ldots, n-1$, and we can therefore recast the problem as the following LP problem:

$$\begin{aligned}
\max_{\mathbf{t},s} \quad & s \\
\text{s.t.} \quad & t_{i+1} - t_i \geq s, \quad i = 1, 2, \ldots, n-1, \\
& \alpha_i \leq t_i \leq \beta_i, \quad i = 1, 2, \ldots, n. \quad \blacksquare
\end{aligned}$$

8.2.2 • Convex Quadratic Problems

Convex quadratic problems are problems consisting of minimizing a convex quadratic function subject to affine constraints. A general form of problems of this class can be written as

$$\begin{aligned}
\min \quad & \mathbf{x}^T \mathbf{Q} \mathbf{x} + 2\mathbf{b}^T \mathbf{x} \\
\text{s.t.} \quad & \mathbf{A}\mathbf{x} \leq \mathbf{c},
\end{aligned}$$

where $\mathbf{Q} \in \mathbb{R}^{n \times n}$ is positive semidefinite, $\mathbf{b} \in \mathbb{R}^n, \mathbf{A} \in \mathbb{R}^{m \times n}$, and $\mathbf{c} \in \mathbb{R}^m$. A well-known example of a convex quadratic problem arises in the area of linear classification and is described in detail next.

8.2.3 • Classification via Linear Separators

Suppose that we are given two types of points in $\mathbb{R}^n$: type A and type B. The type A points are given by

$$\mathbf{x}_1, \mathbf{x}_2, \ldots, \mathbf{x}_m \in \mathbb{R}^n$$

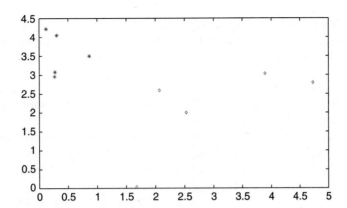

Figure 8.1. *Type A (asterisks) and type B (diamonds) points.*

and the type B points are given by

$$\mathbf{x}_{m+1}, \mathbf{x}_{m+2}, \ldots, \mathbf{x}_{m+p} \in \mathbb{R}^n.$$

For example, Figure 8.1 describes two sets of points in $\mathbb{R}^2$: the type A points are denoted by asterisks and the type B points are denoted by diamonds. The objective is to find a linear separator, which is a hyperplane of the form

$$H(\mathbf{w}, \beta) = \{\mathbf{x} \in \mathbb{R}^n : \mathbf{w}^T \mathbf{x} + \beta = 0\}$$

for which the type A and type B points are in its opposite sides:

$$\mathbf{w}^T \mathbf{x}_i + \beta < 0, \quad i = 1, 2, \ldots, m,$$
$$\mathbf{w}^T \mathbf{x}_i + \beta > 0, \quad i = m+1, m+2, \ldots, m+p.$$

Our underlying assumption is that the two sets of points are *linearly separable*, meaning that the above set of inequalities has a solution. The problem is not well-defined in the sense that there are many linear separators, and what we seek is in fact a separator that is in a sense farthest as possible from all the points. At this juncture we need to define the notion of the margin of the separator, which is the distance of the separator from the closest point, as illustrated in Figure 8.2. The separation problem will thus consist of finding the separator with the largest margin. To compute the margin, we need to have a formula for the distance between a point and a hyperplane. The next lemma provides such a formula, but its proof is postponed to Chapter 10 (see Lemma 10.12), where more general results will be derived.

Lemma 8.6. *Let $H(\mathbf{a}, b) = \{\mathbf{x} \in \mathbb{R}^n : \mathbf{a}^T \mathbf{x} = b\}$, where $0 \neq \mathbf{a} \in \mathbb{R}^n$ and $b \in \mathbb{R}$. Let $\mathbf{y} \in \mathbb{R}^n$. Then the distance between $\mathbf{y}$ and the set H is given by*

$$d(\mathbf{y}, H(\mathbf{a}, b)) = \frac{|\mathbf{a}^T \mathbf{y} - b|}{\|\mathbf{a}\|}.$$

We therefore conclude that the margin corresponding to a hyperplane $H(\mathbf{w}, -\beta)$ ($\mathbf{w} \neq 0$) is

$$\min_{i=1,2,\ldots,m+p} \frac{|\mathbf{w}^T \mathbf{x}_i + \beta|}{\|\mathbf{w}\|}.$$

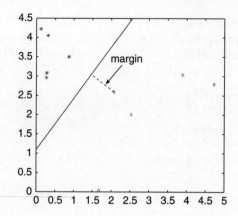

Figure 8.2. *The optimal linear seperator and its margin.*

So far, the problem that we consider is therefore

$$\begin{aligned}
\max \quad & \left\{ \min_{i=1,2,\ldots,m+p} \frac{|\mathbf{w}^T \mathbf{x}_i + \beta|}{\|\mathbf{w}\|} \right\} \\
\text{s.t.} \quad & \mathbf{w}^T \mathbf{x}_i + \beta < 0, \quad i = 1, 2, \ldots, m, \\
& \mathbf{w}^T \mathbf{x}_i + \beta > 0, \quad i = m+1, m+2, \ldots, m+p.
\end{aligned}$$

This is a rather bad formulation of the problem since it is not convex and cannot be easily handled. Our objective is to find a convex reformulation of the problem. For that, note that the problem has a degree of freedom in the sense that if $(\mathbf{w}, \beta)$ is an optimal solution, then so is any nonzero multiplier of it, that is, $(\alpha \mathbf{w}, \alpha \beta)$ for $\alpha \neq 0$. We can therefore decide that

$$\min_{i=1,2,\ldots,m+p} |\mathbf{w}^T \mathbf{x}_i + \beta| = 1,$$

and the problem can then be rewritten as

$$\begin{aligned}
\max \quad & \left\{ \frac{1}{\|\mathbf{w}\|} \right\} \\
\text{s.t.} \quad & \min_{i=1,2,\ldots,m+p} |\mathbf{w}^T \mathbf{x}_i + \beta| = 1, \\
& \mathbf{w}^T \mathbf{x}_i + \beta < 0, \quad i = 1, 2, \ldots, m, \\
& \mathbf{w}^T \mathbf{x}_i + \beta > 0, \quad i = m+1, 2, \ldots, m+p.
\end{aligned}$$

The combination of the first equality and the other inequality constraints implies that a valid reformulation is

$$\begin{aligned}
\min \quad & \tfrac{1}{2} \|\mathbf{w}\|^2 \\
\text{s.t.} \quad & \min_{i=1,2,\ldots,m+p} |\mathbf{w}^T \mathbf{x}_i + \beta| = 1, \\
& \mathbf{w}^T \mathbf{x}_i + \beta \leq -1, \quad i = 1, 2, \ldots, m, \\
& \mathbf{w}^T \mathbf{x}_i + \beta \geq 1, \quad i = m+1, 2, \ldots, m+p,
\end{aligned}$$

where we also used the fact that maximizing $\frac{1}{\|\mathbf{w}\|}$ is the same as minimizing $\|\mathbf{w}\|^2$ in the sense that the optimal set stays the same. Finally, we remove the problematic "min" equality constraint and obtain the following convex quadratic reformulation of the problem:

$$\begin{aligned}
\min \quad & \tfrac{1}{2}\|\mathbf{w}\|^2 \\
\text{s.t.} \quad & \mathbf{w}^T\mathbf{x}_i + \beta \leq -1, \quad i=1,2,\ldots,m, \\
& \mathbf{w}^T\mathbf{x}_i + \beta \geq 1, \quad i=m+1, m+2, \ldots, m+p.
\end{aligned} \quad (8.6)$$

The removal of the "min" constraint is valid since any feasible solution of problem (8.6) surely satisfies $\min_{i=1,2,\ldots,m+p} |\mathbf{w}^T\mathbf{x}_i + \beta| \geq 1$. If $(\mathbf{w}, \beta)$ is in addition optimal, then equality must be satisfied. Otherwise, if $\min_{i=1,2,\ldots,m+p} |\mathbf{w}^T\mathbf{x}_i + \beta| > 1$, then a better solution (i.e., with lower objective function value) will be $\frac{1}{\alpha}(\mathbf{w}, \beta)$, where $\alpha = \min_{i=1,2,\ldots,m+p} |\mathbf{w}^T\mathbf{x}_i + \beta|$.

8.2.4 ▪ Chebyshev Center of a Set of Points

Suppose that we are given m points $\mathbf{a}_1, \mathbf{a}_2, \ldots, \mathbf{a}_m$ in $\mathbb{R}^n$. The objective is to find the center of the minimum radius closed ball containing all the points. This ball is called *the Chebyshev ball* and the corresponding center is *the Chebyshev center*. In mathematical terms, the problem can be written as (r denotes the radius and $\mathbf{x}$ is the center)

$$\begin{aligned}
\min_{\mathbf{x}, r} \quad & r \\
\text{s.t.} \quad & \mathbf{a}_i \in B[\mathbf{x}, r], \quad i=1,2,\ldots, m.
\end{aligned}$$

Of course, recalling that $B[\mathbf{x}, r] = \{\mathbf{y} : \|\mathbf{y} - \mathbf{x}\| \leq r\}$, it follows that the problem can be written as

$$\begin{aligned}
\min_{\mathbf{x}, r} \quad & r \\
\text{s.t.} \quad & \|\mathbf{x} - \mathbf{a}_i\| \leq r, \quad i=1,2,\ldots, m.
\end{aligned} \quad (8.7)$$

This is obviously a convex optimization problem since it consists of minimizing a linear (and hence convex) function subject to convex inequality constraints: the function $\|\mathbf{x} - \mathbf{a}_i\| - r$ is convex as a sum of a translation of the norm function and the linear function $-r$. An illustration of the Chebyshev center and ball is given in Figure 8.3.

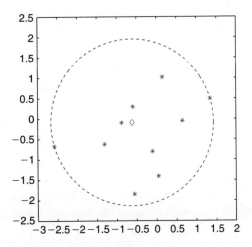

Figure 8.3. *The Chebyshev center (denoted by a diamond marker) of a set of 10 points (asterisks). The boundary of the Chebyshev ball is the dashed circle.*

8.2.5 ▪ Portfolio Selection

Suppose that an investor wishes to construct a portfolio out of n given assets numbered as $1, 2, \ldots, n$. Let $Y_j (j = 1, 2, \ldots, n)$ be the random variable representing the return from asset j. We assume that the expected returns are known,

$$\mu_j = \mathbb{E}(Y_j), \quad j = 1, 2, \ldots, n,$$

and that the covariances of all the pairs of variables are also known,

$$\sigma_{i,j} = COV(Y_i, Y_j), \quad i, j = 1, 2, \ldots, n.$$

There are n decision variables $x_1, x_2, \ldots, x_n$, where x_j denotes the proportion of budget invested in asset j. The decision variables are constrained to be nonnegative and sum up to 1: $\mathbf{x} \in \Delta_n$. The overall return is the random variable,

$$R = \sum_{j=1}^{n} x_j Y_j,$$

whose expectation and variance are given by

$$\mathbb{E}(R) = \mu^T \mathbf{x}, \quad \mathbb{V}(R) = \mathbf{x}^T \mathbf{C} \mathbf{x},$$

where $\mu = (\mu_1, \mu_2, \ldots, \mu_n)^T$ and $\mathbf{C}$ is the *covariance matrix* whose elements are given by $C_{i,j} = \sigma_{i,j}$ for all $1 \leq i, j \leq n$. It is important to note that the covariance matrix is always positive semidefinite. The variance of the portfolio, $\mathbf{x}^T \mathbf{C} \mathbf{x}$, is the *risk* of the suggested portfolio $\mathbf{x}$. There are several formulations of the portfolio optimization problem, which are all referred to as the "Markowitz model" in honor of Harry Markowitz, who first suggested this type of a model in 1952.

One formulation of the problem is to find a portfolio minimizing the risk under the constraint that a minimal return level is guaranteed:

$$\begin{aligned} \min \quad & \mathbf{x}^T \mathbf{C} \mathbf{x} \\ \text{s.t} \quad & \mu^T \mathbf{x} \geq \alpha, \\ & \mathbf{e}^T \mathbf{x} = 1, \\ & \mathbf{x} \geq 0, \end{aligned} \quad (8.8)$$

where $\mathbf{e}$ is the vector of all ones and α is the minimal return value. Another option is to maximize the expected return subject to a bounded risk constraint:

$$\begin{aligned} \max \quad & \mu^T \mathbf{x} \\ \text{s.t} \quad & \mathbf{x}^T \mathbf{C} \mathbf{x} \leq \beta, \\ & \mathbf{e}^T \mathbf{x} = 1, \\ & \mathbf{x} \geq 0, \end{aligned} \quad (8.9)$$

where β is the upper bound on the risk. Finally, a third option is to write an objective function which is a combination of the expected return and the risk:

$$\begin{aligned} \min \quad & -\mu^T \mathbf{x} + \gamma (\mathbf{x}^T \mathbf{C} \mathbf{x}) \\ \text{s.t} \quad & \mathbf{e}^T \mathbf{x} = 1, \\ & \mathbf{x} \geq 0, \end{aligned} \quad (8.10)$$

8.2. Examples

where $\gamma > 0$ is a penalty parameter. Each of the three models (8.8), (8.9), and (8.10) depends on a certain parameter (α, β, or γ) whose value dictates the tradeoff level between profit and risk. Determining the value of each of these parameters is not necessarily an easy task, and it also depends on the subjective preferences of the investors. The three models are all convex optimization problems since $\mathbf{x}^T \mathbf{C} \mathbf{x}$ is a convex function (its associated matrix $\mathbf{C}$ is positive semidefinite). The model (8.10) is a convex quadratic problem.

8.2.6 ▪ Convex QCQPs

A quadratically constrained quadratic problem, or QCQP for short, is a problem consisting of minimizing a quadratic function subject to quadratic inequalities and equalities:

$$\begin{array}{lll} \text{(QCQP)} & \min & \mathbf{x}^T \mathbf{A}_0 \mathbf{x} + 2\mathbf{b}_0^T \mathbf{x} + c_0 \\ & \text{s.t.} & \mathbf{x}^T \mathbf{A}_i \mathbf{x} + 2\mathbf{b}_i^T \mathbf{x} + c_i \leq 0, \quad i = 1, 2, \ldots, m, \\ & & \mathbf{x}^T \mathbf{A}_j \mathbf{x} + 2\mathbf{b}_j^T \mathbf{x} + c_j = 0, \quad j = m+1, m+2, \ldots, m+p. \end{array}$$

Obviously, QCQPs are not necessarily convex problems, but when there are no equality constrainers ($p = 0$) and all the matrices are positive semidefinite, $\mathbf{A}_i \succeq \mathbf{0}$ for $i = 0, 1, \ldots, m$, the problem is convex and is therefore called a *convex QCQP*.

8.2.7 ▪ Hidden Convexity in Trust Region Subproblems

There are several situations in which a certain problem is not convex but nonetheless can be recast as a convex optimization problem. This situation is sometimes called "hidden convexity." Perhaps the most famous nonconvex problem possessing such a hidden convexity property is the *trust region subproblem*, consisting of minimizing a quadratic function (not necessarily convex) subject to an Euclidean norm constraint:

$$\text{(TRS)} \quad \min\{\mathbf{x}^T \mathbf{A} \mathbf{x} + 2\mathbf{b}^T \mathbf{x} + c : \|\mathbf{x}\|^2 \leq 1\}.$$

Here $\mathbf{b} \in \mathbb{R}^n, c \in \mathbb{R}$, and $\mathbf{A}$ is an $n \times n$ symmetric matrix which is not necessarily positive semidefinite. Since the objective function is (possibly) nonconvex, problem (TRS) is (possibly) nonconvex. This is an important class of problems arising, for example, as a subroutine in trust region methods, hence the name of this class of problems. We will now show how to transform (TRS) into a convex optimization problem. First, by the spectral decomposition theorem (Theorem 1.10), there exist an orthogonal matrix $\mathbf{U}$ and a diagonal matrix $\mathbf{D} = \text{diag}(d_1, d_2, \ldots, d_n)$ such that $\mathbf{A} = \mathbf{U} \mathbf{D} \mathbf{U}^T$, and hence (TRS) can be rewritten as

$$\min\{\mathbf{x}^T \mathbf{U} \mathbf{D} \mathbf{U}^T \mathbf{x} + 2\mathbf{b}^T \mathbf{U} \mathbf{U}^T \mathbf{x} + c : \|\mathbf{U}^T \mathbf{x}\|^2 \leq 1\}, \quad (8.11)$$

where we used the fact that $\|\mathbf{U}^T \mathbf{x}\| = \|\mathbf{x}\|$. Making the linear change of variables $\mathbf{y} = \mathbf{U}^T \mathbf{x}$, it follows that (8.11) reduces to

$$\min\{\mathbf{y}^T \mathbf{D} \mathbf{y} + 2\mathbf{b}^T \mathbf{U} \mathbf{y} + c : \|\mathbf{y}\|^2 \leq 1\}.$$

Denoting $\mathbf{f} = \mathbf{U}^T \mathbf{b}$, we obtain the following formulation of the problem:

$$\begin{array}{ll} \min & \sum_{i=1}^{n} d_i y_i^2 + 2 \sum_{i=1}^{n} f_i y_i + c \\ \text{s.t.} & \sum_{i=1}^{n} y_i^2 \leq 1. \end{array} \quad (8.12)$$

The problem is still nonconvex since some of the d_is might be negative. At this point, we will use the following result stating that the signs of the optimal decision variables are actually known in advance.

Lemma 8.7. *Let $\mathbf{y}^*$ be an optimal solution of* (8.12). *Then $f_i y_i^* \leq 0$ for all $i = 1, 2, \ldots, n$.*

Proof. We will denote the objective function of problem (8.12) by $f(\mathbf{y}) \equiv \sum_{i=1}^n d_i y_i^2 + 2\sum_{i=1}^n f_i y_i + c$. Let $i \in \{1, 2, \ldots, n\}$. Define the vector $\tilde{\mathbf{y}}$ to be

$$\tilde{y}_j = \begin{cases} y_j^*, & j \neq i, \\ -y_i^*, & j = i. \end{cases}$$

Then obviously $\tilde{\mathbf{y}}$ is also a feasible solution of (8.12), and since $\mathbf{y}^*$ is an optimal solution of (8.12), it follows that

$$f(\mathbf{y}^*) \leq f(\tilde{\mathbf{y}}),$$

which is the same as

$$\sum_{i=1}^n d_i (y_i^*)^2 + 2\sum_{i=1}^n f_i y_i^* + c \leq \sum_{i=1}^n d_i (\tilde{y}_i)^2 + 2\sum_{i=1}^n f_i \tilde{y}_i + c.$$

Using the definition of $\tilde{\mathbf{y}}$, the above inequality reduces after much cancelation of terms to

$$2 f_i y_i^* \leq 2 f_i (-y_i^*),$$

which implies the desired inequality $f_i y_i^* \leq 0$. □

As a direct result of Lemma 8.7 we have that for any optimal solution $\mathbf{y}^*$, the equality $\operatorname{sgn}(y_i^*) = -\operatorname{sgn}(f_i)$ holds when $f_i \neq 0$ and where the sgn function is defined to be

$$\operatorname{sgn}(x) = \begin{cases} 1, & x \geq 0, \\ -1, & x < 0. \end{cases}$$

When $f_i = 0$, we have the property that both $\mathbf{y}^*$ and $\tilde{\mathbf{y}}$ are optimal (see the proof of Lemma 8.7), and hence the sign of $\mathbf{y}^*$ can be chosen arbitrarily. As a consequence, we can make the change of variables $y_i = -\operatorname{sgn}(f_i)\sqrt{z_i}(z_i \geq 0)$, and problem (8.12) becomes

$$\begin{array}{ll} \min & \sum_{i=1}^n d_i z_i - 2\sum_{i=1}^n |f_i|\sqrt{z_i} + c \\ \text{s.t.} & \sum_{i=1}^n z_i \leq 1, \\ & z_1, z_2, \ldots, z_n \geq 0. \end{array} \quad (8.13)$$

Obviously this is a convex optimization problem since the constraints are linear and the objective function is a sum of linear terms and positive multipliers of the convex functions $-\sqrt{z_i}$. To conclude, we have shown that the nonconvex trust region subproblem (TRS) is equivalent to the convex optimization problem (8.13).

8.3 • The Orthogonal Projection Operator

Given a nonempty closed convex set C, the *orthogonal projection* operator $P_C : \mathbb{R}^n \to C$ is defined by

$$P_C(\mathbf{x}) = \operatorname{argmin}\{\|\mathbf{y} - \mathbf{x}\|^2 : \mathbf{y} \in C\}. \quad (8.14)$$

The orthogonal projection operator with input $\mathbf{x}$ returns the vector in C that is closest to $\mathbf{x}$. Note that the orthogonal projection operator is defined as a solution of a convex optimization problem, specifically, a minimization of a convex quadratic function subject

8.3. The Orthogonal Projection Operator

to a convex feasible set. The first orthogonal projection theorem states that the orthogonal projection operator is in fact well-defined, meaning that the optimization problem in (8.14) has a unique optimal solution.

Theorem 8.8 (first projection theorem). *Let C be a nonempty closed convex set. Then problem (8.14) has a unique optimal solution.*

Proof. Since the objective function in (8.14) is a quadratic function with a positive definite matrix, it follows by Lemma 2.42 that the objective function is coercive and hence, by Theorem 2.32, that the problem has at least one optimal solution. In addition, since the objective function is strictly convex (again, since the objective function is quadratic with positive definite matrix), it follows by Theorem 8.3 that there exists only one optimal solution. $\square$

The distance function was already defined in Example 7.29 as

$$d(\mathbf{x}, C) = \min_{\mathbf{y} \in C} \|\mathbf{x} - \mathbf{y}\|.$$

Evidently, the distance function, in the case where C is a nonempty closed and convex set, can also be written in terms of the orthogonal projection as follows:

$$d(\mathbf{x}, C) = \|\mathbf{x} - P_C(\mathbf{x})\|.$$

Computing the orthogonal projection operator might be a difficult task, but there are some examples of simple sets on which the orthogonal projection can be easily computed.

Example 8.9 (projection on the nonnegative orthant). Let $C = \mathbb{R}^n_+$. To compute the orthogonal projection of $\mathbf{x} \in \mathbb{R}^n$ onto $\mathbb{R}^n_+$, we need to solve the convex optimization problem

$$\begin{aligned} \min \quad & \sum_{i=1}^n (y_i - x_i)^2 \\ \text{s.t.} \quad & y_1, y_2, \ldots, y_n \geq 0. \end{aligned} \quad (8.15)$$

Since this problem is separable, meaning that the objective function is a sum of functions of each of the variables, and the constraints are separable in the sense that each of the variables has its own constraint, it follows that the ith component of the optimal solution $\mathbf{y}^*$ of problem (8.15) is the optimal solution of the univariate problem

$$\min\{(y_i - x_i)^2 : y_i \geq 0\},$$

which is given by $y_i^* = [x_i]_+$, where for a real number $\alpha \in \mathbb{R}$, $[\alpha]_+$ is the *nonnegative part* of α:

$$[\alpha]_+ = \begin{cases} \alpha, & \alpha \geq 0, \\ 0, & \alpha < 0. \end{cases}$$

We will extend the definition of the nonnegative part to vectors, and the nonnegative part of a vector $\mathbf{v} \in \mathbb{R}^n$ is defined by

$$[\mathbf{v}]_+ = ([v_1]_+, [v_2]_+, \ldots, [v_n]_+)^T.$$

To summarize, the orthogonal projection operator onto $\mathbb{R}^n_+$ is given by

$$P_{\mathbb{R}^n_+}(\mathbf{x}) = [\mathbf{x}]_+. \quad \blacksquare$$

Example 8.10 (projection on boxes). A box is a subset of $\mathbb{R}^n$ of the form

$$B = [\ell_1, u_1] \times [\ell_2, u_2] \times \cdots \times [\ell_n, u_n] = \{\mathbf{x} \in \mathbb{R}^n : \ell_i \leq x_i \leq u_i\},$$

where $\ell_i \leq u_i$ for all $i = 1, 2, \ldots, n$. We will also allow some of the u_i's to be equal to ∞ and some of the ℓ_i's to be equal to $-\infty$; in these cases we will assume that ∞ or $-\infty$ are not actually contained in the intervals. A similar separability argument as the one used in the previous example, shows that the orthogonal projection is given by

$$\mathbf{y} = P_B(\mathbf{x}),$$

where

$$y_i = \begin{cases} u_i, & x_i \geq u_i, \\ x_i, & \ell_i < x_i < u_i, \\ \ell_i, & x_i \leq \ell_i, \end{cases}$$

for any $i = 1, 2, \ldots, n$. ∎

Example 8.11 (projection onto balls). Let $C = B[\mathbf{0}, r] = \{\mathbf{y} : \|\mathbf{y}\| \leq r\}$. The optimization problem associated with the computation of $P_C(\mathbf{x})$ is given by

$$\min_{\mathbf{y}} \{\|\mathbf{y} - \mathbf{x}\|^2 : \|\mathbf{y}\|^2 \leq r^2\}. \tag{8.16}$$

If $\|\mathbf{x}\| \leq r$, then obviously $\mathbf{y} = \mathbf{x}$ is the optimal solution of (8.16) since it corresponds to the optimal value 0. When $\|\mathbf{x}\| > r$, then the optimal solution of (8.16) must belong to the boundary of the ball since otherwise, by Theorem 2.6, it would be a stationary point of the objective function, that is, $2(\mathbf{y} - \mathbf{x}) = \mathbf{0}$, and hence $\mathbf{y} = \mathbf{x}$, which is impossible since $\mathbf{x} \notin C$. We thus conclude that the problem in this case is equivalent to

$$\min_{\mathbf{y}} \{\|\mathbf{y} - \mathbf{x}\|^2 : \|\mathbf{y}\|^2 = r^2\},$$

which can be equivalently written as

$$\min_{\mathbf{y}} \{-2\mathbf{x}^T \mathbf{y} + r^2 + \|\mathbf{x}\|^2 : \|\mathbf{y}\|^2 = r^2\},$$

The optimal solution of the above problem is the same as the optimal solution of

$$\min_{\mathbf{y}} \{-2\mathbf{x}^T \mathbf{y} : \|\mathbf{y}\|^2 = r^2\}.$$

By the Cauchy–Schwarz inequality, the objective function can be lower bounded by

$$-2\mathbf{x}^T \mathbf{y} \geq -2\|\mathbf{x}\|\|\mathbf{y}\| = -2r\|\mathbf{x}\|,$$

and on the other hand, this lower bound is attained at $\mathbf{y} = r \frac{\mathbf{x}}{\|\mathbf{x}\|}$, and hence the orthogonal projection is given by

$$P_{B[\mathbf{0}, r]} = \begin{cases} \mathbf{x}, & \|\mathbf{x}\| \leq r, \\ r \frac{\mathbf{x}}{\|\mathbf{x}\|}, & \|\mathbf{x}\| > r. \end{cases} \quad \blacksquare$$

8.4 ▪ CVX

CVX is a MATLAB-based modeling system for convex optimization. It was created by Michael Grant and Stephen Boyd [19]. This MATLAB package is in fact an interface to other convex optimization solvers such as SeDuMi and SDPT3. We will explore here some of the basic features of the software, but a more comprehensive and complete guide can be found at the CVX website (cvxr.com). The basic structure of a CVX program is as follows:

8.4. CVX

```
cvx_begin
{variables declaration}
minimize({objective function}) or maximize({objective function})
subject to
{constraints}
cvx_end
```

Variables Declaration

The variables are declared via the command **variable** or **variables**. Thus, for example,

```
variable x(4);
variable z;
variable Y(2,3);
```

declares three variables:

- **x**, a column vector of length 4,
- **z**, a scalar,
- **Y**, a 2×3 matrix.

The same declaration can be written as

```
variables x(4) z Y(2,3);
```

Atoms

CVX accepts only convex functions as objective and constraint functions. There are several basic convex functions, called "atoms," which are embedded in CVX. Some of these atoms are given in the following table.

function	meaning	attributes
norm(x,p)	$\sqrt[p]{\sum_{i=1}^{n} |x_i|^p} \, (p \geq 1)$	convex
square(x)	x^2	convex
sum_square(x)	$\sum_{i=1}^{n} x_i^2$	convex
square_pos(x)	$[x]_+^2$	convex, nondecreasing
sqrt(x)	$\sqrt{x}$	concave, nondecreasing
inv_pos(x)	$\frac{1}{x} (x > 0)$	convex, nonincreasing
max(x)	$\max\{x_1, x_2, \ldots, x_n\}$	convex, nondecreasing
quad_over_lin(x,y)	$\frac{\|x\|^2}{y} \, (y > 0)$	convex
quad_form(x,P)	$x^T P x \, (P \succeq 0)$	convex

In addition, CVX is aware that the function x^p for an even integer p is a convex function and that affine functions are both convex and concave.

Operations Preserving Convexity

Atoms can be incorporated by several operations which preserve convexity:

- addition,
- multiplication by a nonnegative scalar,

- composition of a nondecreasing convex function with a convex function,
- composition of a convex function with an affine transformation.

CVX is also aware that minus a convex function is a concave function. The constraints that CVX is willing to accept are inequalities of the forms

```
f(x)<=g(x)
g(x)>=f(x)
```

where f is convex and g is concave. Equality constraints must be affine, and the syntax is (h and s are affine functions)

```
h(x)==s(x)
```

Note that the equality must be written in the format ==. Otherwise, it will be interpreted as a substitution operation.

Example 8.12. Suppose that we wish to solve the least squares problem

$$\min \|\mathbf{A}\mathbf{x} - \mathbf{b}\|^2,$$

where

$$\mathbf{A} = \begin{pmatrix} 1 & 2 \\ 3 & 4 \\ 5 & 6 \end{pmatrix}, \qquad \mathbf{b} = \begin{pmatrix} 7 \\ 8 \\ 9 \end{pmatrix}.$$

We can find the solution of this least squares problem by the MATLAB commands

```
>> A=[1,2;3,4;5,6];b=[7;8;9];
>> x=(A'*A)\(A'*b)
x =
   -6.0000
    6.5000
```

To solve this problem via CVX, we can use the function sum_square:

```
cvx_begin
variable x(2)
minimize(sum_square(A*x-b))
cvx_end
```

The obtained solution is as expected:

```
>> x
x =
   -6.0000
    6.5000
```

We can also solve the problem by noting that

$$\|\mathbf{A}\mathbf{x} - \mathbf{b}\|^2 = \mathbf{x}^T \mathbf{A}^T \mathbf{A} \mathbf{x} - 2\mathbf{b}^T \mathbf{A} \mathbf{x} + \|\mathbf{b}\|^2$$

and writing the following commands:

```
cvx_begin
variable x(2)
minimize(quad_form(x,A'*A)-2*b'*A*x)
cvx_end
```

However, the following program is *wrong* and CVX will not accept it:

```
cvx_begin
variable x(2)
minimize(norm(A*x-b)^2)
cvx_end
```

The reason is that the objective function is written as a composition of the square function which is *not* nondecreasing with the function norm(Ax-b). Of course, we know that the image of $\|Ax-b\|$ consists only of nonnegative values and that the square function is nondecreasing over that domain. However, CVX is not aware of that. If we insist on making such a decomposition we can use the function square_pos – the scalar function $\varphi(x) = \max\{x,0\}^2$, which is convex and nondecreasing, and write the legitimate CVX program:

```
cvx_begin
variable x(2)
minimize(square_pos(norm(A*x-b)))
cvx_end
```

It is also worth mentioning that since the problem of minimizing the norm is equivalent to the problem of minimizing the squared norm in the sense that both problems have the same optimal solution, the following CVX program will also find the optimal solution, but the optimal value will be the square root of the optimal value of the original problem:

```
cvx_begin
variable x(2)
minimize(norm(A*x-b))
cvx_end
```

∎

Example 8.13. Suppose that we wish to write a CVX code that solves the convex optimization problem

$$\begin{aligned} \min \quad & \sqrt{x_1^2 + x_2^2 + 1} + 2\max\{x_1, x_2, 0\} \\ \text{s.t.} \quad & |x_1| + |x_2| + \frac{x_1^2}{x_2} \leq 5 \\ & \frac{1}{x_2} + x_1^4 \leq 10 \\ & x_2 \geq 1 \\ & x_1 \geq 0. \end{aligned} \qquad (8.17)$$

In order to write the above problem in CVX, it is important to understand the reason why $\sqrt{x_1^2 + x_2^2 + 1}$ is convex since writing sqrt(x(1)^2+x(2)^2+1) in CVX will result in an error message. Since the expression is written as a composition of an increasing concave function with a convex function, in general it does not result in a convex function. A valid reason why $\sqrt{x_1^2 + x_2^2 + 1}$ is convex is that it can be rewritten as $\|(x_1, x_2, 1)^T\|$. That is, it is a composition of the norm function with an affine transformation. Correspondingly, the correct syntax in CVX will be norm([x;1]). Overall, a CVX program that solves (8.17) is

```
cvx_begin
variable x(2)
minimize(norm([x;1])+2*max(max(x(1),x(2)),0))
subject to
norm(x,1)+quad_over_lin(x(1),x(2))<=5
inv_pos(x(2))+x(1)^4<=10
x(2)>=1
x(1)>=0
cvx_end
```

∎

Example 8.14. Suppose that we wish to find the Chebyshev center of the 5 points

$$(-1,3), \quad (-3,10), \quad (-1,0), \quad (5,0), \quad (-1,-5).$$

Recall that the problem of finding the Chebyshev center of a set of points $a_1, a_2, \ldots, a_m$ is given by (see Section 8.2.4)

$$\begin{aligned} \min_{x,r} \quad & r \\ \text{s.t.} \quad & \|x - a_i\| \leq r, \quad i = 1, 2, \ldots, m, \end{aligned}$$

and thus the following code will solve the problem:

```
A=[-1,-3,-1,5,-1;3,10,0,0,-5];
cvx_begin
variables x(2) r
minimize(r)
subject to
for i=1:5
    norm(x-A(:,i))<=r
end
cvx_end
```

This results in the optimal solution

```
>> x
x =
    -2.0002
     2.5000
>> r
r =
     7.5664
```

To plot the 5 points along with the Chebyshev circle and center we can write

```
plot(A(1,:),A(2,:),'*')
hold on
plot(x(1),x(2),'d')
t=0:0.001:2*pi;
xx=x(1)+r*cos(t);
```

```
yy=x(2)+r*sin(t);
plot(xx,yy)
axis equal
axis([-11,7,-6,11])
hold off
```

The result can be seen in Figure 8.4. ∎

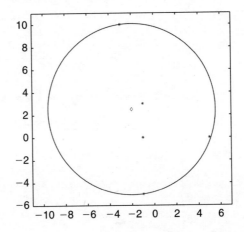

Figure 8.4. *The Chebyshev center (diamond marker) of 5 points (in asterisks).*

Example 8.15 (robust regression). Suppose that we are given 21 points in $\mathbb{R}^2$ generated by the MATLAB commands

```
randn('seed',314);
x=linspace(-5,5,20)';
y=2*x+1+randn(20,1);
x=[x;5];
y=[y;-20];
plot(x,y,'*')
hold on
```

The resulting plot can be seen in Figure 8.5. Note that the point $(5,-20)$ is an outlier; it is far away from all the other points and does not seem to fit into the almost-line structure of the other points. The least squares line, also called *the regression line*, can be found by the commands (see also Chapter 3)

```
A=[x,ones(21,1)];
b=y;
u=A\b;
alpha=u(1);beta=u(2);
plot([-6,6],alpha*[-6,6]+beta);
hold off
```

resulting in the line plotted in Figure 8.6. As can be clearly seen in Figure 8.6, the least squares line is very much affected by the single outlier point, which is a known drawback of the least squares approach. Another option is to replace the l_2-based objective function $\|\mathbf{Ax}-\mathbf{b}\|_2^2$ with an l_1-based objective function; that is, we can consider the optimization problem

$$\min \|\mathbf{Ax}-\mathbf{b}\|_1.$$

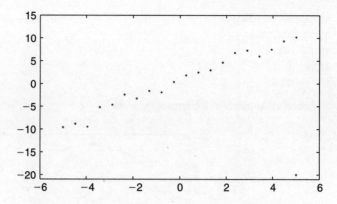

Figure 8.5. 21 *points in the plane. The point* (5,−20) *is an outlier.*

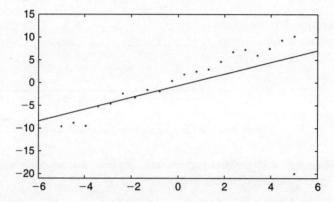

Figure 8.6. 21 *points in the plane along with their least squares (regression) line.*

This approach has the advantage that it is less sensitive to outliers since outliers are not as severely penalized as they are penalized in the least squares objective function. More specifically, in the least squares objective function, the distances to the line are squared, while in the l_1-based function they are not. To find the line using CVX, we can run the commands

```
plot(x,y,'*')
hold on
cvx_begin
variable u_l1(2)
minimize(norm(A*u_l1-b,1))
cvx_end
alpha_l1=u_l1(1);
beta_l1=u_l1(2);
plot([-6,6],alpha_l1*[-6,6]+beta_l1);
axis([-6,6,-21,15])
hold off
```

and the corresponding plot is given in Figure 8.7. Note that the resulting line is insensitive to the outlier. This is why this line is also called *the robust regression line.* ∎

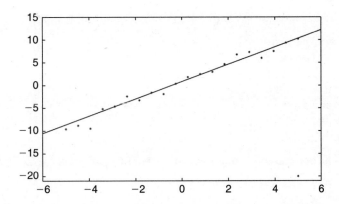

Figure 8.7. 21 *points in the plane along with their robust regression line.*

Example 8.16 (solution of a trust region subproblem). Consider the trust region subproblem (see Section 8.2.7)

$$\begin{aligned} \min \quad & x_1^2 + x_2^2 + 3x_3^2 + 4x_1x_2 + 6x_1x_3 + 8x_2x_3 + x_1 + 2x_2 - x_3 \\ \text{s.t.} \quad & x_1^2 + x_2^2 + x_3^2 \leq 1, \end{aligned}$$

which is the same as

$$\begin{aligned} \min \quad & \mathbf{x}^T \mathbf{A} \mathbf{x} + 2\mathbf{b}^T \mathbf{x} \\ \text{s.t.} \quad & \|\mathbf{x}\|^2 \leq 1, \end{aligned}$$

where

$$\mathbf{A} = \begin{pmatrix} 1 & 2 & 3 \\ 2 & 1 & 4 \\ 3 & 4 & 3 \end{pmatrix}, \qquad \mathbf{b} = \begin{pmatrix} \frac{1}{2} \\ 1 \\ -\frac{1}{2} \end{pmatrix}.$$

The problem is nonconvex since the matrix **A** is not positive definite:

```
>> A=[1,2,3;2,1,4;3,4,3];
>> b=[0.5;1;-0.5];
>> eig(A)

ans =

   -2.1683
   -0.8093
    7.9777
```

It is therefore not possible to solve the problem directly using CVX. Instead, we will use the technique described in Section 8.2.7 to convert the problem into a convex problem, and then we will be able to solve the transformed problem via CVX. We begin by computing the spectral decomposition of **A**,

```
[U,D]=eig(A);
```

and then compute the vectors **d** and **f** in the convex reformulation of the problem:

```
f=U'*b;
d=diag(D);
```

We can now use CVX to solve the equivalent problem (8.13):

```
cvx_begin
variable z(3)
minimize(d'*z-2*abs(f)'*sqrt(z))
subject to
sum(z)<=1
z>=0
cvx_end
```

The optimal solution is then computed by $y_i = -\text{sgn}(f_i)\sqrt{z_i}$ and then $\mathbf{x} = \mathbf{U}\mathbf{y}$:

```
>> y=-sign(f).*sqrt(z);
>> x=U*y
x =

   -0.2300
   -0.7259
    0.6482
```

■

Exercises

8.1. Consider the problem
$$\text{(P)} \quad \begin{array}{ll} \min & f(\mathbf{x}) \\ \text{s.t.} & g(\mathbf{x}) \leq 0 \\ & \mathbf{x} \in X, \end{array}$$

where f and g are convex functions over $\mathbb{R}^n$ and $X \subseteq \mathbb{R}^n$ is a convex set. Suppose that $\mathbf{x}^*$ is an optimal solution of (P) that satisfies $g(\mathbf{x}^*) < 0$. Show that $\mathbf{x}^*$ is also an optimal solution of the problem

$$\begin{array}{ll} \min & f(\mathbf{x}) \\ \text{s.t.} & \mathbf{x} \in X. \end{array}$$

8.2. Let $C = B[\mathbf{x}_0, r]$, where $\mathbf{x}_0 \in \mathbb{R}^n$ and $r > 0$ are given. Find a formula for the orthogonal projection operator P_C.

8.3. Let f be a strictly convex function over $\mathbb{R}^m$ and let g be a convex function over $\mathbb{R}^n$. Define the function
$$h(\mathbf{x}) = f(\mathbf{A}\mathbf{x}) + g(\mathbf{x}),$$
where $\mathbf{A} \in \mathbb{R}^{m \times n}$. Assume that $\mathbf{x}^*$ and $\mathbf{y}^*$ are optimal solutions of the unconstrained problem of minimizing h. Show that $\mathbf{A}\mathbf{x}^* = \mathbf{A}\mathbf{y}^*$.

8.4. For each of the following optimization problems (a) show that it is convex, (b) write a CVX code that solves it, and (c) write down the optimal solution (by running CVX).

(i)
$$\begin{array}{ll} \min & x_1^2 + 2x_1 x_2 + 2x_2^2 + x_3^2 + 3x_1 - 4x_2 \\ \text{s.t.} & \sqrt{2x_1^2 + x_1 x_2 + 4x_2^2 + 4} + \frac{(x_1 - x_2 + x_3 + 1)^2}{x_1 + x_2} \leq 6 \\ & x_1, x_2, x_3 \geq 1. \end{array}$$

(ii)
$$\max \quad x_1 + x_2 + x_3 + x_4$$
$$\text{s.t.} \quad (x_1 - x_2)^2 + (x_3 + 2x_4)^4 \le 5$$
$$x_1 + 2x_2 + 3x_3 + 4x_4 \le 6$$
$$x_1, x_2, x_3, x_4 \ge 0.$$

(iii)
$$\min \quad 5x_1^2 + 4x_2^2 + 7x_3^2 + 4x_1x_2 + 2x_2x_3 + |x_1 - x_2|$$
$$\text{s.t.} \quad \frac{x_1^2 + x_2^2}{x_3} + (x_1^2 + x_2^2 + 1)^4 \le 10$$
$$x_3 \ge 10.$$

(iv)
$$\min \quad \sqrt{x_1^2 + x_2^2 + 2x_1 + 5} + x_1^2 + 2x_1x_2 + x_2^2 + 2x_1 + 3x_2$$
$$\text{s.t.} \quad \frac{x_1^2}{x_1 + x_2} + \left(\frac{x_1^2}{x_2} + 1\right)^8 \le 100$$
$$x_1 + x_2 \ge 4$$
$$x_2 \ge 1.$$

(v)
$$\min \quad |2x_1 + 3x_2 + x_3| + x_1^2 + x_2^2 + x_3^2 + \sqrt{2x_1^2 + 4x_1x_2 + 7x_2^2 + 10x_2 + 6}$$
$$\text{s.t.} \quad \frac{x_1^2 + 1}{x_2} + 2x_1^2 + 5x_2^2 + 10x_3^2 + 4x_1x_2 + 2x_1x_3 + 2x_2x_3 \le 7$$
$$x_1 \ge 0$$
$$x_2 \ge 1.$$

For this problem also show that the expression inside the square root is always nonnegative, i.e., $2x_1^2 + 4x_1x_2 + 7x_2^2 + 10x_2 + 6 \ge 0$ for all x_1, x_2.

(vi)
$$\min \quad \frac{1}{2x_2 + 3x_3} + 5x_1^2 + 4x_2^2 + 7x_3^2 + \frac{x_1^2 + x_1 + 1}{x_2 + x_3}$$
$$\text{s.t.} \quad \max\{x_1 + x_2, x_3^2\} + (x_1^2 + 4x_1x_2 + 5x_2^2 + 1)^2 \le 10$$
$$x_1, x_2, x_3 \ge 0.1.$$

(vii)
$$\min \quad \sqrt{2x_1^2 + 3x_2^2 + x_3^2 + 4x_1x_2 + 7} + (x_1^2 + x_2^2 + x_3^2 + 1)^2$$
$$\text{s.t.} \quad \frac{(x_1 + x_2)^2}{x_3 + 1} + x_1^8 \le 7$$
$$x_1^2 + x_2^2 + 4x_3^2 + 2x_1x_2 + 2x_1x_3 + 2x_2x_3 \le 10$$
$$x_1, x_2, x_3 \ge 0.$$

(viii)
$$\min \quad \frac{x_1^4 + 2x_1^2x_2^2 + x_2^4}{x_1^2 + 2x_1x_2 + x_2^2} + \sqrt{x_3^2 + 1}$$
$$\text{s.t.} \quad x_1^2 + x_2^2 + 2x_3^2 + 2x_1x_2 + 2x_1x_3 + 2x_2x_3 \le 100$$
$$x_1 + x_2 + x_3 = 2$$
$$x_1 + x_2 \ge 1.$$

(ix)
$$\min \quad \frac{x_1^4}{x_2^2} + \frac{x_2^4}{x_1^2} + 2x_1x_2 + |x_1 + 5| + |x_2 + 5| + |x_3 + 5|$$
$$\text{s.t.} \quad \left((x_1^2 + x_2^2 + x_3^2 + 1)^2 + 1\right)^2 + x_1^4 + x_2^4 + x_3^4 \le 200$$
$$\max\{x_1^2 + 4x_1x_2 + 9x_2^2, x_1, x_2\} \le 40$$
$$x_1 \ge 1$$
$$x_2 \ge 1.$$

(x)
$$\begin{aligned} \min \quad & (x_1+x_2+x_3)^8 + x_1^2 + x_2^2 + 3x_3^2 + 2x_1x_2 + 2x_2x_3 + 2x_1x_3 \\ \text{s.t.} \quad & (|x_1-2x_2|+1)^4 + \frac{1}{x_3} \leq 10, \\ & 2x_1+2x_2+x_3 \leq 1, \\ & 0 \leq x_3 \leq 1. \end{aligned}$$

8.5. Suppose that we are given 40 points in the plane. Each of these points belongs to one of two classes. Specifically, there are 19 points of class 1 and 21 points of class 2. The points are generated and plotted by the MATLAB commands

```
rand('seed',314);
x=rand(40,1);
y=rand(40,1);
class=[2*x<y+0.5]+1;
A1=[x(find(class==1)),y(find(class==1))];
A2=[x(find(class==2)),y(find(class==2))];
plot(A1(:,1),A1(:,2),'*','MarkerSize',6)
hold on
plot(A2(:,1),A2(:,2),'d','MarkerSize',6)
hold off
```

The plot of the points is given in Figure 8.8. Note that the rows of $\mathbf{A}_1 \in \mathbb{R}^{19 \times 2}$ are the 19 points of class 1 and the rows of $\mathbf{A}_2 \in \mathbb{R}^{21 \times 2}$ are the 21 points of class 2. Write a CVX-based code for finding the maximum-margin line separating the two classes of points.

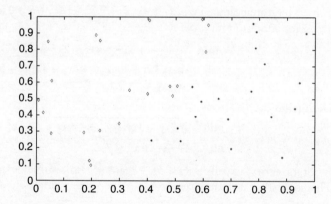

Figure 8.8. *40 points of two classes: class 1 points are denoted by asterisks, and class 2 points are denoted by diamonds.*

Chapter 9

Optimization over a Convex Set

9.1 • Stationarity

Throughout this chapter we will consider the constrained optimization problem (P) given by

$$\text{(P)} \quad \begin{array}{l} \min \ f(\mathbf{x}) \\ \text{s.t.} \ \mathbf{x} \in C, \end{array}$$

where f is a continuously differentiable function and C is a closed and convex set. In Chapter 2 we discussed the notion of *stationary points* of continuously differentiable functions, which are points in which the gradient vanishes. It was shown that stationarity is a necessary condition for a point to be an unconstrained local optimum point. The situation is more complicated when considering constrained problems of the form (P), where instead of looking at stationary points of a function, we need to consider the notion of stationary points of *a problem*.

Definition 9.1 (stationary points of constrained problems). *Let f be a continuously differentiable function over a closed convex set C. Then $\mathbf{x}^* \in C$ is called a* **stationary point** *of (P) if $\nabla f(\mathbf{x}^*)^T(\mathbf{x}-\mathbf{x}^*) \geq 0$ for any $\mathbf{x} \in C$.*

Stationarity actually means that there are no feasible descent directions of f at $\mathbf{x}^*$. This suggests that stationarity is in fact a necessary condition for a local minimum of (P).

Theorem 9.2 (stationarity as a necessary optimality condition). *Let f be a continuously differentiable function over a closed convex set C, and let $\mathbf{x}^*$ be a local minimum of (P). Then $\mathbf{x}^*$ is a stationary point of (P).*

Proof. Let $\mathbf{x}^*$ be a local minimum of (P), and assume in contradiction that $\mathbf{x}^*$ is not a stationary point of (P). Then there exists $\mathbf{x} \in C$ such that $\nabla f(\mathbf{x}^*)^T(\mathbf{x}-\mathbf{x}^*) < 0$. Therefore, $f'(\mathbf{x}^*;\mathbf{d}) < 0$ where $\mathbf{d} = \mathbf{x} - \mathbf{x}^*$, and hence, by Lemma 4.2, it follows that there exists $\varepsilon \in (0,1)$ such that $f(\mathbf{x}^* + t\mathbf{d}) < f(\mathbf{x}^*)$ for all $t \in (0,\varepsilon)$. Since C is convex we have that $\mathbf{x}^* + t\mathbf{d} = (1-t)\mathbf{x}^* + t\mathbf{x} \in C$, leading to the conclusion that $\mathbf{x}^*$ is *not* a local minimum point of (P), in contradiction to the assumption that $\mathbf{x}^*$ is a local minimum point of (P). $\square$

Example 9.3 ($C = \mathbb{R}^n$). If $C = \mathbb{R}^n$, then the stationary points of (P) are the points $\mathbf{x}^*$ satisfying
$$\nabla f(\mathbf{x}^*)^T(\mathbf{x}-\mathbf{x}^*) \geq 0 \qquad (9.1)$$
for all $\mathbf{x} \in \mathbb{R}^n$. Plugging $\mathbf{x} = \mathbf{x}^* - \nabla f(\mathbf{x}^*)$ into (9.1), we obtain that $-\|\nabla f(\mathbf{x}^*)\|^2 \geq 0$, implying that
$$\nabla f(\mathbf{x}^*) = 0.$$
Therefore, it follows that the notion of a stationary point of a function and a stationary point of a *minimization problem* coincide when the problem is unconstrained. ■

Example 9.4 ($C = \mathbb{R}^n_+$). Consider the optimization problem
$$(Q) \quad \begin{array}{l} \min \; f(\mathbf{x}) \\ \text{s.t.} \; x_i \geq 0, \quad i=1,2,\ldots,n, \end{array}$$
where f is a continuously differentiable function over $\mathbb{R}^n_+$. A vector $\mathbf{x}^* \in \mathbb{R}^n_+$ is a stationary point of (Q) if and only if
$$\nabla f(\mathbf{x}^*)^T(\mathbf{x}-\mathbf{x}^*) \geq 0 \text{ for all } \mathbf{x} \geq 0,$$
which is the same as
$$\nabla f(\mathbf{x}^*)^T \mathbf{x} - \nabla f(\mathbf{x}^*)^T \mathbf{x}^* \geq 0 \text{ for all } \mathbf{x} \geq 0. \qquad (9.2)$$
We will now use the following technical result:
$$\mathbf{a}^T\mathbf{x}+b \geq 0 \text{ for all } \mathbf{x} \geq 0 \text{ if and only if } \mathbf{a} \geq 0 \text{ and } b \geq 0.$$
Using this simple result, it follows that (9.2) holds if and only if
$$\nabla f(\mathbf{x}^*) \geq 0 \text{ and } \nabla f(\mathbf{x}^*)^T \mathbf{x}^* \leq 0. \qquad (9.3)$$
Since $\nabla f(\mathbf{x}^*) \geq 0$ and $\mathbf{x}^* \geq 0$, we can conclude that (9.3) holds if and only if
$$\nabla f(\mathbf{x}^*) \geq 0 \text{ and } x_i^* \frac{\partial f}{\partial x_i}(\mathbf{x}^*) = 0, \quad i=1,2,\ldots,n.$$
We can compactly write the above condition as follows:
$$\frac{\partial f}{\partial x_i}(\mathbf{x}^*) \begin{cases} = 0, & x_i^* > 0, \\ \geq 0, & x_i^* = 0. \end{cases} \quad ■$$

Example 9.5 (stationarity over the unit-sum set). Consider the optimization problem
$$(R) \quad \begin{array}{l} \min \; f(\mathbf{x}) \\ \text{s.t.} \; \mathbf{e}^T\mathbf{x}=1, \end{array}$$
where f is a continuously differentiable function over $\mathbb{R}^n$. The feasible set
$$U = \{\mathbf{x} \in \mathbb{R}^n : \mathbf{e}^T\mathbf{x} = 1\} = \left\{\mathbf{x} \in \mathbb{R}^n : \sum_{i=1}^n x_i = 1\right\}$$

9.1. Stationarity

is called *the unit-sum set*. A point $\mathbf{x}^* \in U$ is a stationary point of (R) if and only if

(I) $\quad \nabla f(\mathbf{x}^*)^T(\mathbf{x}-\mathbf{x}^*) \geq 0$ for all $\mathbf{x}$ satisfying $\mathbf{e}^T\mathbf{x} = 1$.

We will show that condition (I) is equivalent to the following simple and explicit condition:

(II) $\quad \dfrac{\partial f}{\partial x_1}(\mathbf{x}^*) = \dfrac{\partial f}{\partial x_2}(\mathbf{x}^*) = \cdots = \dfrac{\partial f}{\partial x_n}(\mathbf{x}^*).$

We begin by showing that (II) implies (I). Indeed, assume that $\mathbf{x}^* \in U$ satisfies (II). Then for any $\mathbf{x} \in U$ one has

$$\nabla f(\mathbf{x}^*)^T(\mathbf{x}-\mathbf{x}^*) = \sum_{i=1}^{n} \dfrac{\partial f}{\partial x_i}(\mathbf{x}^*)(x_i - x_i^*)$$

$$= \dfrac{\partial f}{\partial x_1}(\mathbf{x}^*)\left(\sum_{i=1}^n x_i - \sum_{i=1}^n x_i^*\right) = \dfrac{\partial f}{\partial x_1}(\mathbf{x}^*)(1-1) = 0,$$

and in particular $\nabla f(\mathbf{x}^*)^T(\mathbf{x}-\mathbf{x}^*) \geq 0$. We have thus shown that (I) is satisfied. To show the reverse direction, take $\mathbf{x}^* \in U$ that satisfies (I) and assume in contradiction that (II) does not hold. This means that there exist two different indices $i \neq j$ such that

$$\dfrac{\partial f}{\partial x_i}(\mathbf{x}^*) > \dfrac{\partial f}{\partial x_j}(\mathbf{x}^*).$$

Define the vector $\mathbf{x} \in U$ as

$$x_k = \begin{cases} x_k^*, & k \notin \{i,j\}, \\ x_i^* - 1, & k = i, \\ x_j^* + 1, & k = j. \end{cases}$$

Then

$$\nabla f(\mathbf{x}^*)^T(\mathbf{x}-\mathbf{x}^*) = \dfrac{\partial f}{\partial x_i}(\mathbf{x}^*)(x_i - x_i^*) + \dfrac{\partial f}{\partial x_j}(\mathbf{x}^*)(x_j - x_j^*)$$

$$= -\dfrac{\partial f}{\partial x_i}(\mathbf{x}^*) + \dfrac{\partial f}{\partial x_j}(\mathbf{x}^*)$$

$$< 0,$$

which is a contradiction to the assumption that (I) is satisfied. We thus conclude that (I) implies (II). ∎

Example 9.6 (stationarity over the unit-ball). Consider the optimization problem

(S) $\quad \begin{array}{ll} \min & f(\mathbf{x}) \\ \text{s.t.} & \|\mathbf{x}\| \leq 1, \end{array}$

where f is a continuously differentiable function over $B[\mathbf{0},1]$. A point $\mathbf{x}^* \in B[\mathbf{0},1]$ is a stationary point of problem (S) if and only if

$$\nabla f(\mathbf{x}^*)^T(\mathbf{x}-\mathbf{x}^*) \geq 0 \text{ for all } \mathbf{x} \text{ satisfying } \|\mathbf{x}\| \leq 1, \qquad (9.4)$$

which is equivalent to

$$\min\{\nabla f(\mathbf{x}^*)^T\mathbf{x} - \nabla f(\mathbf{x}^*)^T\mathbf{x}^* : \|\mathbf{x}\| \leq 1\} \geq 0.$$

We will now use the following fact:

[A] for any $\mathbf{a} \in \mathbb{R}^n$ the optimal value of the problem $\min\{\mathbf{a}^T\mathbf{x} : \|\mathbf{x}\| \leq 1\}$ is $-\|\mathbf{a}\|$.

Indeed, if $\mathbf{a} = \mathbf{0}$, then obviously the optimal value is $0 = -\|\mathbf{a}\|$. If $\mathbf{a} \neq \mathbf{0}$, then by the Cauchy–Schwarz inequality we have for any $\mathbf{x} \in B[\mathbf{0}, 1]$

$$\mathbf{a}^T\mathbf{x} \geq -\|\mathbf{a}\| \cdot \|\mathbf{x}\| \geq -\|\mathbf{a}\|,$$

so that

$$\min\{\mathbf{a}^T\mathbf{x} : \|\mathbf{x}\| \leq 1\} \geq -\|\mathbf{a}\|.$$

On the other hand, the lower bound $-\|\mathbf{a}\|$ is attained at $\mathbf{x} = -\frac{\mathbf{a}}{\|\mathbf{a}\|}$, and hence fact [A] is established.

Returing to the characterization of stationary points over the unit ball, by fact [A] it follows that (9.4) is the same as

$$-\nabla f(\mathbf{x}^*)^T\mathbf{x}^* \geq \|\nabla f(\mathbf{x}^*)\|. \tag{9.5}$$

However, we have by the Cauchy–Schwarz inequality that the following inequality holds:

$$-\nabla f(\mathbf{x}^*)^T\mathbf{x}^* \leq \|\nabla f(\mathbf{x}^*)\| \cdot \|\mathbf{x}^*\| \leq \|\nabla f(\mathbf{x}^*)\|,$$

and we conclude that $\mathbf{x}^*$ is a stationary point of (S) if and only if

$$\|\nabla f(\mathbf{x}^*)\| = -\nabla f(\mathbf{x}^*)^T\mathbf{x}^*. \tag{9.6}$$

The above condition is a simple and explicit characterization of the stationarity property. We can, however, find a more informative characterization of stationarity by the following argument: Let $\mathbf{x}^*$ be a point satisfying (9.6). We have two cases:

- If $\nabla f(\mathbf{x}^*) = \mathbf{0}$, then (9.6) holds automatically.
- If $\nabla f(\mathbf{x}^*) \neq \mathbf{0}$, then $\|\mathbf{x}^*\| = 1$ since otherwise, if $\|\mathbf{x}^*\| < 1$, we have by (once again!) the Cauchy–Schwarz inequality that

$$-\nabla f(\mathbf{x}^*)^T\mathbf{x}^* \leq \|\nabla f(\mathbf{x}^*)\| \cdot \|\mathbf{x}^*\| < \|\nabla f(\mathbf{x}^*)\|,$$

which is a contradiction to (9.6). We therefore conclude that when $\nabla f(\mathbf{x}^*) \neq \mathbf{0}$, $\mathbf{x}^*$ is a stationary point if and only if $\|\mathbf{x}^*\| = 1$ and

$$-\nabla f(\mathbf{x}^*)^T\mathbf{x}^* = \|\nabla f(\mathbf{x}^*)\| \cdot \|\mathbf{x}^*\|. \tag{9.7}$$

The latter equality is equivalent to saying that there exists $\lambda \leq 0$ such that $\nabla f(\mathbf{x}^*) = \lambda \mathbf{x}^*$. To show this, note that if there exists such a λ, then indeed

$$-\nabla f(\mathbf{x}^*)^T\mathbf{x}^* = -\lambda\|\mathbf{x}^*\|^2 \stackrel{\lambda \leq 0}{=} \|\lambda\mathbf{x}^*\| \cdot \|\mathbf{x}^*\| = \|\nabla f(\mathbf{x}^*)\| \cdot \|\mathbf{x}^*\|,$$

so that (9.7) is satisfied. In the other direction, if (9.7) holds, then this means that the Cauchy–Schwarz inequality is satisfied as an equality, and hence, by Lemma 1.5, it follows that there exist $\lambda \in \mathbb{R}$ such that $\nabla f(\mathbf{x}^*) = \lambda \mathbf{x}^*$. The parameter λ must be nonpositive since otherwise the left-hand side of (9.7) would be negative, and the right-hand side would be positive, contradicting the equality in (9.7).

In conclusion, $\mathbf{x}^*$ is a stationary point of (S) if and only if either $\nabla f(\mathbf{x}^*) = \mathbf{0}$ or $\|\mathbf{x}^*\| = 1$ and there exists $\lambda \leq 0$ such that $\nabla f(\mathbf{x}^*) = \lambda \mathbf{x}^*$. ∎

To summarize the four examples, we write explicitly each of the stationarity conditions in the following table.

feasible set	explicit stationarity condition
$\mathbb{R}^n$	$\nabla f(\mathbf{x}^*) = \mathbf{0}$
$\mathbb{R}^n_+$	$\frac{\partial f}{\partial x_i}(\mathbf{x}^*) \begin{cases} = 0, & x_i^* > 0 \\ \geq 0, & x_i^* = 0 \end{cases}$
$\{\mathbf{x} \in \mathbb{R}^n : \mathbf{e}^T \mathbf{x} = 1\}$	$\frac{\partial f}{\partial x_1}(\mathbf{x}^*) = \cdots = \frac{\partial f}{\partial x_n}(\mathbf{x}^*)$
$B[\mathbf{0}, 1]$	$\nabla f(\mathbf{x}^*) = \mathbf{0}$ or $\|\mathbf{x}^*\| = 1$ and $\exists \lambda \leq 0 : \nabla f(\mathbf{x}^*) = \lambda \mathbf{x}^*$

9.2 ▪ Stationarity in Convex Problems

Stationarity is a necessary optimality condition for local optimality. However, when the objective function is additionally assumed to be convex, stationarity is a necessary *and sufficient* condition for optimality.

Theorem 9.7. *Let f be a continuously differentiable convex function over a closed and convex set $C \subseteq \mathbb{R}^n$. Then $\mathbf{x}^* \in C$ is a stationary point of*

$$\text{(P)} \quad \begin{array}{ll} \min & f(\mathbf{x}) \\ \text{s.t.} & \mathbf{x} \in C \end{array}$$

if and only if $\mathbf{x}^*$ is an optimal solution of (P).

Proof. If $\mathbf{x}^*$ is an optimal solution of (P), then by Theorem 9.2, it follows that $\mathbf{x}^*$ is a stationary point of (P). To prove the sufficiency of the stationarity condition, assume that $\mathbf{x}^*$ is a stationary point of (P), and let $\mathbf{x} \in C$. Then

$$f(\mathbf{x}) \geq f(\mathbf{x}^*) + \nabla f(\mathbf{x}^*)^T (\mathbf{x} - \mathbf{x}^*) \geq f(\mathbf{x}^*),$$

where the first inequality follows from the gradient inequality for convex functions (Theorem 7.6) and the second inequality follows from the definition of a stationary point. We have thus shown that $\mathbf{x}^*$ is the global minimum point of (P), and the reverse direction is established. □

9.3 ▪ The Orthogonal Projection Revisited

We can use the stationarity property in order to establish an important property of the orthogonal projection operator. This characterization will be called *the second projection theorem*. Geometrically it states that for a given closed and convex set C, $\mathbf{x} \in \mathbb{R}^n$, and $\mathbf{y} \in C$, the angle between $\mathbf{x} - P_C(\mathbf{x})$ and $\mathbf{y} - P_C(\mathbf{x})$ is greater than or equal to 90 degrees. This phenomenon is illustrated in Figure 9.1.

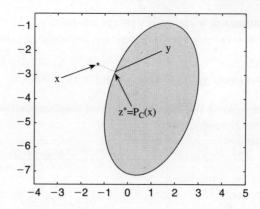

Figure 9.1. *The orthogonal projection operator.*

Theorem 9.8 (second projection theorem). *Let C be a closed convex set and let $\mathbf{x} \in \mathbb{R}^n$. Then $\mathbf{z} = P_C(\mathbf{x})$ if and only if $\mathbf{z} \in C$ and*

$$(\mathbf{x}-\mathbf{z})^T(\mathbf{y}-\mathbf{z}) \leq 0 \text{ for any } \mathbf{y} \in C. \tag{9.8}$$

Proof. $\mathbf{z} = P_C(\mathbf{x})$ if and only if it is the optimal solution of the problem

$$\begin{array}{ll} \min & g(\mathbf{y}) \equiv \|\mathbf{y}-\mathbf{x}\|^2 \\ \text{s.t.} & \mathbf{y} \in C. \end{array}$$

Therefore, by Theorem 9.7 it follows that $\mathbf{z} = P_C(\mathbf{x})$ if and only if

$$\nabla g(\mathbf{z})^T(\mathbf{y}-\mathbf{z}) \geq 0 \text{ for all } \mathbf{y} \in C,$$

which is the same as (9.8). □

Another important property of the orthogonal projection operator is given in the following theorem, which also establishes the so-called *nonexpansiveness* property of P_C.

Theorem 9.9. *Let C be a nonempty closed and convex set. Then*

1. *for any $\mathbf{v}, \mathbf{w} \in \mathbb{R}^n$*

$$(P_C(\mathbf{v}) - P_C(\mathbf{w}))^T(\mathbf{v}-\mathbf{w}) \geq \|P_C(\mathbf{v}) - P_C(\mathbf{w})\|^2, \tag{9.9}$$

2. *(nonexpansiveness) for any $\mathbf{v}, \mathbf{w} \in \mathbb{R}^n$*

$$\|P_C(\mathbf{v}) - P_C(\mathbf{w})\| \leq \|\mathbf{v}-\mathbf{w}\|. \tag{9.10}$$

Proof. Recall that by Theorem 9.8 we have that for any $\mathbf{x} \in \mathbb{R}^n$ and $\mathbf{y} \in C$

$$(\mathbf{x}-P_C(\mathbf{x}))^T(\mathbf{y}-P_C(\mathbf{x})) \leq 0. \tag{9.11}$$

Substituting $\mathbf{x} = \mathbf{v}$ and $\mathbf{y} = P_C(\mathbf{w})$, we have

$$(\mathbf{v}-P_C(\mathbf{v}))^T(P_C(\mathbf{w})-P_C(\mathbf{v})) \leq 0. \tag{9.12}$$

Substituting $\mathbf{x} = \mathbf{w}$ and $\mathbf{y} = P_C(\mathbf{v})$, we obtain

$$(\mathbf{w} - P_C(\mathbf{w}))^T (P_C(\mathbf{v}) - P_C(\mathbf{w})) \leq 0. \tag{9.13}$$

Adding the two inequalities (9.12) and (9.13) yields

$$(P_C(\mathbf{w}) - P_C(\mathbf{v}))^T (\mathbf{v} - \mathbf{w} + P_C(\mathbf{w}) - P_C(\mathbf{v})) \leq 0,$$

and hence,

$$(P_C(\mathbf{v}) - P_C(\mathbf{w}))^T (\mathbf{v} - \mathbf{w}) \geq \|P_C(\mathbf{v}) - P_C(\mathbf{w})\|^2,$$

showing the desired inequality (9.9).

To prove (9.10), note that if $P_C(\mathbf{v}) = P_C(\mathbf{w})$, the inequality is trivial. If $P_C(\mathbf{v}) \neq P_C(\mathbf{w})$, then by the Cauchy–Schwarz inequality we have

$$(P_C(\mathbf{v}) - P_C(\mathbf{w}))^T (\mathbf{v} - \mathbf{w}) \leq \|P_C(\mathbf{v}) - P_C(\mathbf{w})\| \cdot \|\mathbf{v} - \mathbf{w}\|,$$

which combined with (9.9) yields the inequality

$$\|P_C(\mathbf{v}) - P_C(\mathbf{w})\|^2 \leq \|P_C(\mathbf{v}) - P_C(\mathbf{w})\| \cdot \|\mathbf{v} - \mathbf{w}\|.$$

Dividing by $\|P_C(\mathbf{v}) - P_C(\mathbf{w})\|$ implies (9.10). □

Coming back to stationarity, the next result describes an additional useful representation of stationarity in terms of the orthogonal projection operator.

Theorem 9.10. *Let f be a continuously differentiable function defined on the closed and convex set C, and let $s > 0$. Then $\mathbf{x}^* \in C$ is a stationary point of the problem*

$$(\mathrm{P}) \quad \begin{array}{ll} \min & f(\mathbf{x}) \\ s.t. & \mathbf{x} \in C \end{array}$$

if and only if

$$\mathbf{x}^* = P_C(\mathbf{x}^* - s \nabla f(\mathbf{x}^*)). \tag{9.14}$$

Proof. By the second projection theorem (Theorem 9.8), it follows that $\mathbf{x}^* = P_C(\mathbf{x}^* - s\nabla f(\mathbf{x}^*))$ if and only if

$$(\mathbf{x}^* - s\nabla f(\mathbf{x}^*) - \mathbf{x}^*)^T (\mathbf{x} - \mathbf{x}^*) \leq 0 \text{ for any } \mathbf{x} \in C.$$

The above relation is equivalent to

$$\nabla f(\mathbf{x}^*)^T (\mathbf{x} - \mathbf{x}^*) \geq 0 \text{ for any } \mathbf{x} \in C,$$

namely to stationarity. □

Note that condition (9.14) seems to depend on the parameter s, but by its equivalence to stationarity, it is essentially *independent* of s.

9.4 ▪ The Gradient Projection Method

The stationarity condition

$$\mathbf{x}^* = P_C(\mathbf{x}^* - s\nabla f(\mathbf{x}^*)) \tag{9.15}$$

naturally motivates the following algorithm, called *the gradient projection method*, for solving problem (P). This algorithm can be seen as a fixed point method for solving the equation (9.15).

The Gradient Projection Method

Input: $\varepsilon > 0$ - tolerance parameter.

Initialization: Pick $\mathbf{x}_0 \in C$ arbitrarily.
General step: For any $k = 0, 1, 2, \ldots$ execute the following steps:

(a) Pick a stepsize t_k by a line search procedure.

(b) Set $\mathbf{x}_{k+1} = P_C(\mathbf{x}_k - t_k \nabla f(\mathbf{x}_k))$.

(c) If $\|\mathbf{x}_k - \mathbf{x}_{k+1}\| \leq \varepsilon$, then STOP, and $\mathbf{x}_{k+1}$ is the output.

Obviously, in the unconstrained case, that is, when $C = \mathbb{R}^n$, the gradient projection method is just the gradient method studied in Chapter 4.

There are several strategies for choosing the stepsizes t_k. We will consider two choices:

- **constant stepsize** $t_k = \bar{t}$ for all k.

- **backtracking**.

We will elaborate on the specific details of the backtracking procedure in what follows.

9.4.1 ▪ Sufficient Decrease and the Gradient Mapping

To establish the convergence of the method, we will prove a sufficient decrease lemma for $C^{1,1}$ functions, similarly to the sufficient decrease lemma proved for the gradient method (Lemma 4.23).

Lemma 9.11 (sufficient decrease lemma for constrained problems). *Suppose that $f \in C_L^{1,1}(C)$, where C is a closed convex set. Then for any $\mathbf{x} \in C$ and $t \in (0, \frac{2}{L})$ the following inequality holds:*

$$f(\mathbf{x}) - f(P_C(\mathbf{x} - t\nabla f(\mathbf{x}))) \geq t\left(1 - \frac{Lt}{2}\right)\left\|\frac{1}{t}(\mathbf{x} - P_C(\mathbf{x} - t\nabla f(\mathbf{x})))\right\|^2. \tag{9.16}$$

Proof. For the sake of simplicity, we will make the notation $\mathbf{x}^+ = P_C(\mathbf{x} - t\nabla f(\mathbf{x}))$. By the descent lemma (Lemma 4.22) we have that

$$f(\mathbf{x}^+) \leq f(\mathbf{x}) + \langle \nabla f(\mathbf{x}), \mathbf{x}^+ - \mathbf{x} \rangle + \frac{L}{2}\|\mathbf{x} - \mathbf{x}^+\|^2. \tag{9.17}$$

From the second projection theorem (Theorem 9.8) we have

$$\langle \mathbf{x} - t\nabla f(\mathbf{x}) - \mathbf{x}^+, \mathbf{x} - \mathbf{x}^+ \rangle \leq 0,$$

from which it follows that

$$\langle \nabla f(\mathbf{x}), \mathbf{x}^+ - \mathbf{x} \rangle \leq -\frac{1}{t}\|\mathbf{x}^+ - \mathbf{x}\|^2,$$

9.4. The Gradient Projection Method

which combined with (9.17) yields

$$f(\mathbf{x}^+) \leq f(\mathbf{x}) + \left(-\frac{1}{t} + \frac{L}{2}\right) \|\mathbf{x}^+ - \mathbf{x}\|^2.$$

Hence, taking into account the definition of $\mathbf{x}^+$, the desired result follows. □

The result of Lemma 9.11 is a generalization of the sufficient decrease property derived in the unconstrained setting in Lemma 4.23. In fact, when $C = \mathbb{R}^n$ the obtained inequality is exactly the same as the one obtained in Lemma 4.23.

It is convenient to define *the gradient mapping* as

$$G_M(\mathbf{x}) = M\left[\mathbf{x} - P_C\left(\mathbf{x} - \frac{1}{M}\nabla f(\mathbf{x})\right)\right],$$

where $M > 0$. We assume that the identities of f and C are clear from the context. Note that in the unconstrained case $G_M(\mathbf{x}) = \nabla f(\mathbf{x})$, so the gradient mapping is an extension of the usual gradient operation. In addition, by Theorem 9.10, $G_M(\mathbf{x}) = 0$ if and only if $\mathbf{x}$ is a stationary point of (P). This means that we can look at $\|G_L(\mathbf{x})\|$ as an *optimality measure*. The result of Lemma 9.11 essentially states that

$$f(\mathbf{x}) - f(P_C(\mathbf{x} - t\nabla f(\mathbf{x}))) \geq t\left(1 - \frac{Lt}{2}\right)\|G_{\frac{1}{t}}(\mathbf{x})\|^2. \tag{9.18}$$

This generalized sufficient decrease property allows us to prove similar results to those proven in the unconstrained case. Before proceeding to the convergence analysis, we will prove an important monotonicity property of the norm of the gradient mapping $G_M(\mathbf{x})$ with respect to the parameter M.

Lemma 9.12. *Let f be a continuously differentiable function defined on a nonempty closed and convex set C. Suppose that $L_1 \geq L_2 > 0$. Then*

$$\|G_{L_1}(\mathbf{x})\| \geq \|G_{L_2}(\mathbf{x})\| \tag{9.19}$$

and

$$\frac{\|G_{L_1}(\mathbf{x})\|}{L_1} \leq \frac{\|G_{L_2}(\mathbf{x})\|}{L_2} \tag{9.20}$$

for any $\mathbf{x} \in \mathbb{R}^n$.

Proof. Recall that, by the second projection theorem, for any $\mathbf{v} \in \mathbb{R}^n$ and $\mathbf{w} \in C$ the following inequality holds:

$$\langle \mathbf{v} - P_C(\mathbf{v}), P_C(\mathbf{v}) - \mathbf{w} \rangle \geq 0.$$

Plugging $\mathbf{v} = \mathbf{x} - \frac{1}{L_1}\nabla f(\mathbf{x})$ and $\mathbf{w} = P_C(\mathbf{x} - \frac{1}{L_2}\nabla f(\mathbf{x}))$ in the latter inequality it follows that

$$\left\langle \mathbf{x} - \frac{1}{L_1}\nabla f(\mathbf{x}) - P_C\left(\mathbf{x} - \frac{1}{L_1}\nabla f(\mathbf{x})\right), P_C\left(\mathbf{x} - \frac{1}{L_1}\nabla f(\mathbf{x})\right) - P_C\left(\mathbf{x} - \frac{1}{L_2}\nabla f(\mathbf{x})\right) \right\rangle \geq 0,$$

or

$$\left\langle \frac{1}{L_1}G_{L_1}(\mathbf{x}) - \frac{1}{L_1}\nabla f(\mathbf{x}), \frac{1}{L_2}G_{L_2}(\mathbf{x}) - \frac{1}{L_1}G_{L_1}(\mathbf{x}) \right\rangle \geq 0.$$

Exchanging the roles of L_1 and L_2 yields the following inequality:

$$\left\langle \frac{1}{L_2}G_{L_2}(\mathbf{x}) - \frac{1}{L_2}\nabla f(\mathbf{x}), \frac{1}{L_1}G_{L_1}(\mathbf{x}) - \frac{1}{L_2}G_{L_2}(\mathbf{x}) \right\rangle \geq 0.$$

Multiplying the first inequality by L_1 and the second by L_2 and adding them we obtain

$$\left\langle G_{L_1}(\mathbf{x}) - G_{L_2}(\mathbf{x}), \frac{1}{L_2}G_{L_2}(\mathbf{x}) - \frac{1}{L_1}G_{L_1}(\mathbf{x}) \right\rangle \geq 0,$$

which, after some expansion of terms, can be seen to be the same as

$$\frac{1}{L_1}\|G_{L_1}(\mathbf{x})\|^2 + \frac{1}{L_2}\|G_{L_2}(\mathbf{x})\|^2 \leq \left(\frac{1}{L_1} + \frac{1}{L_2}\right) G_{L_1}(\mathbf{x})^T G_{L_2}(\mathbf{x}).$$

Using the Cauchy-Schwarz inequality we obtain that

$$\frac{1}{L_1}\|G_{L_1}(\mathbf{x})\|^2 + \frac{1}{L_2}\|G_{L_2}(\mathbf{x})\|^2 \leq \left(\frac{1}{L_1} + \frac{1}{L_2}\right) \|G_{L_1}(\mathbf{x})\| \cdot \|G_{L_2}(\mathbf{x})\|. \quad (9.21)$$

Note that if $G_{L_2}(\mathbf{x}) = 0$, then by the latter inequality, it follows that $G_{L_1}(\mathbf{x}) = 0$, implying that in this case the inequalities (9.19) and (9.20) hold trivially. Assume then that $G_{L_2}(\mathbf{x}) \neq 0$, and define $t = \frac{\|G_{L_1}(\mathbf{x})\|}{\|G_{L_2}(\mathbf{x})\|}$. Then by (9.21)

$$\frac{1}{L_1}t^2 - \left(\frac{1}{L_1} + \frac{1}{L_2}\right)t + \frac{1}{L_2} \leq 0.$$

Since the roots of the quadratic function are $t = 1, \frac{L_1}{L_2}$, we obtain that

$$1 \leq t \leq \frac{L_1}{L_2},$$

showing that

$$\|G_{L_2}(\mathbf{x})\| \leq \|G_{L_1}(\mathbf{x})\| \leq \frac{L_1}{L_2}\|G_{L_2}(\mathbf{x})\|. \quad \square$$

9.4.2 ▪ Backtracking

Equipped with the definition of the gradient mapping, we are now able to define the backtracking stepsize selection strategy similarly to the definition of the backtracking procedure for descent methods for unconstrained problems (see Chapter 4). The procedure requires three parameters (s, α, β), where $s > 0, \alpha \in (0,1), \beta \in (0,1)$. The choice of t_k is done as follows: First, t_k is set to be equal to the initial guess s. Then while

$$f(\mathbf{x}_k) - f(P_C(\mathbf{x}_k - t_k\nabla f(\mathbf{x}_k))) < \alpha t_k \|G_{\frac{1}{t_k}}(\mathbf{x}_k)\|^2$$

we set $t_k := \beta t_k$. In other words, the stepsize is chosen as $t_k = s\beta^{i_k}$, where i_k is the smallest nonnegative integer for which the condition

$$f(\mathbf{x}_k) - f(P_C(\mathbf{x}_k - s\beta^{i_k}\nabla f(\mathbf{x}_k))) \geq \alpha s\beta^{i_k} \|G_{\frac{1}{s\beta^{i_k}}}(\mathbf{x}_k)\|^2$$

is satisfied.

9.4. The Gradient Projection Method

Note that the backtracking procedure is finite for $C^{1,1}$ functions. Indeed, if $f \in C_L^{1,1}(C)$, then plugging $\mathbf{x} = \mathbf{x}_k$ into (9.18) we obtain

$$f(\mathbf{x}_k) - f(P_C(\mathbf{x}_k - t\nabla f(\mathbf{x}_k))) \geq t\left(1 - \frac{Lt}{2}\right)\left\|G_{\frac{1}{t}}(\mathbf{x}_k)\right\|^2.$$

Hence, since the inequality $t \leq \frac{2(1-\alpha)}{L}$ is the same as $1 - \frac{Lt}{2} \geq \alpha$, we conclude that for any $t \leq \frac{2(1-\alpha)}{L}$ the inequality

$$f(\mathbf{x}_k) - f(P_C(\mathbf{x}_k - t\nabla f(\mathbf{x}_k))) \geq \alpha t \left\|G_{\frac{1}{t}}(\mathbf{x}_k)\right\|^2$$

holds, implying that the backtracking procedure ends when t_k is smaller or equal to $\frac{2(1-\alpha)}{L}$.

We can also compute a lower bound on t_k: either t_k is equal to s or the backtracking procedure is invoked, meaning that the stepsize $\frac{t_k}{\beta}$ did not satisfy the backtracking condition given by

$$f(\mathbf{x}_k) - f(P_C(\mathbf{x}_k - t_k \nabla f(\mathbf{x}_k))) \geq \alpha t_k \|G_{\frac{1}{t_k}}(\mathbf{x}_k)\|^2.$$

In particular, by the above discussion we can conclude that $\frac{t_k}{\beta} > \frac{2(1-\alpha)}{L}$, so that $t_k > \frac{2(1-\alpha)\beta}{L}$. To summarize, in the backtracking procedure, the chosen stepsize t_k satisfies

$$t_k \geq \min\left\{s, \frac{2(1-\alpha)\beta}{L}\right\}. \tag{9.22}$$

9.4.3 ▪ Convergence of the Gradient Projection Method

We will analyze the convergence of the gradient projection method in the case when $f \in C_L^{1,1}(C)$. We begin with the following lemma showing the sufficient decrease of consecutive function values of the sequence generated by the gradient projection method.

Lemma 9.13. *Consider the problem*

$$(P) \quad \begin{array}{l} \min \ f(\mathbf{x}) \\ \text{s.t.} \ \mathbf{x} \in C, \end{array}$$

where $C \subseteq \mathbb{R}^n$ is a nonempty closed and convex set and $f \in C_L^{1,1}(C)$. Let $\{\mathbf{x}_k\}_{k \geq 0}$ be the sequence generated by the gradient projection method for solving problem (P) with either a constant stepsize $t_k = \bar{t} \in (0, \frac{2}{L})$ or with a stepsize chosen by the backtracking procedure with parameters (s, α, β) satisfying $s > 0, \alpha \in (0,1), \beta \in (0,1)$. Then for any $k \geq 0$

$$f(\mathbf{x}_k) - f(\mathbf{x}_{k+1}) \geq M\|G_d(\mathbf{x}_k)\|^2, \tag{9.23}$$

where

$$M = \begin{cases} \bar{t}\left(1 - \frac{\bar{t}L}{2}\right) & \text{constant stepsize,} \\ \alpha \min\left\{s, \frac{2(1-\alpha)\beta}{L}\right\} & \text{backtracking} \end{cases}$$

and

$$d = \begin{cases} 1/\bar{t} & \text{constant stepsize,} \\ 1/s & \text{backtracking.} \end{cases}$$

Proof. The result for the constant stepsize setting follows by plugging $t = \bar{t}$ and $\mathbf{x} = \mathbf{x}_k$ in (9.18). As for the backtracking procedure, by its definition we have

$$f(\mathbf{x}_k) - f(\mathbf{x}_{k+1}) \geq \alpha t_k \|G_{\frac{1}{t_k}}(\mathbf{x}_k)\|^2.$$

The result now follows from the lower bound on t_k given in (9.22) and the fact that $t_k \leq s$, which implies by the monotonicity property of the gradient mapping (Lemma 9.12) that $\|G_{1/t_k}(\mathbf{x}_k)\| \geq \|G_{1/s}(\mathbf{x}_k)\|$. □

We are now ready to prove the convergence of the norm of the gradient mapping to zero.

Theorem 9.14 (convergence of the gradient projection method). *Consider the problem*

$$\text{(P)} \quad \begin{array}{ll} \min & f(\mathbf{x}) \\ \text{s.t.} & \mathbf{x} \in C, \end{array}$$

where C is a nonempty closed and convex set and $f \in C_L^{1,1}(C)$ is bounded below. Let $\{\mathbf{x}_k\}_{k \geq 0}$ be the sequence generated by the gradient projection method for solving problem (P) with either a constant stepsize $t_k = \bar{t} \in (0, \frac{2}{L})$ or a stepsize chosen by the backtracking procedure with parameters (s, α, β) satisfying $s > 0, \alpha, \beta \in (0, 1)$. Then we have the following:

(a) *The sequence $\{f(\mathbf{x}_k)\}$ is nonincreasing. In addition, $f(\mathbf{x}_{k+1}) < f(\mathbf{x}_k)$ unless $\mathbf{x}_k$ is a stationary point of (P).*

(b) $G_d(\mathbf{x}_k) \to 0$ *as* $k \to \infty$, *where d is as defined in Lemma 9.13.*

Proof. (a) By Lemma 9.13 we have that

$$f(\mathbf{x}_k) - f(\mathbf{x}_{k+1}) \geq M \|G_d(\mathbf{x}_k)\|^2. \tag{9.24}$$

Since $M > 0$, it readily follows that $f(\mathbf{x}_k) \geq f(\mathbf{x}_{k+1})$ and equality holds only if $G_d(\mathbf{x}_k) = 0$, which is the same as saying that $\mathbf{x}_k$ is a stationary point of (P).

(b) Since the sequence $\{f(\mathbf{x}_k)\}_{k \geq 0}$ is nonincreasing and bounded below, it converges. Thus, in particular $f(\mathbf{x}_k) - f(\mathbf{x}_{k+1}) \to 0$ as $k \to \infty$, which combined with (9.24) implies that $\|G_d(\mathbf{x}_k)\| \to 0$ as $k \to \infty$. □

We can also get an estimate for the rate of convergence of the gradient projection method, but in the constrained case it will be in terms of the norm of the gradient mapping.

Theorem 9.15 (rate of convergence of gradient mapping norms in the gradient projection method). *Under the setting of Theorem 9.14, let f^* be the limit of the convergent sequence $\{f(\mathbf{x}_k)\}_{k \geq 0}$. Then for any $n = 0, 1, 2, \ldots$*

$$\min_{k=0,1,\ldots,n} \|G_d(\mathbf{x}_k)\| \leq \sqrt{\frac{f(\mathbf{x}_0) - f^*}{M(n+1)}},$$

where

$$M = \begin{cases} \bar{t}\left(1 - \frac{\bar{t}L}{2}\right) & \text{constant stepsize,} \\ \alpha \min\left\{s, \frac{2(1-\alpha)\beta}{L}\right\} & \text{backtracking} \end{cases}$$

9.4. The Gradient Projection Method

and
$$d = \begin{cases} 1/\bar{t} & \text{constant stepsize,} \\ 1/s & \text{backtracking.} \end{cases}$$

Proof. Summing the inequality (9.23) over $k = 0, 1, \ldots, n$ we obtain
$$f(\mathbf{x}_0) - f(\mathbf{x}_{n+1}) \geq M \sum_{k=0}^{n} \|G_d(\mathbf{x}_k)\|^2.$$

Since $f(\mathbf{x}_{n+1}) \geq f^*$, we can thus write
$$f(\mathbf{x}_0) - f^* \geq M \sum_{k=0}^{n} \|G_d(\mathbf{x}_k)\|^2.$$

Finally, using the latter inequality along with the obvious fact that
$$\sum_{k=0}^{n} \|G_d(\mathbf{x}_k)\|^2 \geq (n+1) \min_{k=0,1,\ldots,n} \|G_d(\mathbf{x}_k)\|^2,$$

it follows that
$$f(\mathbf{x}_0) - f^* \geq M(n+1) \min_{k=0,1,\ldots,n} \|G_d(\mathbf{x}_k)\|^2,$$

implying the desired result. $\square$

9.4.4 ▪ The Convex Case

When the objective function is assumed to be convex, the constrained problem (P) becomes convex, and stronger convergence results can be established. We will concentrate on the constant stepsize setting and show that the sequence $\{\mathbf{x}_k\}_{k \geq 0}$ converges to an optimal solution of (P). In addition, we will establish a rate of convergence result for the sequence of function values $\{f(\mathbf{x}_k)\}_{k \geq 0}$ to the optimal value.

Theorem 9.16 (rate of convergence of the sequence of function values). *Consider the problem*
$$\text{(P)} \quad \begin{array}{ll} \min & f(\mathbf{x}) \\ \text{s.t.} & \mathbf{x} \in C, \end{array}$$

where C is a nonempty closed and convex set and $f \in C_L^{1,1}(C)$ is convex over C. Let $\{\mathbf{x}_k\}_{k \geq 0}$ be the sequence generated by the gradient projection method for solving problem (P) with a constant stepsize $t_k = \bar{t} \in (0, \frac{1}{L}]$. Assume that the set of optimal solutions, denoted by X^*, is nonempty, and let f^* be the optimal value of (P). Then

(a) *for any $k \geq 0$ and $\mathbf{x}^* \in X^*$*
$$2\bar{t}(f(\mathbf{x}_{k+1}) - f(\mathbf{x}^*)) \leq \|\mathbf{x}_k - \mathbf{x}^*\|^2 - \|\mathbf{x}_{k+1} - \mathbf{x}^*\|^2, \tag{9.25}$$

(b) *for any $n \geq 0$:*
$$f(\mathbf{x}_n) - f^* \leq \frac{\|\mathbf{x}_0 - \mathbf{x}^*\|^2}{2\bar{t}n}. \tag{9.26}$$

Proof. By the descent lemma we have

$$f(\mathbf{x}_{k+1}) \leq f(\mathbf{x}_k) + \langle \nabla f(\mathbf{x}_k), \mathbf{x}_{k+1} - \mathbf{x}_k \rangle + \frac{L}{2}\|\mathbf{x}_k - \mathbf{x}_{k+1}\|^2. \quad (9.27)$$

Let $\mathbf{x}^*$ be an optimal solution of (P). Then the gradient inequality implies that $f(\mathbf{x}_k) \leq f(\mathbf{x}^*) + \langle \nabla f(\mathbf{x}_k), \mathbf{x}_k - \mathbf{x}^* \rangle$, which combined with (9.27) yields

$$f(\mathbf{x}_{k+1}) \leq f(\mathbf{x}^*) + \langle \nabla f(\mathbf{x}_k), \mathbf{x}_k - \mathbf{x}^* \rangle + \langle \nabla f(\mathbf{x}_k), \mathbf{x}_{k+1} - \mathbf{x}_k \rangle + \frac{L}{2}\|\mathbf{x}_k - \mathbf{x}_{k+1}\|^2. \quad (9.28)$$

By the second projection theorem (Theorem 9.8) we have that

$$\langle \mathbf{x}_k - \bar{t}\nabla f(\mathbf{x}_k) - \mathbf{x}_{k+1}, \mathbf{x}^* - \mathbf{x}_{k+1} \rangle \leq 0,$$

so that

$$\langle \nabla f(\mathbf{x}_k), \mathbf{x}_{k+1} - \mathbf{x}^* \rangle \leq \frac{1}{\bar{t}} \langle \mathbf{x}_k - \mathbf{x}_{k+1}, \mathbf{x}_{k+1} - \mathbf{x}^* \rangle. \quad (9.29)$$

Therefore, combining (9.28), (9.29) and using the inequality $\bar{t} \leq \frac{1}{L}$, we obtain that

$$f(\mathbf{x}_{k+1}) \leq f(\mathbf{x}^*) + \langle \nabla f(\mathbf{x}_k), \mathbf{x}_k - \mathbf{x}^* \rangle + \langle \nabla f(\mathbf{x}_k), \mathbf{x}_{k+1} - \mathbf{x}_k \rangle + \frac{L}{2}\|\mathbf{x}_k - \mathbf{x}_{k+1}\|^2$$

$$= f(\mathbf{x}^*) + \langle \nabla f(\mathbf{x}_k), \mathbf{x}_{k+1} - \mathbf{x}^* \rangle + \frac{L}{2}\|\mathbf{x}_k - \mathbf{x}_{k+1}\|^2$$

$$\leq f(\mathbf{x}^*) + \frac{1}{\bar{t}} \langle \mathbf{x}_k - \mathbf{x}_{k+1}, \mathbf{x}_{k+1} - \mathbf{x}^* \rangle + \frac{L}{2}\|\mathbf{x}_k - \mathbf{x}_{k+1}\|^2$$

$$\leq f(\mathbf{x}^*) + \frac{1}{\bar{t}} \langle \mathbf{x}_k - \mathbf{x}_{k+1}, \mathbf{x}_{k+1} - \mathbf{x}^* \rangle + \frac{1}{2\bar{t}}\|\mathbf{x}_k - \mathbf{x}_{k+1}\|^2$$

$$= f(\mathbf{x}^*) + \frac{1}{2\bar{t}}\left(\|\mathbf{x}_k - \mathbf{x}^*\|^2 - \|\mathbf{x}_{k+1} - \mathbf{x}^*\|^2\right),$$

establishing part (a).

To prove part (b), sum the inequalities (9.25) for $k = 0, 1, 2, \ldots, n-1$ and obtain

$$\|\mathbf{x}_n - \mathbf{x}^*\|^2 - \|\mathbf{x}_0 - \mathbf{x}^*\|^2 \leq 2\bar{t}\sum_{k=0}^{n-1}(f(\mathbf{x}^*) - f(\mathbf{x}_{k+1})) \leq 2\bar{t}n(f(\mathbf{x}^*) - f(\mathbf{x}_n)),$$

where we used in the last inequality the monotonicity of the functions values of the generated sequence. Hence,

$$f(\mathbf{x}_n) - f^* \leq \frac{\|\mathbf{x}_0 - \mathbf{x}^*\|^2 - \|\mathbf{x}_n - \mathbf{x}^*\|^2}{2\bar{t}n} \leq \frac{\|\mathbf{x}_0 - \mathbf{x}^*\|^2}{2\bar{t}n},$$

as required. $\square$

Note that when the constant stepsize is chosen as $\bar{t} = \frac{1}{L}$, the result (9.26) then reads as

$$f(\mathbf{x}_n) - f^* \leq \frac{L\|\mathbf{x}_0 - \mathbf{x}^*\|^2}{2n}.$$

The rate of convergence of $O(1/n)$ obtained in Theorem 9.16 is called a *sublinear rate of convergence*. We can also deduce the convergence of the sequence itself, and for that, we note that by (9.25) the sequence satisfies a property called *Fejér monotonicity*, which we state explicitly in the following lemma.

Lemma 9.17 (Fejér monotonicity of the sequence generated by the gradient projection method). *Under the setting of Theorem 9.16 the sequence $\{\mathbf{x}_k\}_{k\geq 0}$ satisfies the following inequality for any $\mathbf{x}^* \in X^*$ and $k \geq 0$:*

$$\|\mathbf{x}_{k+1} - \mathbf{x}^*\| \leq \|\mathbf{x}_k - \mathbf{x}^*\|.$$

Proof. The proof follows by (9.25) and the fact that $f(\mathbf{x}^*) \leq f(\mathbf{x}_{k+1})$. □

We are now ready to prove the convergence of the sequence $\{\mathbf{x}_k\}_{k\geq 0}$ to an optimal solution.

Theorem 9.18 (convergence of the sequence generated by the gradient projection method). *Under the setting of Theorem 9.16 the sequence $\{\mathbf{x}_k\}_{k\geq 0}$ generated by the gradient projection method with a constant stepsize $t_k = \bar{t} \in (0, \frac{1}{L}]$ converges to an optimal solution.*

Proof. The result follows from the Fejér monotonicity of the sequence. First of all, by part (b) of Theorem 9.16 it follows that $f(\mathbf{x}_k) \to f^*$, where f^* is the optimal value of problem (P), and thus by the continuity of f, any limit point of the sequence $\{\mathbf{x}_k\}_{k\geq 0}$ is an optimal solution. In addition, by Fejér monotonicity property, we have for any $\mathbf{x}^* \in X^*$ that $\|\mathbf{x}_k - \mathbf{x}^*\| \leq \|\mathbf{x}_0 - \mathbf{x}^*\|$, implying that the sequence $\{\mathbf{x}_k\}_{k\geq 0}$ is bounded, establishing the fact that the sequence does have at least one limit point. To prove the convergence of the entire sequence, let $\tilde{\mathbf{x}}$ be a limit point of the sequence $\{\mathbf{x}_k\}_{k\geq 0}$, meaning that there exists a subsequence $\{\mathbf{x}_{k_j}\}_{j\geq 0}$ such that $\mathbf{x}_{k_j} \to \tilde{\mathbf{x}}$. Since $\tilde{\mathbf{x}} \in X^*$, it follows by Fejér monotonicity that for any $k \geq 0$

$$\|\mathbf{x}_{k+1} - \tilde{\mathbf{x}}\| \leq \|\mathbf{x}_k - \tilde{\mathbf{x}}\|.$$

Thus, $\{\|\mathbf{x}_k - \tilde{\mathbf{x}}\|\}_{k\geq 0}$ is a nonincreasing sequence which is bounded below (by zero) and hence convergent. Since $\|\mathbf{x}_{k_j} - \tilde{\mathbf{x}}\| \to 0$ as $j \to \infty$, it follows that the whole sequence $\{\|\mathbf{x}_k - \tilde{\mathbf{x}}\|\}_{k\geq 0}$ converges to zero, and consequently $\mathbf{x}_k \to \tilde{\mathbf{x}}$ as $k \to \infty$. □

9.5 · Sparsity Constrained Problems

So far we considered problems in which the feasible set C is convex. In this section we will study a class of problems in which the constraint set is nonconvex. Therefore, none of the results derived so far apply for this class of problems, and we will see that beside several similarities between the convex and nonconvex cases, there are substantial differences in the type of results that can be obtained.

9.5.1 · Definition and Notation

In this section we consider the problem of minimizing a continuously differentiable objective function subject to a sparsity constraint. More specifically, we consider the problem

$$\text{(S)} \quad \begin{array}{l} \min \ f(\mathbf{x}) \\ \text{s.t.} \ \|\mathbf{x}\|_0 \leq s, \end{array}$$

where $f : \mathbb{R}^n \to \mathbb{R}$ is bounded below and a continuously differentiable function whose gradient is Lipschitz with constant L_f, $s > 0$ is an integer smaller than n, and $\|\mathbf{x}\|_0$ is the so-called ℓ_0 norm of $\mathbf{x}$, which counts the number of nonzero components in $\mathbf{x}$:

$$\|\mathbf{x}\|_0 = |\{i : x_i \neq 0\}|.$$

It is important to note that the l_0 norm is actually *not* a norm. Indeed, it does not even satisfy the homogeneity property since $\|\lambda \mathbf{x}\|_0 = \|\mathbf{x}\|_0$ for $\lambda \neq 0$. We do not assume that f is a convex function. The constraint set is of course nonconvex and most of the analysis of this chapter does not hold for problem (S). Nevertheless, this type of problem is important and has many applications in areas such as compressed sensing, and it is therefore worthwhile to study it and understand how the "classical" results over convex sets can be adjusted in order to be valid for problem (S). First of all, we begin with some notation. For a given vector $\mathbf{x} \in \mathbb{R}^n$, the support set of $\mathbf{x}$ is defined by

$$I_1(\mathbf{x}) \equiv \{i : x_i \neq 0\},$$

and its complement by

$$I_0(\mathbf{x}) \equiv \{i : x_i = 0\}.$$

We denote by C_s the set of vectors $\mathbf{x}$ that are at most s-sparse, meaning that they have at most s nonzero elements:

$$C_s = \{\mathbf{x} : \|\mathbf{x}\|_0 \leq s\}.$$

In this notation problem (S) can be written as

$$\min\{f(\mathbf{x}) : \mathbf{x} \in C_s\}.$$

For a vector $\mathbf{x} \in \mathbb{R}^n$ and $i \in \{1, 2, \ldots, n\}$, the ith largest absolute value component in $\mathbf{x}$ is denoted by $M_i(\mathbf{x})$, so that in particular

$$M_1(\mathbf{x}) \geq M_2(\mathbf{x}) \geq \cdots \geq M_n(\mathbf{x}).$$

Also, $M_1(\mathbf{x}) = \max_{i=1,\ldots,n} |x_i|$ and $M_n(\mathbf{x}) = \min_{i=1,\ldots,n} |x_i|$.

Our objective now is to study necessary optimality conditions for problem (S), which are similar in nature to the stationarity property. As we will see, the nonconvexity of the constraint in (S) prohibits the usual stationarity conditions to hold, but we will see that other optimality conditions can be constructed.

9.5.2 ▪ L-Stationarity

We begin by extending the concept of stationarity to the case of problem (S). We will use the characterization of stationarity in terms of the orthogonal projection. Recall that $\mathbf{x}^*$ is a stationary point of the problem of minimizing a continuously differentiable function over a closed and convex set C if and only if

$$\mathbf{x}^* = P_C(\mathbf{x}^* - s \nabla f(\mathbf{x}^*)), \qquad (9.30)$$

where s is an arbitrary positive number. It is interesting to note (see Theorem 9.10) that condition (9.30)—although expressed in terms of the parameter s—does not actually depend on s. Relation (9.30) is no longer a valid condition when $C = C_s$ since the

9.5. Sparsity Constrained Problems

orthogonal projection is not unique on nonconvex sets and is in fact a multivalued mapping, meaning that its values form a set given by

$$P_{C_s}(\mathbf{x}) = \operatorname{argmin}_{\mathbf{y}} \{\|\mathbf{y} - \mathbf{x}\| : \mathbf{y} \in C_s\}.$$

The existence of vectors in $P_{C_s}(\mathbf{x})$ follows by the closedness of C_s and the coercivity of the function $\|\mathbf{y} - \mathbf{x}\|$ (with respect to $\mathbf{y}$). To extend (9.30) to the sparsity constrained problem (S), we introduce the notion of "L-stationarity."

Definition 9.19. *A vector* $\mathbf{x}^* \in C_s$ *is called an L-stationary point of* (S) *if it satisfies the relation*

$$[NC_L] \quad \mathbf{x}^* \in P_{C_s}\left(\mathbf{x}^* - \frac{1}{L}\nabla f(\mathbf{x}^*)\right). \tag{9.31}$$

As already mentioned, the orthogonal projection operator $P_{C_s}(\cdot)$ is not single-valued. In this case, the orthogonal projection $P_{C_s}(\mathbf{x})$ comprises all vectors consisting of the s components of $\mathbf{x}$ with the largest absolute values and with zeros elsewhere. In general, there can be more than one choice to the s largest components in absolute value, and each of these choices gives rise to another vector in the set $P_{C_s}(\mathbf{x})$. For example

$$P_{C_2}((2,1,1)^T) = \{(2,1,0)^T, (2,0,1)^T\}.$$

Below we will show that under our assumption that $f \in C_{L_f}^{1,1}(\mathbb{R}^n)$, L-stationarity is a necessary condition for optimality whenever $L > L_f$.

Before proving this result, we describe a more explicit representation of $[NC_L]$.

Lemma 9.20. *For any $L > 0$, $\mathbf{x}^* \in C_s$ satisfies $[NC_L]$ if and only if*

$$\left|\frac{\partial f}{\partial x_i}(\mathbf{x}^*)\right| \begin{cases} \leq LM_s(\mathbf{x}^*) & \text{if } i \in I_0(\mathbf{x}^*), \\ = 0 & \text{if } i \in I_1(\mathbf{x}^*). \end{cases} \tag{9.32}$$

Proof. $[NC_L] \Rightarrow$ (9.32). Suppose that $\mathbf{x}^* \in C_s$ satisfies $[NC_L]$. Note that for any index $j \in \{1,2,\ldots,n\}$, the jth component of any vector in $P_{C_s}(\mathbf{x}^* - \frac{1}{L}\nabla f(\mathbf{x}^*))$ is either zero or equal to $x_j^* - \frac{1}{L}\frac{\partial f}{\partial x_j}(\mathbf{x}^*)$. Now, since $\mathbf{x}^* \in P_{C_s}(\mathbf{x}^* - \frac{1}{L}\nabla f(\mathbf{x}^*))$, it follows that if $i \in I_1(\mathbf{x}^*)$, then $x_i^* = x_i^* - \frac{1}{L}\frac{\partial f}{\partial x_i}(\mathbf{x}^*)$, so that $\frac{\partial f}{\partial x_i}(\mathbf{x}^*) = 0$. If $i \in I_0(\mathbf{x}^*)$, then $|x_i^* - \frac{1}{L}\frac{\partial f}{\partial x_i}(\mathbf{x}^*)| \leq M_s(\mathbf{x}^*)$, which combined with the fact that $x_i^* = 0$ implies that $|\frac{\partial f}{\partial x_i}(\mathbf{x}^*)| \leq LM_s(\mathbf{x}^*)$, and consequently (9.32) holds true.

(9.32) $\Rightarrow [NC_L]$. Suppose that $\mathbf{x}^* \in C_s$ satisfies (9.32). If $\|\mathbf{x}^*\|_0 < s$, then $M_s(\mathbf{x}^*) = 0$, and by (9.32) it follows that $\nabla f(\mathbf{x}^*) = 0$; therefore, in this case, $P_{C_s}\left(\mathbf{x}^* - \frac{1}{L}\nabla f(\mathbf{x}^*)\right) = P_{C_s}(\mathbf{x}^*)$ is the set $\{\mathbf{x}^*\}$. If $\|\mathbf{x}^*\|_0 = s$, then $M_s(\mathbf{x}^*) \neq 0$ and $|I_1(\mathbf{x}^*)| = s$. By (9.32)

$$\left|x_i^* - \frac{1}{L}\frac{\partial f}{\partial x_i}(\mathbf{x}^*)\right| \begin{cases} = |x_i^*|, & i \in I_1(\mathbf{x}^*), \\ \leq M_s(\mathbf{x}^*), & i \in I_0(\mathbf{x}^*). \end{cases}$$

Therefore, the vector $\mathbf{x}^* - \frac{1}{L}\nabla f(\mathbf{x}^*)$ contains the s components of $\mathbf{x}^*$ with the largest absolute value and all other components are smaller or equal (in absolute value) to them, and consequently $[NC_L]$ holds. □

Evidently, the condition (9.32) depends on the constant L; the condition is more restrictive for smaller values of L. This shows that the state of affairs in the nonconvex case is different. Before proceeding to show under which conditions is L-stationarity a necessary optimality condition, we will prove the following technical and useful result. The analysis depends heavily on the descent lemma (Lemma 4.22). We recall that the lemma states that if $f \in C_{L_f}^{1,1}(\mathbb{R}^n)$ and $L \geq L_f$, then

$$f(\mathbf{y}) \leq h_L(\mathbf{y}, \mathbf{x}),$$

where

$$h_L(\mathbf{y}, \mathbf{x}) \equiv f(\mathbf{x}) + \langle \nabla f(\mathbf{x}), \mathbf{y} - \mathbf{x} \rangle + \frac{L}{2}\|\mathbf{x} - \mathbf{y}\|^2. \tag{9.33}$$

Lemma 9.21. *Suppose that $f \in C_{L_f}^{1,1}(\mathbb{R}^n)$ and that $L > L_f$. Then for any $\mathbf{x} \in C_s$ and $\mathbf{y} \in \mathbb{R}^n$ satisfying*

$$\mathbf{y} \in P_{C_s}\left(\mathbf{x} - \frac{1}{L}\nabla f(\mathbf{x})\right), \tag{9.34}$$

we have

$$f(\mathbf{x}) - f(\mathbf{y}) \geq \frac{L - L_f}{2}\|\mathbf{x} - \mathbf{y}\|^2. \tag{9.35}$$

Proof. Note that (9.34) can be written as

$$\mathbf{y} \in \mathrm{argmin}_{\mathbf{z} \in C_s} \left\| \mathbf{z} - \left(\mathbf{x} - \frac{1}{L}\nabla f(\mathbf{x})\right) \right\|^2. \tag{9.36}$$

Since

$$h_L(\mathbf{z}, \mathbf{x}) = f(\mathbf{x}) + \langle \nabla f(\mathbf{x}), \mathbf{z} - \mathbf{x} \rangle + \frac{L}{2}\|\mathbf{z} - \mathbf{x}\|^2$$

$$= \frac{L}{2}\left\|\mathbf{z} - \left(\mathbf{x} - \frac{1}{L}\nabla f(\mathbf{x})\right)\right\|^2 + \underbrace{f(\mathbf{x}) - \frac{1}{2L}\|\nabla f(\mathbf{x})\|^2}_{\text{constant w.r.t. } \mathbf{z}},$$

it follows that the minimization problem (9.36) is equivalent to

$$\mathbf{y} \in \mathrm{argmin}_{\mathbf{z} \in C_s} h_L(\mathbf{z}, \mathbf{x}).$$

This implies that

$$h_L(\mathbf{y}, \mathbf{x}) \leq h_L(\mathbf{x}, \mathbf{x}) = f(\mathbf{x}). \tag{9.37}$$

Now by the descent lemma we have

$$f(\mathbf{x}) - f(\mathbf{y}) \geq f(\mathbf{x}) - h_{L_f}(\mathbf{y}, \mathbf{x}),$$

which combined with (9.37) and the identity

$$h_{L_f}(\mathbf{x}, \mathbf{y}) = h_L(\mathbf{x}, \mathbf{y}) - \frac{L - L_f}{2}\|\mathbf{x} - \mathbf{y}\|^2$$

yields (9.35). □

9.5. Sparsity Constrained Problems

We are now ready to prove the necessity of the L-stationarity property under the condition $L > L_f$.

Theorem 9.22 (**L-stationarity as a necessary optimality condition**). *Suppose that $f \in C^{1,1}_{L_f}(\mathbb{R}^n)$ and that $L > L_f$. Let $\mathbf{x}^*$ be an optimal solution of (S). Then*

(i) $\mathbf{x}^*$ *is an L-stationary point,*

(ii) *the set* $P_{C_s}(\mathbf{x}^* - \frac{1}{L}\nabla f(\mathbf{x}^*))$ *is a singleton.*[3]

Proof. We will prove both parts simultaneously. Suppose to the contrary that there exists a vector

$$\mathbf{y} \in P_{C_s}\left(\mathbf{x}^* - \frac{1}{L}\nabla f(\mathbf{x}^*)\right), \tag{9.38}$$

which is different from $\mathbf{x}^*$ ($\mathbf{y} \neq \mathbf{x}^*$). Invoking Lemma 9.21 with $\mathbf{x} = \mathbf{x}^*$, we have

$$f(\mathbf{x}^*) - f(\mathbf{y}) \geq \frac{L - L_f}{2}\|\mathbf{x}^* - \mathbf{y}\|^2,$$

contradicting the optimality of $\mathbf{x}^*$. We conclude that $\mathbf{x}^*$ is the only vector in the set $P_{C_s}(\mathbf{x}^* - \frac{1}{L}\nabla f(\mathbf{x}^*))$. $\square$

To summarize, we have shown that under a Lipschitz condition on ∇f, L-stationarity for any $L > L_f$ is a necessary optimality condition.

9.5.3 ▪ The Iterative Hard-Thresholding Method

One approach for solving problem (S) is to employ the following natural generalization of the gradient projection algorithm. The method can be seen as a fixed point method aimed at "enforcing" the L-stationary condition (9.31):

$$\mathbf{x}^{k+1} \in P_{C_s}\left(\mathbf{x}^k - \frac{1}{L}\nabla f(\mathbf{x}^k)\right), \quad k = 0,1,2,\ldots. \tag{9.39}$$

The method is known in the literature as the *iterative hard-thresholding (IHT)* method, and we will adopt this terminology.

The IHT Method
Input: a constant $L > L_f$.

- **Initialization:** Choose $\mathbf{x}_0 \in C_s$.
- **General step:** $\mathbf{x}^{k+1} \in P_{C_s}\left(\mathbf{x}^k - \frac{1}{L}\nabla f(\mathbf{x}^k)\right), \quad k = 0,1,2,\ldots.$

It can be shown that the general step of the IHT method is equivalent to the relation

$$\mathbf{x}^{k+1} \in \operatorname{argmin}_{\mathbf{x} \in C_s} h_L(\mathbf{x}, \mathbf{x}^k), \tag{9.40}$$

where $h_L(\mathbf{x}, \mathbf{y})$ is defined by (9.33). (See also the proof of Lemma 9.21.)

---

[3] A set is called a *singleton* if it contains exactly one element.

Several basic properties of the IHT method are summarized in the following lemma.

Lemma 9.23. *Suppose that $f \in C_{L_f}^{1,1}(\mathbb{R}^n)$ and that f is lower bounded. Let $\{\mathbf{x}^k\}_{k \geq 0}$ be the sequence generated by the IHT method with a constant stepsize $\frac{1}{L}$, where $L > L_f$. Then*

(a) $f(\mathbf{x}^k) - f(\mathbf{x}^{k+1}) \geq \frac{L - L_f}{2} \|\mathbf{x}^k - \mathbf{x}^{k+1}\|^2$,

(b) $\{f(\mathbf{x}^k)\}_{k \geq 0}$ *is a nonincreasing sequence,*

(c) $\|\mathbf{x}^k - \mathbf{x}^{k+1}\| \to 0$,

(d) *for every $k = 0, 1, 2, \ldots$, if $\mathbf{x}^k \neq \mathbf{x}^{k+1}$, then $f(\mathbf{x}^{k+1}) < f(\mathbf{x}^k)$.*

Proof. Part (a) follows by substituting $\mathbf{x} = \mathbf{x}^k, \mathbf{y} = \mathbf{x}^{k+1}$ into (9.35). Part (b) follows immediately from part (a). To prove (c), note that since $\{f(\mathbf{x}_k)\}_{k \geq 0}$ is a nonincreasing sequence, which is also lower bounded (by the fact that f is bounded below), it follows that it converges to some limit ℓ and hence $f(\mathbf{x}_k) - f(\mathbf{x}_{k+1}) \to \ell - \ell = 0$ as $k \to \infty$. Therefore, by part (a) and the fact that $L > L_f$, the limit $\|\mathbf{x}^k - \mathbf{x}^{k+1}\| \to 0$ holds. Finally, (d) is a direct consequence of (a). $\square$

As already mentioned, the IHT algorithm can be viewed as a fixed point method for solving the condition for L-stationarity. The following theorem states that all accumulation points of the sequence generated by the IHT method with constant stepsize $\frac{1}{L}$ are indeed L-stationary points.

Theorem 9.24. *Suppose that $f \in C_{L_f}^{1,1}(\mathbb{R}^n)$ and that f is lower bounded. Let $\{\mathbf{x}^k\}_{k \geq 0}$ be the sequence generated by the IHT method with stepsize $\frac{1}{L}$, where $L > L_f$. Then any accumulation point of $\{\mathbf{x}^k\}_{k \geq 0}$ is an L-stationary point.*

Proof. Suppose that $\mathbf{x}^*$ is an accumulation point of the sequence. Then there exists a subsequence $\{\mathbf{x}^{k_n}\}_{n \geq 0}$ that converges to $\mathbf{x}^*$. By Lemma 9.23

$$f(\mathbf{x}^{k_n}) - f(\mathbf{x}^{k_n+1}) \geq \frac{L - L_f}{2} \|\mathbf{x}^{k_n} - \mathbf{x}^{k_n+1}\|^2. \tag{9.41}$$

Since $\{f(\mathbf{x}^{k_n})\}_{n \geq 0}$ and $\{f(\mathbf{x}^{k_n+1})\}_{n \geq 0}$, as nonincreasing and lower bounded sequences, both converge to the same limit, it follows that $f(\mathbf{x}^{k_n}) - f(\mathbf{x}^{k_n+1}) \to 0$ as $n \to \infty$, which combined with (9.41) yields that $\mathbf{x}^{k_n+1} \to \mathbf{x}^*$ as $n \to \infty$. Recall that for all $n \geq 0$

$$\mathbf{x}^{k_n+1} \in P_{C_s}\left(\mathbf{x}^{k_n} - \frac{1}{L}\nabla f(\mathbf{x}^{k_n})\right).$$

Let $i \in I_1(\mathbf{x}^*)$. By the convergence of $\mathbf{x}^{k_n}$ and $\mathbf{x}^{k_n+1}$ to $\mathbf{x}^*$, it follows that there exists an N such that

$$x_i^{k_n}, x_i^{k_n+1} \neq 0 \text{ for all } n > N,$$

and therefore, for $n > N$,

$$x_i^{k_n+1} = x_i^{k_n} - \frac{\partial f}{\partial x_i}(\mathbf{x}^{k_n}).$$

Taking n to ∞ and using the continuity of f, we obtain that

$$\frac{\partial f}{\partial x_i}(\mathbf{x}^*) = 0.$$

Now let $i \in I_0(\mathbf{x}^*)$. If there exist an infinite number of indices k_n for which $x_i^{k_n+1} \neq 0$, then as in the previous case, we obtain that $x_i^{k_n+1} = x_i^{k_n} - \frac{\partial f}{\partial x_i}(\mathbf{x}^{k_n})$ for these indices, implying (by taking the limit) that $\frac{\partial f}{\partial x_i}(\mathbf{x}^*) = 0$. In particular, $|\frac{\partial f}{\partial x_i}(\mathbf{x}^*)| \leq LM_s(\mathbf{x}^*)$. On the other hand, if there exists an $M > 0$ such that for all $n > M$ $x_i^{k_n+1} = 0$, then

$$\left| x_i^{k_n} - \frac{1}{L}\frac{\partial f}{\partial x_i}(\mathbf{x}^{k_n}) \right| \leq M_s\left(\mathbf{x}^{k_n} - \frac{1}{L}\nabla f(\mathbf{x}^{k_n})\right) = M_s(\mathbf{x}^{k_n+1}).$$

Thus, taking n to infinity, while exploiting the continuity of the function M_s, we obtain that

$$\left|\frac{\partial f}{\partial x_i}(\mathbf{x}^*)\right| \leq LM_s(\mathbf{x}^*),$$

and hence, by Lemma 9.20, the desired result is established. $\square$

Exercises

9.1. Let f be a continuously differentiable convex function over a closed and convex set $C \subseteq \mathbb{R}^n$. Show that $\mathbf{x}^* \in C$ is an optimal solution of the problem

$$(\text{P}) \quad \min\{f(\mathbf{x}) : \mathbf{x} \in C\}$$

if and only if

$$\langle \nabla f(\mathbf{x}), \mathbf{x}^* - \mathbf{x} \rangle \leq 0 \text{ for all } \mathbf{x} \in C.$$

9.2. Consider the Huber function

$$H_\mu(\mathbf{x}) = \begin{cases} \frac{\|\mathbf{x}\|^2}{2\mu}, & \|\mathbf{x}\| \leq \mu, \\ \|\mathbf{x}\| - \frac{\mu}{2} & \text{else,} \end{cases}$$

where $\mu > 0$ is a given parameter. Show that $H_\mu \in C^{1,1}_{\frac{1}{\mu}}$.

9.3. Consider the minimization problem

$$(\text{Q}) \quad \begin{array}{ll} \min & 2x_1^2 + 3x_2^2 + 4x_3^2 + 2x_1 x_2 - 2x_1 x_3 - 8x_1 - 4x_2 - 2x_3 \\ \text{s.t.} & x_1, x_2, x_3 \geq 0. \end{array}$$

(i) Show that the vector $(\frac{17}{7}, 0, \frac{6}{7})^T$ is an optimal solution of (Q).

(ii) Employ the gradient projection method via MATLAB with constant stepsize $\frac{1}{L}$ (L being the Lipschitz constant of the gradient of the objective function). Show the function values of the first 100 iterations and the produced solution.

9.4. Consider the minimization problem

$$(\text{P}) \quad \min\{f(\mathbf{x}): \mathbf{a}^T\mathbf{x} = 1, \mathbf{x} \in \mathbb{R}^n\},$$

where f is a continuously differentiable function over $\mathbb{R}^n$ and $\mathbf{a} \in \mathbb{R}^n_{++}$. Show that $\mathbf{x}^*$ satisfying $\mathbf{a}^T\mathbf{x}^* = 1$ is a stationary point of (P) if and only if

$$\frac{\frac{\partial f}{\partial x_1}(\mathbf{x}^*)}{a_1} = \frac{\frac{\partial f}{\partial x_2}(\mathbf{x}^*)}{a_2} = \cdots = \frac{\frac{\partial f}{\partial x_n}(\mathbf{x}^*)}{a_n}.$$

9.5. Consider the minimization problem

$$(\text{P}) \quad \min\{f(\mathbf{x}): \mathbf{x} \in \Delta_n\},$$

where f is a continuously differentiable function over Δ_n. Prove that $\mathbf{x}^* \in \Delta_n$ is a stationary point of (P) if and only if there exists $\mu \in \mathbb{R}$ such that

$$\frac{\partial f}{\partial x_i}(\mathbf{x}^*) \begin{cases} = \mu, & x_i^* > 0, \\ \geq \mu, & x_i^* = 0. \end{cases}$$

9.6. Let $S \subseteq \mathbb{R}^n$ be a nonempty closed and convex set, and let $f \in C_L^{1,1}(S)$ be a convex function over S. Assume that the optimal value of the problem

$$(\text{P}) \quad \min\{f(\mathbf{x}): \mathbf{x} \in S\},$$

denoted by f^* is real. Prove that for any $\mathbf{x} \in S$ the following inequality holds:

$$f(\mathbf{x}) - f^* \geq \frac{1}{2L}\|G_L(\mathbf{x})\|^2,$$

where $G_L(\mathbf{x}) = L[\mathbf{x} - P_S(\mathbf{x} - \frac{1}{L}\nabla f(\mathbf{x}))]$.

9.7. Let $f \in C_L^{1,1}(S)$ be a strongly convex function over a nonempty closed convex set $S \subseteq \mathbb{R}^n$ with strong convexity parameter $\sigma > 0$, that is,

$$f(\mathbf{y}) \geq f(\mathbf{x}) + \langle \nabla f(\mathbf{x}), \mathbf{y} - \mathbf{x}\rangle + \frac{\sigma}{2}\|\mathbf{x}-\mathbf{y}\|^2 \text{ for all } \mathbf{x},\mathbf{y} \in S.$$

Consider the problem

$$(\text{P}) \quad \begin{array}{l}\min \ f(\mathbf{x}) \\ \text{s.t.} \ \mathbf{x} \in S,\end{array}$$

where S is a closed and convex set. Let $\{\mathbf{x}_k\}_{k \geq 0}$ be the sequence generated by the gradient projection method for solving problem (P) with a constant stepsize $t_k = \frac{1}{L}$. Let $\mathbf{x}^*$ be the optimal solution of (P), and let f^* be the optimal value of (P). Prove that there exists a constant $c \in (0,1)$ such that for any $k \geq 0$

$$\|\mathbf{x}_{k+1} - \mathbf{x}^*\| \leq c\|\mathbf{x}_k - \mathbf{x}^*\|.$$

Find an explicit expression for c.

Chapter 10
Optimality Conditions for Linearly Constrained Problems

In the previous chapter we discussed the notion of *stationarity*, which is a necessary optimality condition for problems with differentiable objective functions and closed convex feasible sets. One of the main drawbacks of this concept is that for most feasible sets, it is rather difficult to validate whether this condition is satisfied or not, and it is even more difficult to use it in order to actually solve the underlying optimization problem. Our main objective in this chapter is to derive an equivalent optimality condition that is much easier to handle. We will establish the so-called KKT conditions for the special case of linearly constrained problems.

10.1 ▪ Separation and Alternative Theorems

We begin with a very simple yet powerful result on convex sets, namely the separation theorem between a point and a closed convex set. This result will be the basis for all the optimality conditions that will be discussed later on. Given a set $S \subseteq \mathbb{R}^n$, a hyperplane $H = \{\mathbf{x} \in \mathbb{R}^n : \mathbf{a}^T \mathbf{x} = b\}$ ($\mathbf{a} \in \mathbb{R}^n \setminus \{\mathbf{0}\}, b \in \mathbb{R}$) is said to *strictly separate* a point $\mathbf{y} \notin S$ from S if

$$\mathbf{a}^T \mathbf{y} > b$$

and

$$\mathbf{a}^T \mathbf{x} \leq b \text{ for all } \mathbf{x} \in S.$$

An illustration of a separation between a point and a closed and convex set can be seen in Figure 10.1. Our next result shows that a point can always be strictly separated from a closed convex set, as long as it does not belong to it.

Theorem 10.1 (strict separation theorem). *Let $C \subseteq \mathbb{R}^n$ be a nonempty closed and convex set, and let $\mathbf{y} \notin C$. Then there exist $\mathbf{p} \in \mathbb{R}^n \setminus \{\mathbf{0}\}$ and $\alpha \in \mathbb{R}$ such that*

$$\mathbf{p}^T \mathbf{y} > \alpha$$

and

$$\mathbf{p}^T \mathbf{x} \leq \alpha \text{ for all } \mathbf{x} \in C.$$

Proof. By the second projection theorem (Theorem 9.8), the vector $\bar{\mathbf{x}} = P_C(\mathbf{y}) \in C$ satisfies

$$(\mathbf{y} - \bar{\mathbf{x}})^T (\mathbf{x} - \bar{\mathbf{x}}) \leq 0 \text{ for all } \mathbf{x} \in C,$$

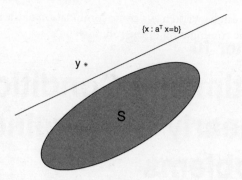

Figure 10.1. *Strict separation of point from a closed and convex set.*

which is the same as
$$(\mathbf{y}-\bar{\mathbf{x}})^T\mathbf{x} \leq (\mathbf{y}-\bar{\mathbf{x}})^T\bar{\mathbf{x}} \text{ for all } \mathbf{x} \in C.$$
Denote $\mathbf{p} = \mathbf{y}-\bar{\mathbf{x}} \neq \mathbf{0}$ (since $\mathbf{y} \notin C$) and $\alpha = (\mathbf{y}-\bar{\mathbf{x}})^T\bar{\mathbf{x}}$. Then we have that $\mathbf{p}^T\mathbf{x} \leq \alpha$ for all $\mathbf{x} \in C$. On the other hand,
$$\mathbf{p}^T\mathbf{y} = (\mathbf{y}-\bar{\mathbf{x}})^T\mathbf{y} = (\mathbf{y}-\bar{\mathbf{x}})^T(\mathbf{y}-\bar{\mathbf{x}}) + (\mathbf{y}-\bar{\mathbf{x}})^T\bar{\mathbf{x}} = \|\mathbf{y}-\bar{\mathbf{x}}\|^2 + \alpha > \alpha,$$
and the result is established. $\square$

As was already mentioned, the latter separation theorem is extremely important since it is the basis for many optimality conditions. We begin by using it in order to prove an *alternative theorem*, which is known in the literature as *Farkas' lemma*. We refer to it as an alternative theorem since it essentially states that exactly one of two systems ("alternatives") is feasible.

Lemma 10.2 (Farkas' lemma). *Let* $\mathbf{c} \in \mathbb{R}^n$ *and* $\mathbf{A} \in \mathbb{R}^{m \times n}$. *Then* **exactly** *one of the following systems has a solution:*

I. $\mathbf{A}\mathbf{x} \leq \mathbf{0}, \mathbf{c}^T\mathbf{x} > 0.$

II. $\mathbf{A}^T\mathbf{y} = \mathbf{c}, \mathbf{y} \geq \mathbf{0}.$

Before proceeding to the proof of the lemma, let us begin with an illustration. For that, consider the following example:
$$\mathbf{A} = \begin{pmatrix} 1 & 5 \\ -1 & 2 \end{pmatrix}, \quad \mathbf{c} = \begin{pmatrix} -1 \\ 9 \end{pmatrix},$$

Stating that system I is *infeasible* means that the system $\mathbf{A}\mathbf{x} \leq \mathbf{0}$ implies the inequality $\mathbf{c}^T\mathbf{x} \leq 0$. Thus, the relevant question is whether the inequality $-x_1 + 9x_2 \leq 0$ holds whenever the two inequalities
$$x_1 + 5x_2 \leq 0,$$
$$-x_1 + 2x_2 \leq 0$$
are satisfied. The answer to this question is affirmative. Indeed, we can see the implication by noting that adding twice the second inequality to the first inequality yields the desired inequality $-x_1 + 9x_2 \leq 0$. Thus, the argument for showing the implication is that the row

10.1. Separation and Alternative Theorems

vector $\mathbf{c}^T$ can be written as a conic combination of the rows of $\mathbf{A}$, or in other words, that $\mathbf{c}$ is a conic combination of the columns of $\mathbf{A}^T$:

$$\begin{pmatrix} 1 \\ 5 \end{pmatrix} + 2 \begin{pmatrix} -1 \\ 2 \end{pmatrix} = \begin{pmatrix} -1 \\ 9 \end{pmatrix}$$

or

$$\underbrace{\begin{pmatrix} 1 & -1 \\ 5 & 2 \end{pmatrix}}_{\mathbf{A}^T} \begin{pmatrix} 1 \\ 2 \end{pmatrix} = \underbrace{\begin{pmatrix} -1 \\ 9 \end{pmatrix}}_{\mathbf{c}}.$$

The interesting question is whether it is always correct that a system of linear inequalities ("base system") implies another linear inequality ("new inequality") if and only if the new inequality can be written as a conic combination of the inequalities in the base system. The answer to this question, according to Farkas' lemma, is yes! We will actually prefer to state Farkas' lemma in the spirit of this discussion. This can be done since an alternative theorem stating that exactly one of two statements A and B is true is equivalent to a result stating that B is equivalent to the denial of A.

Lemma 10.3 (Farkas' lemma, second formulation). *Let $\mathbf{c} \in \mathbb{R}^n$ and $\mathbf{A} \in \mathbb{R}^{m \times n}$. Then the following two claims are equivalent:*

A. *The implication $\mathbf{Ax} \leq \mathbf{0} \Rightarrow \mathbf{c}^T \mathbf{x} \leq 0$ holds true.*

B. *There exists $\mathbf{y} \in \mathbb{R}^m_+$ such that $\mathbf{A}^T \mathbf{y} = \mathbf{c}$.*

Proof. Suppose that system B is feasible, meaning that there exists $\mathbf{y} \in \mathbb{R}^m_+$ such that $\mathbf{A}^T \mathbf{y} = \mathbf{c}$. To see that the implication A holds, suppose that $\mathbf{Ax} \leq \mathbf{0}$ for some $\mathbf{x} \in \mathbb{R}^n$. Then multiplying this inequality from the left by $\mathbf{y}^T$ (a valid operation since $\mathbf{y} \geq \mathbf{0}$) yields

$$\mathbf{y}^T \mathbf{A} \mathbf{x} \leq 0.$$

Finally, using the fact that $\mathbf{c}^T = \mathbf{y}^T \mathbf{A}$, we obtain the desired inequality

$$\mathbf{c}^T \mathbf{x} \leq 0.$$

The reverse direction is much less obvious. Suppose that the implication A is satisfied, and let us show that system B is feasible. Suppose in contradiction that system B is infeasible, and consider the following closed and convex set:

$$S = \{\mathbf{x} \in \mathbb{R}^n : \mathbf{x} = \mathbf{A}^T \mathbf{y} \text{ for some } \mathbf{y} \in \mathbb{R}^m_+\}.$$

The closedness of the above set follows from Lemma 6.32. The infeasibility of B means that $\mathbf{c} \notin S$. By Theorem 10.1, it follows that there exists a vector $\mathbf{p} \in \mathbb{R}^n \setminus \{\mathbf{0}\}$ and $\alpha \in \mathbb{R}$ such that $\mathbf{p}^T \mathbf{c} > \alpha$ and

$$\mathbf{p}^T \mathbf{x} \leq \alpha \text{ for all } \mathbf{x} \in S. \tag{10.1}$$

Since $\mathbf{0} \in S$, we can conclude that $\alpha \geq 0$, and hence also that $\mathbf{p}^T \mathbf{c} > 0$. In addition, (10.1) is equivalent to

$$\mathbf{p}^T \mathbf{A}^T \mathbf{y} \leq \alpha \text{ for all } \mathbf{y} \geq \mathbf{0}$$

or to
$$(\mathbf{Ap})^T \mathbf{y} \leq \alpha \text{ for all } \mathbf{y} \geq 0, \tag{10.2}$$

which implies that $\mathbf{Ap} \leq 0$. Indeed, if there was an index $i \in \{1, 2, \ldots, m\}$ such that $[\mathbf{Ap}]_i > 0$, then for $\mathbf{y} = \beta \mathbf{e}_i$, we would have $(\mathbf{Ap})^T \mathbf{y} = \beta [\mathbf{Ap}]_i$, which is an expression that goes to ∞ as $\beta \to \infty$. Taking a large enough β will contradict (10.2). We have thus arrived at a contradiction to the assumption that the implication A holds (using the vector $\mathbf{p}$), and consequently B is satisfied. $\square$

The next alternative theorem, called "Gordan's theorem," is heavily based on Farkas' lemma.

Theorem 10.4 (Gordan's alternative theorem). *Let $\mathbf{A} \in \mathbb{R}^{m \times n}$. Then exactly one of the following two systems has a solution:*

A. $\mathbf{Ax} < 0$.

B. $\mathbf{p} \neq 0, \mathbf{A}^T \mathbf{p} = 0, \mathbf{p} \geq 0$.

Proof. Suppose that system A has a solution. We will prove that system B is infeasible. Assume in contradiction that B is feasible, meaning that there exists $\mathbf{p} \neq 0$ satisfying $\mathbf{A}^T \mathbf{p} = 0, \mathbf{p} \geq 0$. Multiplying the equality $\mathbf{A}^T \mathbf{p} = 0$ from the left by $\mathbf{x}^T$ yields

$$(\mathbf{Ax})^T \mathbf{p} = 0,$$

which is impossible since $\mathbf{Ax} < 0$ and $0 \neq \mathbf{p} \geq 0$.

Now suppose that system A does not have a solution. Note that system A is equivalent to (s is a scalar)

$$\mathbf{Ax} + s\mathbf{e} \leq 0,$$
$$s > 0.$$

The latter system can be rewritten as

$$\tilde{\mathbf{A}} \begin{pmatrix} \mathbf{x} \\ s \end{pmatrix} \leq 0, \quad \mathbf{c}^T \begin{pmatrix} \mathbf{x} \\ s \end{pmatrix} > 0,$$

where $\tilde{\mathbf{A}} = (\mathbf{A} \quad \mathbf{e})$ and $\mathbf{c} = \mathbf{e}_{n+1}$. The infeasibility of A is thus equivalent to the infeasibility of the system

$$\tilde{\mathbf{A}} \mathbf{w} \leq 0, \quad \mathbf{c}^T \mathbf{w} > 0, \quad \mathbf{w} \in \mathbb{R}^{n+1}.$$

By Farkas' lemma, there exists $\mathbf{z} \in \mathbb{R}_+^m$ such that

$$\begin{pmatrix} \mathbf{A}^T \\ \mathbf{e}^T \end{pmatrix} \mathbf{z} = \mathbf{c};$$

that is, there exists $\mathbf{z} \in \mathbb{R}_+^m$ such that

$$\mathbf{A}^T \mathbf{z} = 0, \quad \mathbf{e}^T \mathbf{z} = 1.$$

Since $\mathbf{e}^T \mathbf{z} = 1$, it follows in particular that $\mathbf{z} \neq 0$, and we have thus shown the existence of $0 \neq \mathbf{z} \in \mathbb{R}_+^m$ such that $\mathbf{A}^T \mathbf{z} = 0$; that is, that system B is feasible. $\square$

10.2 • The KKT conditions

We will now show how Gordan's alternative theorem can be used to establish a very useful optimality criterion that is in fact a special case of the so-called Karush–Kuhn–Tucker (abbreviated KKT) conditions, which will be discussed later on in Chapter 11. The new optimality condition follows from the stationarity condition already discussed in Chapter 9.

Theorem 10.5 (KKT conditions for linearly constrained problems; necessary optimality conditions). *Consider the minimization problem*

$$\text{(P)} \quad \begin{array}{l} \min \ f(\mathbf{x}) \\ \text{s.t.} \ \mathbf{a}_i^T \mathbf{x} \leq b_i, \ i = 1, 2, \ldots, m, \end{array}$$

where f is continuously differentiable over $\mathbb{R}^n$, $\mathbf{a}_1, \mathbf{a}_2, \ldots, \mathbf{a}_m \in \mathbb{R}^n$, $b_1, b_2, \ldots, b_m \in \mathbb{R}$, and let $\mathbf{x}^*$ be a local minimum point of (P). Then there exist $\lambda_1, \lambda_2, \ldots, \lambda_m \geq 0$ such that

$$\nabla f(\mathbf{x}^*) + \sum_{i=1}^m \lambda_i \mathbf{a}_i = 0 \tag{10.3}$$

and

$$\lambda_i(\mathbf{a}_i^T \mathbf{x}^* - b_i) = 0, \quad i = 1, 2, \ldots, m. \tag{10.4}$$

Proof. Since $\mathbf{x}^*$ is a local minimum point of (P), it follows by Theorem 9.2 that $\mathbf{x}^*$ is a stationary point, meaning that $\nabla f(\mathbf{x}^*)^T(\mathbf{x} - \mathbf{x}^*) \geq 0$ for every $\mathbf{x} \in \mathbb{R}^n$ satisfying $\mathbf{a}_i^T \mathbf{x} \leq b_i$ for any $i = 1, 2, \ldots, m$. Let us denote the set of *active constraints* by

$$I(\mathbf{x}^*) = \{i : \mathbf{a}_i^T \mathbf{x}^* = b_i\}.$$

Making the change of variables $\mathbf{y} = \mathbf{x} - \mathbf{x}^*$, we obtain that $\nabla f(\mathbf{x}^*)^T \mathbf{y} \geq 0$ for any $\mathbf{y} \in \mathbb{R}^n$ satisfying $\mathbf{a}_i^T(\mathbf{y} + \mathbf{x}^*) \leq b_i$ for any $i = 1, 2, \ldots, m$, that is, for any $\mathbf{y} \in \mathbb{R}^n$ satisfying

$$\begin{array}{ll} \mathbf{a}_i^T \mathbf{y} \leq 0, & i \in I(\mathbf{x}^*), \\ \mathbf{a}_i^T \mathbf{y} \leq b_i - \mathbf{a}_i^T \mathbf{x}^*, & i \notin I(\mathbf{x}^*). \end{array}$$

We will show that in fact the second set of inequalities in the latter system can be removed, that is, that the following implication is valid:

$$\mathbf{a}_i^T \mathbf{y} \leq 0 \text{ for all } i \in I(\mathbf{x}^*) \Rightarrow \nabla f(\mathbf{x}^*)^T \mathbf{y} \geq 0.$$

Suppose then that $\mathbf{y}$ satisfies $\mathbf{a}_i^T \mathbf{y} \leq 0$ for all $i \in I(\mathbf{x}^*)$. Since $b_i - \mathbf{a}_i^T \mathbf{x}^* > 0$ for all $i \notin I(\mathbf{x}^*)$, it follows that there exists a small enough $\alpha > 0$ for which[4] $\mathbf{a}_i^T(\alpha \mathbf{y}) \leq b_i - \mathbf{a}_i^T \mathbf{x}^*$. Thus, since in addition $\mathbf{a}_i^T(\alpha \mathbf{y}) \leq 0$ for any $i \in I(\mathbf{x}^*)$ it follows by the stationarity condition that $\nabla f(\mathbf{x}^*)^T(\alpha \mathbf{y}) \geq 0$, and hence that $\nabla f(\mathbf{x}^*)^T \mathbf{y} \geq 0$. We have thus shown that

$$\mathbf{a}_i^T \mathbf{y} \leq 0 \text{ for all } i \in I(\mathbf{x}^*) \Rightarrow \nabla f(\mathbf{x}^*)^T \mathbf{y} \geq 0.$$

Thus, by Farkas' lemma it follows that there exist $\lambda_i \geq 0, i \in I(\mathbf{x}^*)$, such that

$$-\nabla f(\mathbf{x}^*) = \sum_{i \in I(\mathbf{x}^*)} \lambda_i \mathbf{a}_i.$$

---

[4] Indeed, if $\mathbf{a}_i^T \mathbf{y} \leq 0$ for all $i \notin I(\mathbf{x}^*)$, then we can take $\alpha = 1$. Otherwise, denote $J = \{i \notin I(\mathbf{x}^*) : \mathbf{a}_i^T \mathbf{y} > 0\}$ and take $\alpha = \min_{i \in J} \frac{b_i - \mathbf{a}_i^T \mathbf{x}^*}{\mathbf{a}_i^T \mathbf{y}}$.

Defining $\lambda_i = 0$ for all $i \notin I(\mathbf{x}^*)$, we get that $\lambda_i(\mathbf{a}_i^T\mathbf{x}^* - b_i) = 0$ for all $i = 1, 2, \ldots, m$ and that

$$\nabla f(\mathbf{x}^*) + \sum_{i=1}^m \lambda_i \mathbf{a}_i = 0$$

as required. □

The KKT conditions are necessary optimality conditions, but when the objective function is convex, they are both necessary and sufficient global optimality conditions.

Theorem 10.6 (KKT conditions for convex linearly constrained problems; necessary and sufficient optimality conditions). *Consider the minimization problem*

$$\text{(P)} \quad \begin{array}{ll} \min & f(\mathbf{x}) \\ \text{s.t.} & \mathbf{a}_i^T\mathbf{x} \leq b_i, \; i = 1, 2, \ldots, m, \end{array}$$

where f is a convex continuously differentiable function over $\mathbb{R}^n$, $\mathbf{a}_1, \mathbf{a}_2, \ldots, \mathbf{a}_m \in \mathbb{R}^n$, $b_1, b_2, \ldots, b_m \in \mathbb{R}$, and let $\mathbf{x}^*$ be a feasible solution of (P). Then $\mathbf{x}^*$ is an optimal solution of (P) if and only if there exist $\lambda_1, \lambda_2, \ldots, \lambda_m \geq 0$ such that

$$\nabla f(\mathbf{x}^*) + \sum_{i=1}^m \lambda_i \mathbf{a}_i = 0 \tag{10.5}$$

and

$$\lambda_i(\mathbf{a}_i^T\mathbf{x}^* - b_i) = 0, \quad i = 1, 2, \ldots, m. \tag{10.6}$$

Proof. If $\mathbf{x}^*$ is an optimal solution of (P), then by Theorem 10.5 there exist $\lambda_1, \lambda_2, \ldots, \lambda_m \geq 0$ such that (10.5) and (10.6) are satisfied. To prove the sufficiency, suppose that $\mathbf{x}^*$ is a feasible solution of (P) satisfying (10.5) and (10.6). Let $\mathbf{x}$ be any feasible solution of (P). Define the function

$$h(\mathbf{x}) = f(\mathbf{x}) + \sum_{i=1}^m \lambda_i(\mathbf{a}_i^T\mathbf{x} - b_i).$$

Then by (10.5) it follows that $\nabla h(\mathbf{x}^*) = 0$, and since h is convex, it follows by Proposition 7.8 that $\mathbf{x}^*$ is a minimizer of h over $\mathbb{R}^n$, which combined with (10.6) implies that

$$f(\mathbf{x}^*) = f(\mathbf{x}^*) + \sum_{i=1}^m \lambda_i(\mathbf{a}_i^T\mathbf{x}^* - b_i) = h(\mathbf{x}^*) \leq h(\mathbf{x}) = f(\mathbf{x}) + \sum_{i=1}^m \lambda_i(\mathbf{a}_i^T\mathbf{x} - b_i) \leq f(\mathbf{x}),$$

where the last inequality follows from the fact that $\lambda_i \geq 0$ and $\mathbf{a}_i^T\mathbf{x} - b_i \leq 0$ for $i = 1, 2, \ldots, m$. We have thus proven that $\mathbf{x}^*$ is a global optimal solution of (P). □

The scalars $\lambda_1, \ldots, \lambda_m$ that appear in the KKT conditions are also called *Lagrange multipliers*, and each of the multipliers is associated with a corresponding constraint: λ_i is the multiplier associated with the ith constraint $\mathbf{a}_i^T\mathbf{x} \leq b_i$. Note that the multipliers associated with inequality constraints are nonnegative. The conditions (10.6) are known in the literature as the *complementary slackness conditions*. We can also generalize Theorems 10.5 and 10.6 to the case where linear *equality* constraints are also present. The main difference is that the multipliers associated with equality constraints are not restricted to be nonnegative. The proof of the variant that also incorporates equality constraints is based on the simple observation that a linear equality constraint $\mathbf{a}^T\mathbf{x} = b$ can be written as two inequality constraints, $\mathbf{a}^T\mathbf{x} \leq b$ and $-\mathbf{a}^T\mathbf{x} \leq -b$.

10.2. The KKT conditions

Theorem 10.7 (KKT conditions for linearly constrained problems). *Consider the minimization problem*

$$\text{(Q)} \quad \begin{array}{ll} \min & f(\mathbf{x}) \\ \text{s.t.} & \mathbf{a}_i^T \mathbf{x} \leq b_i, \quad i = 1, 2, \ldots, m, \\ & \mathbf{c}_j^T \mathbf{x} = d_j, \quad j = 1, 2, \ldots, p, \end{array}$$

where f is a continuously differentiable function over $\mathbb{R}^n$, $\mathbf{a}_1, \mathbf{a}_2, \ldots, \mathbf{a}_m, \mathbf{c}_1, \mathbf{c}_2, \ldots, \mathbf{c}_p \in \mathbb{R}^n$, $b_1, b_2, \ldots, b_m, d_1, d_2, \ldots, d_p \in \mathbb{R}$. Then we have the following:

(a) **(necessity of the KKT conditions)** *If $\mathbf{x}^*$ is a local minimum point of (Q), then there exist $\lambda_1, \lambda_2, \ldots, \lambda_m \geq 0$ and $\mu_1, \mu_2, \ldots, \mu_p \in \mathbb{R}$ such that*

$$\nabla f(\mathbf{x}^*) + \sum_{i=1}^{m} \lambda_i \mathbf{a}_i + \sum_{j=1}^{p} \mu_j \mathbf{c}_j = 0 \quad (10.7)$$

and

$$\lambda_i (\mathbf{a}_i^T \mathbf{x}^* - b_i) = 0, \quad i = 1, 2, \ldots, m. \quad (10.8)$$

(b) **(sufficiency in the convex case)** *If in addition f is convex over $\mathbb{R}^n$ and $\mathbf{x}^*$ is a feasible solution of (Q) for which there exist $\lambda_1, \lambda_2, \ldots, \lambda_m \geq 0$ and $\mu_1, \mu_2, \ldots, \mu_p \in \mathbb{R}$ such that (10.7) and (10.8) are satisfied, then $\mathbf{x}^*$ is an optimal solution of (Q).*

Proof. (a). Consider the equivalent problem

$$\text{(Q')} \quad \begin{array}{ll} \min & f(\mathbf{x}) \\ \text{s.t.} & \mathbf{a}_i^T \mathbf{x} \leq b_i, \quad i = 1, 2, \ldots, m, \\ & \mathbf{c}_j^T \mathbf{x} \leq d_j, \quad j = 1, 2, \ldots, p, \\ & -\mathbf{c}_j^T \mathbf{x} \leq -d_j, \quad j = 1, 2, \ldots, p. \end{array}$$

Then since $\mathbf{x}^*$ is an optimal solution of (Q), it is also an optimal solution of (Q'), and thus, by Theorem 10.5, it follows that there exist multipliers $\lambda_1, \lambda_2, \ldots, \lambda_m \geq 0$ and $\mu_1^+, \mu_1^-, \mu_2^+, \mu_2^-, \ldots, \mu_p^+, \mu_p^- \geq 0$ such that

$$\nabla f(\mathbf{x}^*) + \sum_{i=1}^{m} \lambda_i \mathbf{a}_i + \sum_{j=1}^{p} \mu_j^+ \mathbf{c}_j - \sum_{j=1}^{p} \mu_j^- \mathbf{c}_j = 0 \quad (10.9)$$

and

$$\lambda_i (\mathbf{a}_i^T \mathbf{x}^* - b_i) = 0, \quad i = 1, 2, \ldots, m, \quad (10.10)$$
$$\mu_j^+ (\mathbf{c}_j^T \mathbf{x}^* - d_j) = 0, \quad j = 1, 2, \ldots, p, \quad (10.11)$$
$$\mu_j^- (-\mathbf{c}_j^T \mathbf{x}^* + d_j) = 0, \quad j = 1, 2, \ldots, p. \quad (10.12)$$

We thus obtain that (10.7) and (10.8) are satisfied with $\mu_j = \mu_j^+ - \mu_j^-, j = 1, 2, \ldots, p$.

(b) To prove the second part, suppose that $\mathbf{x}^*$ satisfies (10.7) and (10.8). Then it also satisfies (10.9), (10.10), (10.11), and (10.12) with

$$\mu_j^+ = [\mu_j]_+, \mu_j^- = [\mu_j]_- = -\min\{\mu_j, 0\},$$

which by Theorem 10.6 implies that $\mathbf{x}^*$ is an optimal solution of (Q') and thus also an optimal solution of (Q). $\square$

We note that a feasible point $\mathbf{x}^*$ is called *a KKT point* if there exist multipliers for which (10.7) and (10.8) are satisfied.

A very popular representation of the KKT conditions is via the Lagrangian function, which we will present in the setting of general nonlinear programming problems:

$$\text{(NLP)} \quad \begin{array}{ll} \min & f(\mathbf{x}) \\ \text{s.t.} & g_i(\mathbf{x}) \leq 0, \quad i = 1, 2, \ldots, m, \\ & h_j(\mathbf{x}) = 0, \quad j = 1, 2, \ldots, p. \end{array}$$

Here $f, g_1, g_2, \ldots, g_m, h_1, h_2, \ldots, h_p$ are all continuously differentiable functions over $\mathbb{R}^n$. The associated Lagrangian function takes the form

$$L(\mathbf{x}, \lambda, \mu) = f(\mathbf{x}) + \sum_{i=1}^{m} \lambda_i g_i(\mathbf{x}) + \sum_{j=1}^{p} \mu_j h_j(\mathbf{x}).$$

In the linearly constrained case of problem (Q), the condition (10.7) is the same as

$$\nabla_{\mathbf{x}} L(\mathbf{x}, \lambda, \mu) = \nabla f(\mathbf{x}) + \sum_{i=1}^{m} \lambda_i \nabla g_i(\mathbf{x}) + \sum_{j=1}^{p} \mu_j \nabla h_j(\mathbf{x}) = \mathbf{0}.$$

Back to problem (Q), if in addition we define the matrices $\mathbf{A}$ and $\mathbf{C}$ and the vectors $\mathbf{b}$ and $\mathbf{d}$ by

$$\mathbf{A} = \begin{pmatrix} \mathbf{a}_1^T \\ \mathbf{a}_2^T \\ \vdots \\ \mathbf{a}_m^T \end{pmatrix}, \quad \mathbf{C} = \begin{pmatrix} \mathbf{c}_1^T \\ \mathbf{c}_2^T \\ \vdots \\ \mathbf{c}_p^T \end{pmatrix}, \quad \mathbf{b} = \begin{pmatrix} b_1 \\ b_2 \\ \vdots \\ b_m \end{pmatrix}, \quad \mathbf{d} = \begin{pmatrix} d_1 \\ d_2 \\ \vdots \\ d_p \end{pmatrix},$$

then the constraints of problem (Q) can be written as

$$\mathbf{A}\mathbf{x} \leq \mathbf{b}, \quad \mathbf{C}\mathbf{x} = \mathbf{d}.$$

The Lagrangian can be also written as

$$L(\mathbf{x}, \lambda, \mu) = f(\mathbf{x}) + \lambda^T (\mathbf{A}\mathbf{x} - \mathbf{b}) + \mu^T (\mathbf{C}\mathbf{x} - \mathbf{d}),$$

and condition (10.7) takes the form

$$\nabla_{\mathbf{x}} L(\mathbf{x}, \lambda, \mu) = \nabla f(\mathbf{x}) + \mathbf{A}^T \lambda + \mathbf{C}^T \mu = \mathbf{0}.$$

Example 10.8. Consider the problem

$$\begin{array}{ll} \min & \frac{1}{2}(x_1^2 + x_2^2 + x_3^2) \\ \text{s.t.} & x_1 + x_2 + x_3 = 3. \end{array}$$

Since the problem is convex, the KKT conditions are necessary and sufficient. The Lagrangian of the problem is

$$L(x_1, x_2, x_3, \mu) = \frac{1}{2}(x_1^2 + x_2^2 + x_3^2) + \mu(x_1 + x_2 + x_3 - 3).$$

10.2. The KKT conditions

The KKT conditions are (we also incorporate feasibility within the KKT system)

$$\frac{\partial L}{\partial x_1} = x_1 + \mu = 0,$$

$$\frac{\partial L}{\partial x_2} = x_2 + \mu = 0,$$

$$\frac{\partial L}{\partial x_3} = x_3 + \mu = 0,$$

$$x_1 + x_2 + x_3 = 3.$$

By the first three equalities we obtain that $x_1 = x_2 = x_3 = -\mu$. Substituting this in the last equation yields $\mu = -1$, and we obtained that the unique solution of the KKT system is $x_1 = x_2 = x_3 = 1, \mu = -1$. Hence, the unique optimal solution of the problem is $(x_1, x_2, x_3) = (1, 1, 1)$. ∎

Example 10.9. Consider the problem

$$\begin{aligned} \min \quad & x_1^2 + 2x_2^2 + 4x_1 x_2 \\ \text{s.t.} \quad & x_1 + x_2 = 1, \\ & x_1, x_2 \geq 0. \end{aligned}$$

The problem is nonconvex, since the matrix associated with the quadratic objective function $\mathbf{A} = \begin{pmatrix} 1 & 2 \\ 2 & 2 \end{pmatrix}$ is indefinite. However, the KKT conditions are still necessary optimality conditions. The Lagrangian of the problem is

$$L(x_1, x_2, \mu, \lambda_1, \lambda_2) = x_1^2 + 2x_2^2 + 4x_1 x_2 + \mu(x_1 + x_2 - 1) - \lambda_1 x_1 - \lambda_2 x_2, \quad \lambda_1, \lambda_2 \in \mathbb{R}_+, \mu \in \mathbb{R}.$$

The KKT conditions are

$$\frac{\partial L}{\partial x_1} = 2x_1 + 4x_2 + \mu - \lambda_1 = 0,$$

$$\frac{\partial L}{\partial x_2} = 4x_2 + 4x_1 + \mu - \lambda_2 = 0,$$

$$\lambda_1 x_1 = 0,$$

$$\lambda_2 x_2 = 0,$$

$$x_1 + x_2 = 1,$$

$$x_1, x_2 \geq 0,$$

$$\lambda_1, \lambda_2 \geq 0.$$

We will split the analysis into 4 cases.

- $\lambda_1 = \lambda_2 = 0$. In this case we obtain the three equations

$$2x_1 + 4x_2 + \mu = 0,$$
$$4x_2 + 4x_1 + \mu = 0,$$
$$x_1 + x_2 = 1,$$

whose solution is $x_1 = 0, x_2 = 1, \mu = -4$. We thus obtain that $(x_1, x_2) = (0, 1)$ is a KKT point.

- $\lambda_1, \lambda_2 > 0$. In this case, by the complementary slackness conditions we obtain that $x_1 = x_2 = 0$, which contradicts the constraint $x_1 + x_2 = 1$.

- $\lambda_1 > 0, \lambda_2 = 0$. In this case, by the complementary slackness conditions we have $x_1 = 0$ and consequently $x_2 = 1$, which was already shown to be a KKT point.

- $\lambda_1 = 0, \lambda_2 > 0$. Here $x_2 = 0$ and $x_1 = 1$. The first two equations in the KKT system reduce to

$$2 + \mu = 0,$$
$$4 + \mu - \lambda_2 = 0.$$

The solution of this system is $\mu = -2, \lambda_2 = 2$. We thus obtain that $(1,0)$ is also a KKT point.

To summarize, there are two KKT points: $(0,1)$ and $(1,0)$. Since the problem consists of minimizing a continuous function over a compact set it follows by the Weierstrass theorem (Theorem 2.30) that it has a global optimal solution. The KKT conditions are necessary optimality conditions, and hence the optimal solution is either $(0,1)$ or $(1,0)$. Since the respective objective function values are 2 and 1, it follows that $(1,0)$ is the global optimal solution of the problem. ∎

Example 10.10 (orthogonal projection onto an affine space). Let C be the affine space

$$C = \{\mathbf{x} \in \mathbb{R}^n : \mathbf{A}\mathbf{x} = \mathbf{b}\},$$

where $\mathbf{A} \in \mathbb{R}^{m \times n}$ and $\mathbf{b} \in \mathbb{R}^m$. We assume that the rows of $\mathbf{A}$ are linearly independent. Given $\mathbf{y} \in \mathbb{R}^n$, the optimization problem associated with the problem of finding $P_C(\mathbf{y})$ is

$$\begin{aligned} \min \quad & \|\mathbf{x} - \mathbf{y}\|^2 \\ \text{s.t.} \quad & \mathbf{A}\mathbf{x} = \mathbf{b}. \end{aligned}$$

This is a convex optimization problem, so the KKT conditions are necessary and sufficient. The Lagrangian function is

$$L(\mathbf{x}, \lambda) = \|\mathbf{x} - \mathbf{y}\|^2 + 2\lambda^T(\mathbf{A}\mathbf{x} - \mathbf{b}) = \|\mathbf{x}\|^2 - 2(\mathbf{y} - \mathbf{A}^T\lambda)^T\mathbf{x} - 2\lambda^T\mathbf{b} + \|\mathbf{y}\|^2, \quad \lambda \in \mathbb{R}^m.$$

Therefore, the KKT conditions are

$$2\mathbf{x} - 2(\mathbf{y} - \mathbf{A}^T\lambda) = 0,$$
$$\mathbf{A}\mathbf{x} = \mathbf{b}.$$

The first equation can be written as

$$\mathbf{x} = \mathbf{y} - \mathbf{A}^T\lambda. \tag{10.13}$$

Substituting this expression for $\mathbf{x}$ in the second equation yields the equation

$$\mathbf{A}(\mathbf{y} - \mathbf{A}^T\lambda) = \mathbf{b},$$

which is the same as

$$\mathbf{A}\mathbf{A}^T\lambda = \mathbf{A}\mathbf{y} - \mathbf{b}.$$

Thus,
$$\lambda = (\mathbf{AA}^T)^{-1}(\mathbf{Ay}-\mathbf{b}),$$
where here we used the fact that $\mathbf{AA}^T$ is nonsingular since the rows of $\mathbf{A}$ are linearly independent. Plugging the latter expression for λ into (10.13), we obtain that
$$P_C(\mathbf{y}) = \mathbf{y} - \mathbf{A}^T(\mathbf{AA}^T)^{-1}(\mathbf{Ay}-\mathbf{b}).$$
Note that the projection onto an affine space is by itself an affine transformation. ∎

In the last example we multiplied the Lagrange multiplier in the Lagrangian function by 2. This was done to simplify the computations, and it is always allowed to multiply the Lagrange multipliers by a positive constant.

Example 10.11 (orthogonal projection onto hyperplanes). Consider the hyperplane
$$H = \{\mathbf{x} \in \mathbb{R}^n : \mathbf{a}^T\mathbf{x} = b\}.$$
where $0 \neq \mathbf{a} \in \mathbb{R}^n$ and $b \in \mathbb{R}$. Since a hyperplane is a spacial case of an affine space, we can use the formula obtained in the last example in order to derive an explicit expression for the projection onto H:
$$P_H(\mathbf{y}) = \mathbf{y} - \mathbf{a}(\mathbf{a}^T\mathbf{a})^{-1}(\mathbf{a}^T\mathbf{y}-b) = \mathbf{y} - \frac{\mathbf{a}^T\mathbf{y}-b}{\|\mathbf{a}\|^2}\mathbf{a}. \quad \blacksquare$$

As a consequence of the last example, we can write a result providing an explicit expression for the distance between a point and a hyperplane. (The result was already stated and not proved in Lemma 8.6.)

Lemma 10.12 (distance of a point from a hyperplane). *Let $H = \{\mathbf{x} \in \mathbb{R}^n : \mathbf{a}^T\mathbf{x} = b\}$, where $0 \neq \mathbf{a} \in \mathbb{R}^n$ and $b \in \mathbb{R}$. Then*
$$d(\mathbf{y}, H) = \frac{|\mathbf{a}^T\mathbf{y}-b|}{\|\mathbf{a}\|}.$$

Proof.
$$d(\mathbf{y},H) = \|\mathbf{y}-P_H(\mathbf{y})\| = \left\|\mathbf{y} - \left(\mathbf{y} - \frac{\mathbf{a}^T\mathbf{y}-b}{\|\mathbf{a}\|^2}\mathbf{a}\right)\right\| = \frac{|\mathbf{a}^T\mathbf{y}-b|}{\|\mathbf{a}\|}. \quad \square$$

Example 10.13 (orthogonal projection onto half-spaces). Let
$$H^- = \{\mathbf{x} \in \mathbb{R}^n : \mathbf{a}^T\mathbf{x} \leq b\},$$
where $0 \neq \mathbf{a} \in \mathbb{R}^n$ and $b \in \mathbb{R}$. The corresponding optimization problem is
$$\begin{aligned}\min_{\mathbf{x}} \quad & \|\mathbf{x}-\mathbf{y}\|^2 \\ \text{s.t.} \quad & \mathbf{a}^T\mathbf{x} \leq b.\end{aligned}$$

The Lagrangian of the problem is
$$L(\mathbf{x}, \lambda) = \|\mathbf{x}-\mathbf{y}\|^2 + 2\lambda(\mathbf{a}^T\mathbf{x}-b), \quad \lambda \geq 0,$$

and the KKT conditions are

$$2(\mathbf{x}-\mathbf{y})+2\lambda\mathbf{a}=0,$$
$$\lambda(\mathbf{a}^T\mathbf{x}-b)=0,$$
$$\mathbf{a}^T\mathbf{x}\leq b,$$
$$\lambda\geq 0.$$

If $\lambda = 0$, then $\mathbf{x} = \mathbf{y}$ and the KKT conditions are satisfied when $\mathbf{a}^T\mathbf{y} \leq b$. That is, when $\mathbf{a}^T\mathbf{y} \leq b$, the optimal solution is $\mathbf{x} = \mathbf{y}$, which is not a surprise since *for any* set C, $P_C(\mathbf{y}) = \mathbf{y}$ whenever $\mathbf{y} \in C$. Now assume that $\lambda > 0$; then by the complementary slackness condition we have that

$$\mathbf{a}^T\mathbf{x}=b. \tag{10.14}$$

Plugging the first equation $\mathbf{x} = \mathbf{y} - \lambda\mathbf{a}$ into (10.14), we obtain that

$$\mathbf{a}^T(\mathbf{y}-\lambda\mathbf{a})=b,$$

so that $\lambda = \frac{\mathbf{a}^T\mathbf{y}-b}{\|\mathbf{a}\|^2}$. The multiplier λ is indeed positive when $\mathbf{a}^T\mathbf{y} > b$. We have thus obtained that when $\mathbf{a}^T\mathbf{y} > b$, the optimal solution is

$$\mathbf{x}=\mathbf{y}-\frac{\mathbf{a}^T\mathbf{y}-b}{\|\mathbf{a}\|^2}\mathbf{a}.$$

To summarize,

$$P_H(\mathbf{y})=\begin{cases}\mathbf{y}, & \mathbf{a}^T\mathbf{y}\leq b,\\ \mathbf{y}-\frac{\mathbf{a}^T\mathbf{y}-b}{\|\mathbf{a}\|^2}\mathbf{a}, & \mathbf{a}^T\mathbf{y}> b.\end{cases}$$

The latter expression can also be compactly written as

$$P_H(\mathbf{y})=\mathbf{y}-\frac{[\mathbf{a}^T\mathbf{y}-b]_+}{\|\mathbf{a}\|^2}\mathbf{a}.$$

An illustration of the orthogonal projection onto a hyperplane can be found in Figure 10.2. ∎

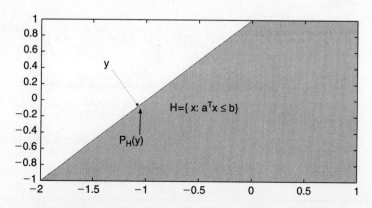

Figure 10.2. *A vector* $\mathbf{y}$ *and its orthogonal projection onto a half-space.*

10.3. Orthogonal Regression

Example 10.14. Consider the optimization problem

$$\min\{\mathbf{x}^T\mathbf{Q}\mathbf{x} + 2\mathbf{c}^T\mathbf{x} : \mathbf{A}\mathbf{x} = \mathbf{b}\},$$

where $\mathbf{Q} \in \mathbb{R}^{n \times n}$ is a positive definite matrix, $\mathbf{c} \in \mathbb{R}^n, \mathbf{b} \in \mathbb{R}^m$, and $\mathbf{A}$ is an $m \times n$ matrix with linearly independent rows. The Lagrangian of the problem is

$$L(\mathbf{x}, \lambda) = \mathbf{x}^T\mathbf{Q}\mathbf{x} + 2\mathbf{c}^T\mathbf{x} + 2\lambda^T(\mathbf{A}\mathbf{x} - \mathbf{b}),$$

and the KKT conditions are

$$\nabla_\mathbf{x} L(\mathbf{x}, \lambda) = 2[\mathbf{Q}\mathbf{x} + \mathbf{c} + \mathbf{A}^T\lambda] = 0,$$
$$\mathbf{A}\mathbf{x} = \mathbf{b}.$$

The first equation implies that

$$\mathbf{x} = -\mathbf{Q}^{-1}(\mathbf{c} + \mathbf{A}^T\lambda). \quad (10.15)$$

Plugging this expression of $\mathbf{x}$ into the feasibility constraint we obtain

$$-\mathbf{A}\mathbf{Q}^{-1}(\mathbf{c} + \mathbf{A}^T\lambda) = \mathbf{b},$$

so that

$$\lambda = -(\mathbf{A}\mathbf{Q}^{-1}\mathbf{A}^T)^{-1}(\mathbf{b} + \mathbf{A}\mathbf{Q}^{-1}\mathbf{c}). \quad (10.16)$$

The optimal solution of the problem is given by (10.15) with λ as in (10.16). ∎

10.3 ▪ Orthogonal Regression

An interesting application to the formula for the distance between a point and a hyperplane given in Lemma 10.12 is in the orthogonal regression problem, which we now recall. Consider the points $\mathbf{a}_1, \ldots, \mathbf{a}_m$ in $\mathbb{R}^n$. For a given $\mathbf{0} \neq \mathbf{x} \in \mathbb{R}^n$ and $y \in \mathbb{R}$, we define the hyperplane

$$H_{\mathbf{x}, y} := \{\mathbf{a} \in \mathbb{R}^n : \mathbf{x}^T\mathbf{a} = y\}.$$

In the orthogonal regression problem we seek to find a nonzero vector $\mathbf{x} \in \mathbb{R}^n$ and $y \in \mathbb{R}$ such that the sum of squared Euclidean distances between the points $\mathbf{a}_1, \ldots, \mathbf{a}_m$ to $H_{\mathbf{x}, y}$ is minimal; that is, the problem is given by

$$\min_{\mathbf{x}, y}\left\{\sum_{i=1}^m d(\mathbf{a}_i, H_{\mathbf{x}, y})^2 : \mathbf{0} \neq \mathbf{x} \in \mathbb{R}^n, y \in \mathbb{R}\right\}. \quad (10.17)$$

An illustration of the solution to the orthogonal regression problem is given in Figure 10.3.

The optimal solution of the orthogonal regression problem is described in the next result whose proof strongly relies on the formula of the distance between a point and a hyperplane.

Proposition 10.15. *Let $\mathbf{a}_1, \ldots, \mathbf{a}_m \in \mathbb{R}^n$ and let $\mathbf{A}$ be the matrix given by*

$$\mathbf{A} = \begin{pmatrix} \mathbf{a}_1^T \\ \mathbf{a}_2^T \\ \vdots \\ \mathbf{a}_m^T \end{pmatrix}.$$

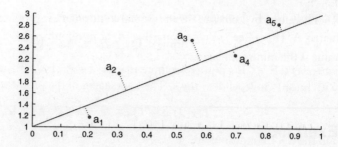

Figure 10.3. *A two-dimensional example: given 5 points* $\mathbf{a}_1, \ldots, \mathbf{a}_5$ *in the plane, the orthogonal regression problem seeks to find the line for which the sum of squared norms of the dashed lines is minimal.*

Then an optimal solution of problem (10.17) *is given by* $\mathbf{x}$ *that is an eigenvector of the matrix* $\mathbf{A}^T(\mathbf{I}_m - \frac{1}{m}\mathbf{e}\mathbf{e}^T)\mathbf{A}$ *associated with the minimum eigenvalue and* $y = \frac{1}{m}\sum_{i=1}^m \mathbf{a}_i^T \mathbf{x}$. *Here* $\mathbf{e}$ *is the m-length vector of ones. The optimal function value of problem* (10.17) *is* $\lambda_{\min}[\mathbf{A}^T(\mathbf{I}_m - \frac{1}{m}\mathbf{e}\mathbf{e}^T)\mathbf{A}]$.

Proof. By Lemma 10.12, the squared Euclidean distance between the point $\mathbf{a}_i$ to $H_{\mathbf{x},y}$ is given by

$$d(\mathbf{a}_i, H_{\mathbf{x},y})^2 = \frac{(\mathbf{a}_i^T \mathbf{x} - y)^2}{\|\mathbf{x}\|^2}, \quad i = 1, \ldots, m.$$

It follows that (10.17) is the same as

$$\min\left\{ \sum_{i=1}^m \frac{(\mathbf{a}_i^T \mathbf{x} - y)^2}{\|\mathbf{x}\|^2} : 0 \neq \mathbf{x} \in \mathbb{R}^n, y \in \mathbb{R} \right\}. \tag{10.18}$$

Fixing $\mathbf{x}$ and minimizing first with respect to y we obtain that the optimal y is given by

$$y = \frac{1}{m}\sum_{i=1}^m \mathbf{a}_i^T \mathbf{x} = \frac{1}{m}\mathbf{e}^T \mathbf{A}\mathbf{x}.$$

Using the latter expression for y we obtain that

$$\sum_{i=1}^m (\mathbf{a}_i^T \mathbf{x} - y)^2 = \sum_{i=1}^m \left(\mathbf{a}_i^T \mathbf{x} - \frac{1}{m}\mathbf{e}^T \mathbf{A}\mathbf{x}\right)^2$$

$$= \sum_{i=1}^m (\mathbf{a}_i^T \mathbf{x})^2 - \frac{2}{m}\sum_{i=1}^m (\mathbf{e}^T \mathbf{A}\mathbf{x})(\mathbf{a}_i^T \mathbf{x}) + \frac{1}{m}(\mathbf{e}^T \mathbf{A}\mathbf{x})^2$$

$$= \sum_{i=1}^m (\mathbf{a}_i^T \mathbf{x})^2 - \frac{1}{m}(\mathbf{e}^T \mathbf{A}\mathbf{x})^2 = \|\mathbf{A}\mathbf{x}\|^2 - \frac{1}{m}(\mathbf{e}^T \mathbf{A}\mathbf{x})^2$$

$$= \mathbf{x}^T \mathbf{A}^T \left(\mathbf{I}_m - \frac{1}{m}\mathbf{e}\mathbf{e}^T\right)\mathbf{A}\mathbf{x}.$$

Therefore, we arrive at the following reformulation of (10.17) as a problem consisting of minimizing a Rayleigh quotient:

$$\min_{\mathbf{x}} \left\{ \frac{\mathbf{x}^T[\mathbf{A}^T(\mathbf{I}_m - \frac{1}{m}\mathbf{e}\mathbf{e}^T)\mathbf{A}]\mathbf{x}}{\|\mathbf{x}\|^2} : \mathbf{x} \neq 0 \right\}.$$

Therefore, by Lemma 1.12 an optimal solution of the problem is an eigenvector of the matrix $\mathbf{A}^T(\mathbf{I}_m - \frac{1}{m}\mathbf{e}\mathbf{e}^T)\mathbf{A}$ corresponding to the minimum eigenvalue; the optimal function value is the minimum eigenvalue $\lambda_{\min}[\mathbf{A}^T(\mathbf{I}_m - \frac{1}{m}\mathbf{e}\mathbf{e}^T)\mathbf{A}]$. □

Exercises

10.1. Show that the dual cone of
$$M = \{\mathbf{x} \in \mathbb{R}^n : \mathbf{A}\mathbf{x} \geq 0\} \quad (\mathbf{A} \in \mathbb{R}^{m \times n})$$
is
$$M^* = \{\mathbf{A}^T\mathbf{v} : \mathbf{v} \in \mathbb{R}^m_+\}.$$

10.2. (**nonhomogenous Farkas' lemma**) Let $\mathbf{A} \in \mathbb{R}^{m \times n}, \mathbf{c} \in \mathbb{R}^n, \mathbf{b} \in \mathbb{R}^m$, and $d \in \mathbb{R}$. Suppose that there exists $\mathbf{y} \geq 0$ such that $\mathbf{A}^T\mathbf{y} = \mathbf{c}$. Prove that exactly one of the following two systems is feasible:

A. $\mathbf{A}\mathbf{x} \leq \mathbf{b}, \mathbf{c}^T\mathbf{x} > d$.

B. $\mathbf{A}^T\mathbf{y} = \mathbf{c}, \mathbf{b}^T\mathbf{y} \leq d, \mathbf{y} \geq 0$.

10.3. Let $\mathbf{A} \in \mathbb{R}^{m \times n}$ and $\mathbf{c} \in \mathbb{R}^n$. Show that exactly one of the following two systems is feasible:

A. $\mathbf{A}\mathbf{x} \geq 0, \mathbf{x} \geq 0, \mathbf{c}^T\mathbf{x} > 0$.

B. $\mathbf{A}^T\mathbf{y} \geq \mathbf{c}, \mathbf{y} \leq 0$.

10.4. Prove Motzkin's theorem of the alternative: the system
$$\text{(I)} \quad \begin{array}{rl} \mathbf{A}\mathbf{d} < & 0, \\ \mathbf{B}\mathbf{d} \leq & 0 \end{array}$$
has a solution if and only if the system
$$\text{(II)} \quad \begin{array}{l} \mathbf{A}^T\mathbf{u} + \mathbf{B}^T\mathbf{y} = 0, \\ \mathbf{u}, \mathbf{y} \geq 0, \quad \mathbf{u} \neq 0 \end{array}$$
does not have a solution (here $\mathbf{A} \in \mathbb{R}^{m \times n}, \mathbf{B} \in \mathbb{R}^{k \times n}$).

10.5. Prove the following nonhomogenous version of Gordan's alternative theorem: Given $\mathbf{A} \in \mathbb{R}^{m \times n}$, exactly one of these two systems is feasible.

A. $\mathbf{A}\mathbf{z} < \mathbf{b}$.

B. $\mathbf{A}^T\mathbf{y} = 0, \mathbf{b}^T\mathbf{y} \leq 0, \mathbf{y} \geq 0, \mathbf{y} \neq 0$.

10.6. Consider the maximization problem
$$\begin{array}{ll} \max & x_1^2 + 2x_1 x_2 + 2x_2^2 - 3x_1 + x_2 \\ \text{s.t.} & x_1 + x_2 = 1 \\ & x_1, x_2 \geq 0. \end{array}$$

(i) Is the problem convex?

(ii) Find all the KKT points of the problem.

(iii) Find the optimal solution of the problem.

10.7. Consider the problem

$$\min \quad -x_1 x_2 x_3$$
$$\text{s.t.} \quad x_1 + 3x_2 + 6x_3 \leq 48,$$
$$x_1, x_2, x_3 \geq 0.$$

(i) Write the KKT conditions for the problem.

(ii) Find the optimal solution of the problem.

10.8. Consider the problem

$$\min \quad x_1^2 + 2x_2^2 + x_1$$
$$\text{s.t.} \quad x_1 + x_2 \leq a,$$

where $a \in \mathbb{R}$ is a parameter.

(i) Prove that for any $a \in \mathbb{R}$, the problem has a unique optimal solution (without actually solving it).

(ii) Solve the problem (the solution will be in terms of the parameter a).

(iii) Let $f(a)$ be the optimal value of the problem with parameter a. Write an explicit expression for f and prove that it is a convex function.

10.9. Consider the problem

$$\min \quad x_1^2 + x_2^2 + x_3^2 + x_1 x_2 + x_2 x_3 - 2x_1 - 4x_2 - 6x_3$$
$$\text{s.t.} \quad x_1 + x_2 + x_3 \leq 1.$$

(i) Is the problem convex?

(ii) Find all the KKT points of the problem.

(iii) Find the optimal solution of the problem.

10.10. Consider the problem

$$\min \quad x_1^2 + x_2^2 + x_3^2$$
$$\text{s.t.} \quad x_1 + 2x_2 + 3x_3 \geq 4$$
$$x_3 \leq 1.$$

(i) Write down the KKT conditions.

(ii) Without solving the KKT system, prove that the problem has a unique optimal solution and that this solution satisfies the KKT conditions.

(iii) Find the optimal solution of the problem using the KKT system.

10.11. Use the KKT conditions in order to solve the problem

$$\min \quad x_1^2 + x_2^2$$
$$\text{s.t.} \quad -2x_1 - x_2 + 10 \leq 0$$
$$x_2 \geq 0.$$

Chapter 11
The KKT Conditions

In this chapter we will further develop the KKT conditions discussed in Chapter 10, but we will consider general constraints and not restrict ourselves to linear constraints. The Karush-Kuhn-Tucker (KKT) conditions were originally named after Harold Kuhn and Albert Tucker, who first published the conditions in 1951. Later on it was discovered that William Karush developed the necessary conditions in his master's thesis back in 1939, and the conditions were thus named after the three researchers.

11.1 ▪ Inequality Constrained Problems

We will begin our exploration into the KKT conditions by analyzing the inequality constrained problem

$$\text{(P)} \quad \begin{array}{ll} \min & f(\mathbf{x}) \\ \text{s.t.} & g_i(\mathbf{x}) \leq 0, \quad i = 1, 2, \ldots, m, \end{array} \tag{11.1}$$

where $f, g_1, \ldots, g_m$ are continuously differentiable functions over $\mathbb{R}^n$. Our first task is to develop necessary optimality conditions. For that, we will define the concept of a feasible descent direction.

Definition 11.1 (feasible descent directions). *Consider the problem*

$$\begin{array}{ll} \min & h(\mathbf{x}) \\ \text{s.t.} & \mathbf{x} \in C, \end{array}$$

*where h is continuously differentiable over the set $C \subseteq \mathbb{R}^n$. Then a vector $\mathbf{d} \neq \mathbf{0}$ is called a **feasible descent direction** at $\mathbf{x} \in C$ if $\nabla h(\mathbf{x})^T \mathbf{d} < 0$, and there exists $\varepsilon > 0$ such that $\mathbf{x} + t\mathbf{d} \in C$ for all $t \in [0, \varepsilon]$.*

Obviously, a necessary local optimality condition of a point $\mathbf{x}$ is that it does not have any feasible descent directions.

Lemma 11.2. *Consider the problem*

$$\text{(G)} \quad \begin{array}{ll} \min & f(\mathbf{x}) \\ \text{s.t.} & \mathbf{x} \in C, \end{array}$$

where f is a continuously differentiable function over the set $C \subseteq \mathbb{R}^n$. If $\mathbf{x}^*$ is a local optimal solution of (G), then there are no feasible descent directions at $\mathbf{x}^*$.

Proof. The proof is by contradiction. If there is a feasible descent direction, that is, a vector $\mathbf{d}$ and $\varepsilon_1 > 0$ such that $\mathbf{x}^* + t\mathbf{d} \in C$ for all $t \in [0, \varepsilon_1]$ and $\nabla f(\mathbf{x}^*)^T \mathbf{d} < 0$, then by the definition of the directional derivative (see also Lemma 4.2) there is an $\varepsilon_2 < \varepsilon_1$ such that $f(\mathbf{x}^* + t\mathbf{d}) < f(\mathbf{x}^*)$ for all $t \in [0, \varepsilon_2]$, which is a contradiction to the local optimality of $\mathbf{x}^*$. $\square$

We can now write a necessary condition for local optimality in the form of an infeasibility of a set of strict linear inequalities. Before doing that, we mention the following terminology: Given a set of inequalities

$$g_i(\mathbf{x}) \leq 0, \quad i = 1, 2, \ldots, m,$$

where $g_i : \mathbb{R}^n \to \mathbb{R}$ are functions, and a vector $\tilde{\mathbf{x}} \in \mathbb{R}^n$, the *active constraints* at $\tilde{\mathbf{x}}$ are the constraints satisfied as equalities at $\tilde{\mathbf{x}}$. The set of active constraints is denoted by

$$I(\tilde{\mathbf{x}}) = \{i : g_i(\tilde{\mathbf{x}}) = 0\}.$$

Lemma 11.3. *Let $\mathbf{x}^*$ be a local minimum of the problem*

$$\begin{aligned} \min \quad & f(\mathbf{x}) \\ \text{s.t.} \quad & g_i(\mathbf{x}) \leq 0, \quad i = 1, 2, \ldots, m, \end{aligned}$$

where $f, g_1, \ldots, g_m$ are continuously differentiable functions over $\mathbb{R}^n$. Let $I(\mathbf{x}^*)$ be the set of active constraints at $\mathbf{x}^*$:

$$I(\mathbf{x}^*) = \{i : g_i(\mathbf{x}^*) = 0\}.$$

Then there does not exist a vector $\mathbf{d} \in \mathbb{R}^n$ such that

$$\begin{aligned} \nabla f(\mathbf{x}^*)^T \mathbf{d} &< 0, \\ \nabla g_i(\mathbf{x}^*)^T \mathbf{d} &< 0, \quad i \in I(\mathbf{x}^*). \end{aligned} \tag{11.2}$$

Proof. Suppose by contradiction that $\mathbf{d}$ satisfies the system of inequalities (11.2). Then by Lemma 4.2, it follows that there exists $\varepsilon_1 > 0$ such that $f(\mathbf{x}^* + t\mathbf{d}) < f(\mathbf{x}^*)$ and $g_i(\mathbf{x}^* + t\mathbf{d}) < g_i(\mathbf{x}^*) = 0$ for any $t \in (0, \varepsilon_1)$ and $i \in I(\mathbf{x}^*)$. For any $i \notin I(\mathbf{x}^*)$ we have that $g_i(\mathbf{x}^*) < 0$, and hence, by the continuity of g_i for all i, it follows that there exists $\varepsilon_2 > 0$ such that $g_i(\mathbf{x}^* + t\mathbf{d}) < 0$ for any $t \in (0, \varepsilon_2)$ and $i \notin I(\mathbf{x}^*)$. We can thus conclude that

$$\begin{aligned} f(\mathbf{x}^* + t\mathbf{d}) &< f(\mathbf{x}^*), \\ g_i(\mathbf{x}^* + t\mathbf{d}) &< 0, \quad i = 1, 2, \ldots, m, \end{aligned}$$

for all $t \in (0, \min\{\varepsilon_1, \varepsilon_2\})$, which is a contradiction to the local optimality of $\mathbf{x}^*$. $\square$

We have thus shown that a necessary optimality condition for local optimality is the infeasibility of a certain system of strict inequalities. We can now invoke Gordan's theorem of the alternative (Theorem 10.4) in order to obtain the so-called Fritz-John conditions.

Theorem 11.4 (Fritz-John conditions for inequality constrained problems). *Let $\mathbf{x}^*$ be a local minimum of the problem*

$$\begin{aligned} \min \quad & f(\mathbf{x}) \\ \text{s.t.} \quad & g_i(\mathbf{x}) \leq 0, \quad i = 1, 2, \ldots, m, \end{aligned}$$

where $f, g_1, \ldots, g_m$ are continuously differentiable functions over $\mathbb{R}^n$. Then there exist multipliers $\lambda_0, \lambda_1, \ldots, \lambda_m \geq 0$, which are not all zeros, such that

$$\lambda_0 \nabla f(\mathbf{x}^*) + \sum_{i=1}^{m} \lambda_i \nabla g_i(\mathbf{x}^*) = 0, \tag{11.3}$$

$$\lambda_i g_i(\mathbf{x}^*) = 0, \quad i = 1, 2, \ldots, m.$$

Proof. By Lemma 11.3 it follows that the following system of inequalities does not have a solution:

$$(S) \quad \begin{aligned} \nabla f(\mathbf{x}^*)^T \mathbf{d} &< 0, \\ \nabla g_i(\mathbf{x}^*)^T \mathbf{d} &< 0, \quad i \in I(\mathbf{x}^*), \end{aligned} \tag{11.4}$$

where $I(\mathbf{x}^*) = \{i : g_i(\mathbf{x}^*) = 0\} = \{i_1, i_2, \ldots, i_k\}$. System (S) can be rewritten as

$$\mathbf{A}\mathbf{d} < 0,$$

where

$$\mathbf{A} = \begin{pmatrix} \nabla f(\mathbf{x}^*)^T \\ \nabla g_{i_1}(\mathbf{x}^*)^T \\ \vdots \\ \nabla g_{i_k}(\mathbf{x}^*)^T \end{pmatrix}.$$

By Gordan's theorem of the alternative (Theorem 10.4), system (S) is infeasible if and only if there exists a vector $\boldsymbol{\eta} = (\lambda_0, \lambda_{i_1}, \ldots, \lambda_{i_k})^T \neq 0$ such that

$$\mathbf{A}^T \boldsymbol{\eta} = 0, \quad \boldsymbol{\eta} \geq 0,$$

which is the same as

$$\lambda_0 \nabla f(\mathbf{x}^*) + \sum_{i \in I(\mathbf{x}^*)} \lambda_i \nabla g_i(\mathbf{x}^*) = 0,$$

$$\lambda_i \geq 0, \quad i \in I(\mathbf{x}^*).$$

Define $\lambda_i = 0$ for any $i \notin I(\mathbf{x}^*)$, and we obtain that

$$\lambda_0 \nabla f(\mathbf{x}^*) + \sum_{i=1}^{m} \lambda_i \nabla g_i(\mathbf{x}^*) = 0$$

and that $\lambda_i g_i(\mathbf{x}^*) = 0$ for any $i \in \{1, 2, \ldots, m\}$ as required. $\square$

A major drawback of the Fritz-John conditions is in the fact that they allow λ_0 to be zero. The case $\lambda_0 = 0$ is not particularly informative since condition (11.3) then becomes

$$\sum_{i=1}^{m} \lambda_i \nabla g_i(\mathbf{x}^*) = 0,$$

which means that the gradients of the active constraints $\{\nabla g_i(\mathbf{x}^*)\}_{i \in I(\mathbf{x}^*)}$ are linearly dependent. This condition has nothing to do with the objective function, implying that there might be a lot of points satisfying the Fritz-John conditions which are not local minimum points. If we add an assumption that the gradients of the active constraints are linearly independent at $\mathbf{x}^*$, then we can establish the KKT conditions, which are the same as the Fritz-John conditions with $\lambda_0 = 1$.

Theorem 11.5 (KKT conditions for inequality constrained problems). *Let $\mathbf{x}^*$ be a local minimum of the problem*

$$\begin{aligned}\min\quad & f(\mathbf{x})\\ \text{s.t.}\quad & g_i(\mathbf{x}) \leq 0, \quad i = 1, 2, \ldots, m,\end{aligned}$$

where $f, g_1, \ldots, g_m$ are continuously differentiable functions over $\mathbb{R}^n$. Let

$$I(\mathbf{x}^*) = \{i : g_i(\mathbf{x}^*) = 0\}$$

be the set of active constraints. Suppose that the gradients of the active constraints $\{\nabla g_i(\mathbf{x}^*)\}_{i \in I(\mathbf{x}^*)}$ are linearly independent. Then there exist multipliers $\lambda_1, \lambda_2, \ldots, \lambda_m \geq 0$ such that

$$\nabla f(\mathbf{x}^*) + \sum_{i=1}^{m} \lambda_i \nabla g_i(\mathbf{x}^*) = 0, \tag{11.5}$$

$$\lambda_i g_i(\mathbf{x}^*) = 0, \quad i = 1, 2, \ldots, m. \tag{11.6}$$

Proof. By the Fritz-John conditions it follows that there exist $\tilde{\lambda}_0, \tilde{\lambda}_1, \ldots, \tilde{\lambda}_m \geq 0$, not all zeros, such that

$$\tilde{\lambda}_0 \nabla f(\mathbf{x}^*) + \sum_{i=1}^{m} \tilde{\lambda}_i \nabla g_i(\mathbf{x}^*) = 0, \tag{11.7}$$

$$\tilde{\lambda}_i g_i(\mathbf{x}^*) = 0, \quad i = 1, 2, \ldots, m. \tag{11.8}$$

We have that $\tilde{\lambda}_0 \neq 0$ since otherwise, if $\tilde{\lambda}_0 = 0$, by (11.7) and (11.8) it follows that

$$\sum_{i \in I(\mathbf{x}^*)} \tilde{\lambda}_i \nabla g_i(\mathbf{x}^*) = 0,$$

where not all the scalars $\tilde{\lambda}_i, i \in I(\mathbf{x}^*)$ are zeros, leading to a contradiction to the basic assumption that $\{\nabla g_i(\mathbf{x}^*)\}_{i \in I(\mathbf{x}^*)}$ are linearly independent, and hence $\tilde{\lambda}_0 > 0$. Defining $\lambda_i = \frac{\tilde{\lambda}_i}{\tilde{\lambda}_0}$, the result directly follows from (11.7) and (11.8). $\square$

The condition that the gradients of the active constraints are linearly independent is one of many types of assumptions that are referred to in the literature as "constraint qualifications."

11.2 ▪ Inequality and Equality Constrained Problems

By using the implicit function theorem, it is possible to generalize the KKT conditions for problems involving also equality constraints. We will state this generalization without a proof.

Theorem 11.6 (KKT conditions for inequality/equality constrained problems). *Let $\mathbf{x}^*$ be a local minimum of the problem*

$$\begin{aligned}\min\quad & f(\mathbf{x})\\ \text{s.t.}\quad & g_i(\mathbf{x}) \leq 0, \quad i = 1, 2, \ldots, m,\\ & h_j(\mathbf{x}) = 0, \quad j = 1, 2, \ldots, p.\end{aligned} \tag{11.9}$$

where $f, g_1, \ldots, g_m, h_1, h_2, \ldots, h_p$ are continuously differentiable functions over $\mathbb{R}^n$. Suppose that the gradients of the active constraints and the equality constraints

$$\{\nabla g_i(\mathbf{x}^*) : i \in I(\mathbf{x}^*)\} \cup \{\nabla h_j(\mathbf{x}^*) : j = 1, 2, \ldots, p\}$$

are linearly independent (where as before $I(\mathbf{x}^*) = \{i : g_i(\mathbf{x}^*) = 0\}$). Then there exist multipliers $\lambda_1, \lambda_2, \ldots, \lambda_m \geq 0$ and $\mu_1, \mu_2, \ldots, \mu_p \in \mathbb{R}$ such that

$$\nabla f(\mathbf{x}^*) + \sum_{i=1}^m \lambda_i \nabla g_i(\mathbf{x}^*) + \sum_{j=1}^p \mu_j \nabla h_j(\mathbf{x}^*) = \mathbf{0},$$

$$\lambda_i g_i(\mathbf{x}^*) = 0, \quad i = 1, 2, \ldots, m.$$

We will now add to our terminology two concepts: KKT points and regularity. The first was already discussed in the previous chapter in the context of linearly constrained problems and is now extended.

Definition 11.7 (KKT points). *Consider the minimization problem (11.9), where $f, g_1, \ldots, g_m, h_1, h_2, \ldots, h_p$ are continuously differentiable functions over $\mathbb{R}^n$. A feasible point $\mathbf{x}^*$ is called a **KKT point** if there exist $\lambda_1, \lambda_2, \ldots, \lambda_m \geq 0$ and $\mu_1, \mu_2, \ldots, \mu_p \in \mathbb{R}$ such that*

$$\nabla f(\mathbf{x}^*) + \sum_{i=1}^m \lambda_i \nabla g_i(\mathbf{x}^*) + \sum_{j=1}^p \mu_j \nabla h_j(\mathbf{x}^*) = \mathbf{0},$$

$$\lambda_i g_i(\mathbf{x}^*) = 0, \quad i = 1, 2, \ldots, m.$$

Definition 11.8 (regularity). *Consider the minimization problem (11.9), where $f, g_1, \ldots, g_m, h_1, h_2, \ldots, h_p$ are continuously differentiable functions over $\mathbb{R}^n$. A feasible point $\mathbf{x}^*$ is called **regular** if the gradients of the active constraints among the inequality constraints and of the equality constraints*

$$\{\nabla g_i(\mathbf{x}^*) : i \in I(\mathbf{x}^*)\} \cup \{\nabla h_j(\mathbf{x}^*) : j = 1, 2, \ldots, p\}$$

are linearly independent.

In the terminology of the above definitions, Theorem 11.6 states that a necessary optimality condition for local optimality of a regular point is that it is a KKT point. The additional requirement of regularity is not required in the linearly constrained case in which no such assumption is needed; see Theorem 10.7.

Example 11.9. Consider the problem

$$\begin{array}{ll} \min & x_1 + x_2 \\ \text{s.t.} & x_1^2 + x_2^2 = 1. \end{array}$$

Note that this is not a convex optimization problem due to the fact that the equality constraint is nonlinear. In addition, since the problem consists of minimizing a continuous function over a nonempty compact set, it follows that the minimizer exists (by the Weierstrass theorem, Theorem 2.30).

Let us first address the issue of whether the KKT conditions are necessary for this problem. Since by Theorem 11.6 we know that the KKT conditions are necessary optimality conditions for regular points, we will find the irregular points of the problem. These are exactly the points $\mathbf{x}^*$ for which the set of gradients of active constraints, which is given here by $\{2\binom{x_1^*}{x_2^*}\}$, is a dependent set of vectors; this can happen only when $x_1^* = x_2^* = 0$, that is, at a nonfeasible point. The conclusion is that the problem does not have irregular points, and hence the KKT conditions are necessary optimality conditions.

In order to write the KKT conditions, we will form the Lagrangian:

$$L(x_1, x_2, \lambda) = x_1 + x_2 + \lambda(x_1^2 + x_2^2 - 1).$$

The KKT conditions are

$$\frac{\partial L}{\partial x_1} = 1 + 2\lambda x_1 = 0,$$

$$\frac{\partial L}{\partial x_2} = 1 + 2\lambda x_2 = 0,$$

$$x_1^2 + x_2^2 = 1.$$

By the first two conditions, $\lambda \neq 0$, and hence $x_1 = x_2 = -\frac{1}{2\lambda}$. Plugging this expression of x_1, x_2 into the last equation yields

$$\left(-\frac{1}{2\lambda}\right)^2 + \left(-\frac{1}{2\lambda}\right)^2 = 1,$$

so that $\lambda = \pm\frac{1}{\sqrt{2}}$. The problem thus has two KKT points: $(\frac{1}{\sqrt{2}}, \frac{1}{\sqrt{2}})$ and $(-\frac{1}{\sqrt{2}}, -\frac{1}{\sqrt{2}})$. Since an optimal solution does exist, and the KKT conditions are necessary for this problem, it follows that at least one of these two points is optimal, and obviously the point $(-\frac{1}{\sqrt{2}}, -\frac{1}{\sqrt{2}})$ which has the smaller objective value $(-\sqrt{2})$ is the optimal solution. ∎

Example 11.10. Consider a problem which is equivalent to the one given in the previous example:

$$\begin{array}{ll} \min & x_1 + x_2 \\ \text{s.t.} & (x_1^2 + x_2^2 - 1)^2 = 0. \end{array}$$

Writing the KKT conditions yields

$$1 + 4\lambda x_1(x_1^2 + x_2^2 - 1) = 0,$$
$$1 + 4\lambda x_2(x_1^2 + x_2^2 - 1) = 0,$$
$$(x_1^2 + x_2^2 - 1)^2 = 0.$$

This system is of course infeasible since the combination of the first and third equations gives the impossible equation $1 = 0$. It is not a surprise that the KKT conditions are not satisfied here, since all the feasible points are in fact irregular. Indeed, the gradient of the constraint is $\binom{4x_1(x_1^2+x_2^2-1)}{4x_2(x_1^2+x_2^2-1)}$, which is the zeros vector for any feasible point. ∎

Example 11.11. Consider the optimization problem

$$\begin{array}{ll} \min & 2x_1 + 3x_2 - x_3 \\ \text{s.t.} & x_1^2 + x_2^2 + x_3^2 = 1, \\ & x_1^2 + 2x_2^2 + 2x_3^2 = 2. \end{array}$$

The problem does have an optimal solution since it consists of minimizing a continuous function over a nonempty and compact set. The KKT system is

$$2 + 2(\lambda + \mu)x_1 = 0,$$
$$3 + 2(\lambda + 2\mu)x_2 = 0,$$
$$-1 + 2(\lambda + 2\mu)x_3 = 0,$$
$$x_1^2 + x_2^2 + x_3^2 = 1,$$
$$x_1^2 + 2x_2^2 + 2x_3^2 = 2.$$

Obviously, by the first three equations $\lambda + \mu \neq 0, \lambda + 2\mu \neq 0$, and in addition

$$x_1 = -\frac{1}{\lambda + \mu}, \quad x_2 = -\frac{3}{2(\lambda + 2\mu)}, \quad x_3 = \frac{1}{2(\lambda + 2\mu)}.$$

Denoting $t_1 = \frac{1}{\lambda+\mu}, t_2 = \frac{1}{2(\lambda+2\mu)}$, we obtain that $x_1 = -t_1, x_2 = -3t_2, x_3 = t_2$. Substituting these expressions into the constraints we obtain that

$$t_1^2 + 10t_2^2 = 1,$$
$$t_1^2 + 20t_2^2 = 2.$$

implying that $t_1^2 = 0$, which is impossible. We obtain that there are no KKT points. Thus, since as was already observed, an optimal solution does exist, it follows that it is an irregular point. To find the irregular points, note that the gradients of the constraints are given by

$$\begin{pmatrix} 2x_1 \\ 2x_2 \\ 2x_3 \end{pmatrix}, \quad \begin{pmatrix} 2x_1 \\ 4x_2 \\ 4x_3 \end{pmatrix}.$$

The two gradients are linearly dependent in the following two cases. (A) $x_1 = 0$. In this case, the problem becomes

$$\begin{aligned} \min \quad & 3x_2 - x_3 \\ \text{s.t.} \quad & x_2^2 + x_3^2 = 1, \end{aligned}$$

whose optimal solution is $(x_2, x_3) = (-\frac{3}{\sqrt{10}}, \frac{1}{\sqrt{10}})$, and the obtained objective function value is $-\sqrt{10}$.

(B) $x_2 = x_3 = 0$. However, the constraints in this case reduce to $x_1^2 = 1, x_1^2 = 2$, which is impossible.

The conclusion is that the optimal solution of the problem is

$$(x_1, x_2, x_3) = \left(0, -\frac{3}{\sqrt{10}}, \frac{1}{\sqrt{10}}\right)$$

with optimal value $-\sqrt{10}$. ∎

11.3 ▪ The Convex Case

The KKT conditions are necessary optimality condition under the regularity condition. When the problem is convex, the KKT conditions are *always* sufficient and no further conditions are required.

Theorem 11.12 (sufficiency of the KKT conditions for convex optimization problems). *Let $\mathbf{x}^*$ be a feasible solution of*

$$\begin{aligned} \min \quad & f(\mathbf{x}) \\ \text{s.t.} \quad & g_i(\mathbf{x}) \leq 0, \quad i = 1, 2, \ldots, m, \\ & h_j(\mathbf{x}) = 0, \quad j = 1, 2, \ldots, p, \end{aligned} \quad (11.10)$$

where $f, g_1, \ldots, g_m$ are continuously differentiable convex functions over $\mathbb{R}^n$ and $h_1, h_2, \ldots, h_p$ are affine functions. Suppose that there exist multipliers $\lambda_1, \lambda_2, \ldots, \lambda_m \geq 0$ and $\mu_1, \mu_2, \ldots, \mu_p \in \mathbb{R}$ such that

$$\nabla f(\mathbf{x}^*) + \sum_{i=1}^m \lambda_i \nabla g_i(\mathbf{x}^*) + \sum_{j=1}^p \mu_j \nabla h_j(\mathbf{x}^*) = 0,$$

$$\lambda_i g_i(\mathbf{x}^*) = 0, \quad i = 1, 2, \ldots, m.$$

Then $\mathbf{x}^*$ is an optimal solution of (11.10).

Proof. Let $\mathbf{x}$ be a feasible solution of (11.10). We will show that $f(\mathbf{x}) \geq f(\mathbf{x}^*)$. Note that the function

$$s(\mathbf{x}) = f(\mathbf{x}) + \sum_{i=1}^m \lambda_i g_i(\mathbf{x}) + \sum_{j=1}^p \mu_j h_j(\mathbf{x})$$

is convex, and since $\nabla s(\mathbf{x}^*) = \nabla f(\mathbf{x}^*) + \sum_{i=1}^m \lambda_i \nabla g_i(\mathbf{x}^*) + \sum_{j=1}^p \mu_j \nabla h_j(\mathbf{x}^*) = 0$, it follows by Proposition 7.8 that $\mathbf{x}^*$ is a minimizer of $s(\cdot)$ over $\mathbb{R}^n$, and in particular $s(\mathbf{x}^*) \leq s(\mathbf{x})$. We can thus conclude that

$$\begin{aligned} f(\mathbf{x}^*) &= f(\mathbf{x}^*) + \sum_{i=1}^m \lambda_i g_i(\mathbf{x}^*) + \sum_{j=1}^p \mu_j h_j(\mathbf{x}^*) \quad (\lambda_i g_i(\mathbf{x}^*) = 0, h_j(\mathbf{x}^*) = 0) \\ &= s(\mathbf{x}^*) \\ &\leq s(\mathbf{x}) \\ &= f(\mathbf{x}) + \sum_{i=1}^m \lambda_i g_i(\mathbf{x}) + \sum_{j=1}^p \mu_j h_j(\mathbf{x}) \\ &\leq f(\mathbf{x}) \quad (\lambda_i \geq 0, g_i(\mathbf{x}) \leq 0, h_j(\mathbf{x}) = 0), \end{aligned}$$

showing that $\mathbf{x}^*$ is the optimal solution of (11.10). $\square$

In the convex case we can find a different condition than regularity that guarantees the necessity of the KKT condition. This condition is called *Slater's condition*. Slater's condition, like regularity, is a condition on the constraints of the problem. We will say that Slater's condition is satisfied for a set of convex inequalities

$$g_i(\mathbf{x}) \leq 0, \quad i = 1, 2, \ldots, m,$$

where $g_1, g_2, \ldots, g_m$ are given convex functions if there exists $\hat{\mathbf{x}} \in \mathbb{R}^n$ such that

$$g_i(\hat{\mathbf{x}}) < 0, \quad i = 1, 2, \ldots, m.$$

Note that Slater's condition requires that there exists a point that strictly satisfies the constraints, and does not require, like in the regularity condition, an a priori knowledge on the point that is a candidate to be an optimal solution. This is the reason why checking the validity of Slater's condition is usually a much easier task than checking regularity.

Next, the necessity of the KKT conditions for problems with convex inequalities under Slater's condition is stated and proved.

11.3. The Convex Case

Theorem 11.13 (necessity of the KKT conditions under Slater's condition). *Let $\mathbf{x}^*$ be an optimal solution of the problem*

$$\begin{array}{ll} \min & f(\mathbf{x}) \\ \text{s.t.} & g_i(\mathbf{x}) \leq 0, \quad i = 1, 2, \ldots, m, \end{array} \quad (11.11)$$

where $f, g_1, \ldots, g_m$ are continuously differentiable functions over $\mathbb{R}^n$. In addition, $g_1, g_2, \ldots, g_m$ are convex functions over $\mathbb{R}^n$. Suppose that there exists $\hat{\mathbf{x}} \in \mathbb{R}^n$ such that

$$g_i(\hat{\mathbf{x}}) < 0, \quad i = 1, 2, \ldots, m.$$

Then there exist multipliers $\lambda_1, \lambda_2, \ldots, \lambda_m \geq 0$ such that

$$\nabla f(\mathbf{x}^*) + \sum_{i=1}^{m} \lambda_i \nabla g_i(\mathbf{x}^*) = 0, \quad (11.12)$$

$$\lambda_i g_i(\mathbf{x}^*) = 0, \quad i = 1, 2, \ldots, m. \quad (11.13)$$

Proof. By Theorem 11.4, since $\mathbf{x}^*$ is an optimal solution of (11.11), then the Fritz-John conditions are satisfied. That is, there exist $\tilde{\lambda}_0, \tilde{\lambda}_1, \ldots, \tilde{\lambda}_m \geq 0$, which are not all zeros, such that

$$\tilde{\lambda}_0 \nabla f(\mathbf{x}^*) + \sum_{i=1}^{m} \tilde{\lambda}_i \nabla g_i(\mathbf{x}^*) = 0,$$

$$\tilde{\lambda}_i g_i(\mathbf{x}^*) = 0, \quad i = 1, 2, \ldots, m. \quad (11.14)$$

All that we need to show is that $\tilde{\lambda}_0 > 0$, and then the conditions (11.12) and (11.13) will be satisfied with $\lambda_i = \frac{\tilde{\lambda}_i}{\tilde{\lambda}_0}, i = 1, 2, \ldots, m$. To prove that $\tilde{\lambda}_0 > 0$, assume in contradiction that it is zero; then

$$\sum_{i=1}^{m} \tilde{\lambda}_i \nabla g_i(\mathbf{x}^*) = 0. \quad (11.15)$$

By the gradient inequality we have that for all $i = 1, 2, \ldots, m$

$$0 > g_i(\hat{\mathbf{x}}) \geq g_i(\mathbf{x}^*) + \nabla g_i(\mathbf{x}^*)^T (\hat{\mathbf{x}} - \mathbf{x}^*).$$

Multiplying the ith inequality by $\tilde{\lambda}_i$ and summing over $i = 1, 2, \ldots, m$, we obtain

$$0 > \sum_{i=1}^{m} \tilde{\lambda}_i g_i(\mathbf{x}^*) + \left[\sum_{i=1}^{m} \tilde{\lambda}_i \nabla g_i(\mathbf{x}^*) \right]^T (\hat{\mathbf{x}} - \mathbf{x}^*), \quad (11.16)$$

where the inequality is strict since not all the $\tilde{\lambda}_i$ are zero. Plugging the identities (11.14) and (11.15) into (11.16), we obtain the impossible statement that $0 > 0$, thus establishing the result. $\square$

Example 11.14. Consider the convex optimization problem

$$\begin{array}{ll} \min & x_1^2 - x_2 \\ \text{s.t.} & x_2 = 0. \end{array}$$

The Lagrangian is

$$L(x_1, x_2, \lambda) = x_1^2 - x_2 + \lambda x_2.$$

Since the problem is a linearly constrained convex problem, the KKT conditions are necessary and sufficient (Theorem 10.7). The conditions are

$$2x_1 = 0,$$
$$-1 + \lambda = 0,$$
$$x_2 = 0,$$

and they are satisfied for $(x_1, x_2) = (0, 0)$, which is the optimal solution.

Now, consider a different reformulation of the problem:

$$\begin{aligned} \min \quad & x_1^2 - x_2 \\ \text{s.t.} \quad & x_2^2 \leq 0. \end{aligned}$$

Slater's condition is not satisfied since the constraint cannot be satisfied strictly, and therefore the KKT conditions are not guaranteed to hold at the optimal solution. The KKT conditions in this case are

$$2x_1 = 0,$$
$$-1 + 2\lambda x_2 = 0,$$
$$\lambda x_2^2 = 0,$$
$$x_2^2 \leq 0,$$
$$\lambda \geq 0.$$

The above system is infeasible since $x_2 = 0$, and hence the equality $-1 + 2\lambda x_2 = 0$ is impossible. ∎

A slightly more refined analysis can show that in the presence of affine constraints, one can prove the necessity of the KKT conditions under a generalized Slater's condition which states that there exists a point that strictly satisfies all the nonlinear inequality constraints as well as satisfies the affine equality and inequality constraints.

Definition 11.15 (generalized Slater's condition). *Consider the system*

$$g_i(\mathbf{x}) \leq 0, \quad i = 1, 2, \ldots, m,$$
$$h_j(\mathbf{x}) \leq 0, \quad j = 1, 2, \ldots, p,$$
$$s_k(\mathbf{x}) = 0, \quad k = 1, 2, \ldots, q,$$

where $g_i, i = 1, 2, \ldots, m$, are convex functions and $h_j, s_k, j = 1, 2, \ldots, p, k = 1, 2, \ldots, q$, are affine functions. Then we say that the **generalized Slater's condition** *is satisfied if there exists $\hat{\mathbf{x}} \in \mathbb{R}^n$ for which*

$$g_i(\hat{\mathbf{x}}) < 0, \quad i = 1, 2, \ldots, m,$$
$$h_j(\hat{\mathbf{x}}) \leq 0, \quad j = 1, 2, \ldots, p,$$
$$s_k(\hat{\mathbf{x}}) = 0, \quad k = 1, 2, \ldots, q.$$

The necessity of the KKT conditions under the generalized Slater's condition is now stated.

11.3. The Convex Case

Theorem 11.16 (necessity of the KKT conditions under the generalized Slater's condition). *Let $\mathbf{x}^*$ be an optimal solution of the problem*

$$\begin{aligned}
\min \quad & f(\mathbf{x}) \\
\text{s.t.} \quad & g_i(\mathbf{x}) \leq 0, \quad i=1,2,\ldots,m, \\
& h_j(\mathbf{x}) \leq 0, \quad j=1,2,\ldots,p, \\
& s_k(\mathbf{x}) = 0, \quad k=1,2,\ldots,q,
\end{aligned}$$

where $f, g_1, \ldots, g_m$ are continuously differentiable convex functions over $\mathbb{R}^n$, and $h_j, s_k, j = 1,2,\ldots,p, k = 1,2,\ldots,q$, are affine. Suppose that there exists $\hat{\mathbf{x}} \in \mathbb{R}^n$ such that

$$\begin{aligned}
g_i(\hat{\mathbf{x}}) &< 0, \quad i=1,2,\ldots,m, \\
h_j(\hat{\mathbf{x}}) &\leq 0, \quad j=1,2,\ldots,p, \\
s_k(\hat{\mathbf{x}}) &= 0, \quad k=1,2,\ldots,q.
\end{aligned}$$

Then there exist multipliers $\lambda_1, \lambda_2, \ldots, \lambda_m, \eta_1, \eta_2, \ldots, \eta_p \geq 0, \mu_1, \mu_2, \ldots, \mu_q \in \mathbb{R}$ such that

$$\nabla f(\mathbf{x}^*) + \sum_{i=1}^m \lambda_i \nabla g_i(\mathbf{x}^*) + \sum_{j=1}^p \eta_j \nabla h_j(\mathbf{x}^*) + \sum_{k=1}^q \mu_k \nabla s_k(\mathbf{x}^*) = 0,$$

$$\begin{aligned}
\lambda_i g_i(\mathbf{x}^*) &= 0, \quad i=1,2,\ldots,m, \\
\eta_j h_j(\mathbf{x}^*) &= 0, \quad j=1,2,\ldots,p.
\end{aligned}$$

Example 11.17. Consider the convex optimization problem

$$\begin{aligned}
\min \quad & 4x_1^2 + x_2^2 - x_1 - 2x_2 \\
\text{s.t.} \quad & 2x_1 + x_2 \leq 1, \\
& x_1^2 \leq 1.
\end{aligned}$$

Slater's condition is satisfied with $(\hat{x}_1, \hat{x}_2) = (0,0)$ ($2 \cdot 0 + 0 < 1, 0^2 - 1 < 0$), so the KKT conditions are necessary and sufficient. The Lagrangian function is

$$L(x_1, x_2, \lambda_1, \lambda_2) = 4x_1^2 + x_2^2 - x_1 - 2x_2 + \lambda_1(2x_1 + x_2 - 1) + \lambda_2(x_1^2 - 1),$$

and the KKT system is

$$\frac{\partial L}{\partial x_1} = 8x_1 - 1 + 2\lambda_1 + 2\lambda_2 x_1 = 0, \quad (11.17)$$

$$\frac{\partial L}{\partial x_2} = 2x_2 - 2 + \lambda_1 = 0, \quad (11.18)$$

$$\begin{aligned}
\lambda_1(2x_1 + x_2 - 1) &= 0, \\
\lambda_2(x_1^2 - 1) &= 0, \\
2x_1 + x_2 &\leq 1, \\
x_1^2 &\leq 1, \\
\lambda_1, \lambda_2 &\geq 0.
\end{aligned}$$

We will consider four cases.
Case I: If $\lambda_1 = \lambda_2 = 0$, then by the first two equations, $x_1 = \frac{1}{8}, x_2 = 1$, which is not a feasible solution.

Case II: If $\lambda_1 > 0, \lambda_2 > 0$, then by the complementary slackness conditions

$$2x_1 + x_2 = 1,$$
$$x_1^2 = 1.$$

The two solutions of this system are $(1,-1),(-1,3)$. Plugging the first solution into the first equation of the KKT system (equation (11.17)) yields

$$7 + 2\lambda_1 + 2\lambda_2 = 0,$$

which is impossible since $\lambda_1, \lambda_2 > 0$. Plugging the second solution into the second equation of the KKT system (equation (11.18)) results in the equation

$$4 + \lambda_1 = 0,$$

which has no solution since $\lambda_1 > 0$.

Case III: If $\lambda_1 > 0, \lambda_2 = 0$, then by the complementary slackness conditions we have that

$$2x_1 + x_2 = 1, \qquad (11.19)$$

which combined with (11.17) and (11.18) yields the set of equations (recalling that $\lambda_2 = 0$)

$$8x_1 - 1 + 2\lambda_1 = 0,$$
$$2x_2 - 2 + \lambda_1 = 0,$$
$$2x_1 + x_2 = 1,$$

whose unique solution is $(x_1, x_2, \lambda_1) = (\frac{1}{16}, \frac{7}{8}, \frac{1}{4})$, and we obtain that $(x_1, x_2, \lambda_1, \lambda_2) = (\frac{1}{16}, \frac{7}{8}, \frac{1}{4}, 0)$ satisfies the KKT system. Hence, $(x_1, x_2) = (\frac{1}{16}, \frac{7}{8})$ is a KKT point, and since the problem is convex, it is an optimal solution. In principle, we do not have to check the fourth case since we already found an optimal solution, but it might be that there exist additional optimal solutions, and if our objective is to find *all* the optimal solutions, then all the cases should be covered.

Case IV: If $\lambda_1 = 0, \lambda_2 > 0$, then by the complementary slackness conditions we have $x_1^2 = 1$. By (11.18) we have that $x_2 = 1$. The two candidate solutions in this case are therefore $(1,1)$ and $(-1,1)$. The point $(1,1)$ does not satisfy the first constraint of the problem and is therefore infeasible. Plugging $(x_1, x_2) = (-1, 1)$ and $\lambda_1 = 0$ into (11.17) yields

$$-9 - 2\lambda_2 = 0,$$

which is a contradiction to the positivity of λ_2.

To conclude, the unique optimal solution of the problem is $(x_1, x_2) = (\frac{1}{16}, \frac{7}{8})$. ∎

11.4 ▪ Constrained Least Squares

Consider the problem

$$\text{(CLS)} \quad \begin{array}{ll} \min & \|\mathbf{A}\mathbf{x} - \mathbf{b}\|^2 \\ \text{s.t.} & \|\mathbf{x}\|^2 \leq \alpha, \end{array}$$

where $\mathbf{A} \in \mathbb{R}^{m \times n}$ is assumed to be of full column rank, $\mathbf{b} \in \mathbb{R}^m$, and $\alpha > 0$. We will refer to this problem as a *constrained least squares (CLS)* problem. In Section 3.3 we considered

11.4. Constrained Least Squares

the regularized least squares (RLS) problem which has the form[5] $\min\{\|\mathbf{Ax}-\mathbf{b}\|^2+\mu\|\mathbf{x}\|^2\}$. The two problems are related in the sense that they both regularize the least squares solution by a quadratic regularization term. In the RLS problem, the regularization is done by a penalty function, while in the CLS problem the regularization is performed by incorporating it as a constraint.

Problem (CLS) is a convex problem and satisfies Slater's condition since $\hat{\mathbf{x}}=0$ strictly satisfies the constraint of the problem. To solve the problem, we begin by forming the Lagrangian:

$$L(\mathbf{x},\lambda) = \|\mathbf{Ax}-\mathbf{b}\|^2 + \lambda(\|\mathbf{x}\|^2-\alpha) \quad (\lambda \geq 0).$$

The KKT conditions are

$$\nabla_{\mathbf{x}} L = 2\mathbf{A}^T(\mathbf{Ax}-\mathbf{b}) + 2\lambda\mathbf{x} = 0,$$
$$\lambda(\|\mathbf{x}\|^2-\alpha) = 0,$$
$$\|\mathbf{x}\|^2 \leq \alpha,$$
$$\lambda \geq 0.$$

There are two options. In the first, $\lambda = 0$, and then by the first equation we have that

$$\mathbf{x} = \mathbf{x}_{LS} \equiv (\mathbf{A}^T\mathbf{A})^{-1}\mathbf{A}^T\mathbf{b}.$$

This is a KKT point and hence the optimal solution if and only if $\mathbf{x}_{LS}$ is feasible, that is, if $\|\mathbf{x}_{LS}\|^2 \leq \alpha$. This is not a surprising result since it is clear that when the unconstrained minimizer ($\mathbf{x}_{LS}$) satisfies the constraint, it is also the optimal solution of the constrained problem.

On the other hand, if $\|\mathbf{x}_{LS}\|^2 > \alpha$, then $\lambda > 0$. By the complementary slackness condition we have that $\|\mathbf{x}\|^2 = \alpha$, and the first equation implies that

$$\mathbf{x} = \mathbf{x}_\lambda \equiv (\mathbf{A}^T\mathbf{A} + \lambda\mathbf{I})^{-1}\mathbf{A}^T\mathbf{b}.$$

The multiplier $\lambda > 0$ should be thus chosen to satisfy $\|\mathbf{x}_\lambda\|^2 = \alpha$; that is, λ is the solution of

$$f(\lambda) \equiv \|(\mathbf{A}^T\mathbf{A} + \lambda\mathbf{I})^{-1}\mathbf{A}^T\mathbf{b}\|^2 - \alpha = 0. \tag{11.20}$$

We have $f(0) = \|(\mathbf{A}^T\mathbf{A})^{-1}\mathbf{A}^T\mathbf{b}\|^2 - \alpha = \|\mathbf{x}_{LS}\|^2 - \alpha > 0$, and it is not difficult to show that f is a strictly decreasing function satisfying $f(\lambda) \to -\alpha$ as $\lambda \to \infty$. Thus, there exists a unique λ for which $f(\lambda) = 0$, and this λ can be found for example by a simple bisection procedure. To conclude, the optimal solution of the CLS problem is given by

$$\mathbf{x} = \begin{cases} \mathbf{x}_{LS}, & \|\mathbf{x}_{LS}\|^2 \leq \alpha, \\ (\mathbf{A}^T\mathbf{A} + \lambda\mathbf{I})^{-1}\mathbf{A}^T\mathbf{b} & \|\mathbf{x}_{LS}\|^2 > \alpha, \end{cases}$$

where λ is the unique root of f over $[0,\infty)$. We will now construct a MATLAB function for solving the CLS problem. For that, we need to write a MATLAB function that performs a bisection algorithm. The bisection method for finding the root of a scalar equation $f(x) = 0$ is described below.

---

[5] In Section 3.3 we actually considered a more general model in which the regularizer had the form $\mu\|\mathbf{Dx}\|^2$, where $\mathbf{D}$ is a given matrix.

> **Bisection**
>
> **Input:** $\varepsilon > 0$ - tolerance parameter. $a < b$ - two numbers satisfying $f(a)f(b) < 0$.
>
> **Initialization:** Take $l_0 = a, u_0 = b$.
> **General step:** For any $k = 0, 1, 2, \ldots$ execute the following steps:
>
> (a) Take $x_k = \frac{u_k + l_k}{2}$.
>
> (b) If $f(l_k) \cdot f(x_k) > 0$, define $l_{k+1} = x_k, u_{k+1} = u_k$. Otherwise, define $l_{k+1} = l_k, u_{k+1} = x_k$.
>
> (c) if $u_{k+1} - l_{k+1} \leq \varepsilon$, then STOP, and x_k is the output.

A MATLAB function implementing the bisection method is given below.

```
function z=bisection(f,lb,ub,eps)
%INPUT
%=================
%f .................. a scalar function
%lb ................. the initial lower bound
%ub ................. the initial upper bound
%eps ................ tolerance parameter
%OUTPUT
%=================
% z .................. a root of the equation f(x)=0
if (f(lb)*f(ub)>0)
    error('f(lb)*f(ub)>0')
end

iter=0;
while (ub-lb>eps)
    z=(lb+ub)/2;
    iter=iter+1;
    if(f(lb)*f(z)>0)
        lb=z;
    else
        ub=z;
    end
    fprintf('iter_numer = %3d current_sol = %2.6f \n',iter,z);
end
```

Therefore, for example, if we wish to find the square root of 2 with an accuracy of 10^{-4}, then we can solve the equation $x^2 - 2 = 0$ by the following MATLAB command:

```
>> bisection(@(x)x^2-2,1,2,1e-4);
iter_numer =    1 current_sol = 1.500000
iter_numer =    2 current_sol = 1.250000
iter_numer =    3 current_sol = 1.375000
iter_numer =    4 current_sol = 1.437500
```

11.4. Constrained Least Squares

```
iter_numer =    5 current_sol = 1.406250
iter_numer =    6 current_sol = 1.421875
iter_numer =    7 current_sol = 1.414063
iter_numer =    8 current_sol = 1.417969
iter_numer =    9 current_sol = 1.416016
iter_numer =   10 current_sol = 1.415039
iter_numer =   11 current_sol = 1.414551
iter_numer =   12 current_sol = 1.414307
iter_numer =   13 current_sol = 1.414185
iter_numer =   14 current_sol = 1.414246
```

As for the CLS problem, note that the scalar function f given in (11.20) satisfies $f(0) > 0$. Therefore, all that is left is to find a point $u > 0$ satisfying $f(u) < 0$. For that, we will start with guessing $u = 1$ and then make the update $u \leftarrow 2u$ until $f(u) < 0$. The MATLAB function implementing these ideas is given below.

```
function x_cls=cls(A,b,alpha)
%INPUT
%=================
%A .................. an mxn matrix
%b .................. an m-length vector
%alpha .............. positive scalar
%OUTPUT
%=================
% x_cls ............ an optimal solution of
%                    min{||A*x-b||:||x||^2<=alpha}
d=size(A);
n=d(2);
x_ls=A\b;
if (norm(x_ls)^2<=alpha)
    x_cls=x_ls;
else
    f=@(lam) norm((A'*A+lam*eye(n))\(A'*b))^2-alpha;
    u=1;
    while (f(u)>0)
        u=2*u;
    end
    lam=bisection(f,0,u,1e-7);
    x_cls=(A'*A+lam*eye(n))\(A'*b);
end
```

For example, assume that we pick **A** and **b** as

```
A=[1,2;3,1;2,3];
b=[2;3;4];
```

The least squares solution and its squared norm can be easily found:

```
>> x_ls=A\b
x_ls =

    0.7600
    0.7600
```

```
>> norm(x_ls)^2

ans =

    1.1552
```

If we use the `cls` function with an α which is greater than 1.1552, then we will obviously get back the least squares solution:

```
>> cls(A,b,1.5)
ans =

    0.7600
    0.7600
```

On the other hand, taking an α with a smaller value than 1.552, will result in a different solution (the bisection output was suppressed):

```
>> cls(A,b,0.5)
ans =

    0.5000
    0.5000
```

To double check the result, we can run CVX,

```
cvx_begin
variable x_cvx(2)
minimize(norm(A*x_cvx-b))
norm(x_cvx)<=sqrt(0.5)
cvx_end
```

and get the same result:

```
>> x_cvx
x_cvx =

    0.5000
    0.5000
```

11.5 ▪ Second Order Optimality Conditions

11.5.1 ▪ Necessary Conditions for Inequality Constrained Problems

We can also establish necessary second order optimality conditions in the general nonconvex case. We will begin by stating and proving the result for inequality constrained problem.

Theorem 11.18 (second order necessary conditions for inequality constrained problems). *Consider the problem*

$$\begin{aligned} \min \quad & f_0(\mathbf{x}) \\ \text{s.t.} \quad & f_i(\mathbf{x}) \leq 0, \quad i = 1, 2, \ldots, m, \end{aligned} \quad (11.21)$$

11.5. Second Order Optimality Conditions

where $f_0, f_1, \ldots, f_m$ are twice continuously differentiable over $\mathbb{R}^n$. Let $\mathbf{x}^*$ be a local minimum of problem (11.21), and suppose that $\mathbf{x}^*$ is regular, meaning that the set $\{\nabla f_i(\mathbf{x}^*)\}_{i \in I(\mathbf{x}^*)}$ is linearly independent, where

$$I(\mathbf{x}^*) = \{i \in \{1, 2, \ldots, m\} : f_i(\mathbf{x}^*) = 0\}.$$

Denote the Lagrangian by

$$L(\mathbf{x}, \lambda) = f_0(\mathbf{x}) + \sum_{i=1}^{m} \lambda_i f_i(\mathbf{x}).$$

Then there exist $\lambda_1, \lambda_2, \ldots, \lambda_m \geq 0$ such that

$$\nabla_\mathbf{x} L(\mathbf{x}^*, \lambda) = 0,$$
$$\lambda_i f_i(\mathbf{x}^*) = 0, \quad i = 1, 2, \ldots, m,$$

and

$$\mathbf{y}^T \nabla_{\mathbf{xx}}^2 L(\mathbf{x}^*, \lambda) \mathbf{y} = \mathbf{y}^T \left[\nabla^2 f_0(\mathbf{x}^*) + \sum_{i=1}^{m} \lambda_i \nabla^2 f_i(\mathbf{x}^*) \right] \mathbf{y} \geq 0$$

for all $\mathbf{y} \in \Lambda(\mathbf{x}^*)$, where

$$\Lambda(\mathbf{x}^*) \equiv \{\mathbf{d} \in \mathbb{R}^n : \nabla f_i(\mathbf{x}^*)^T \mathbf{d} = 0, i \in I(\mathbf{x}^*)\}.$$

Proof. Let $\mathbf{d} \in D(\mathbf{x}^*)$, where

$$D(\mathbf{x}^*) = \{\mathbf{d} \in \mathbb{R}^n : \nabla f_i(\mathbf{x}^*)^T \mathbf{d} \leq 0, i \in I(\mathbf{x}^*) \cup \{0\}\}.$$

From this point until further notice we will assume that $\mathbf{d}$ is fixed. Let $\mathbf{z} \in \mathbb{R}^n$, and $i \in \{0, 1, 2, \ldots, m\}$. Define

$$\mathbf{x}(t) \equiv \mathbf{x}^* + t\mathbf{d} + \frac{t^2}{2}\mathbf{z}$$

and the one-dimensional functions

$$g_i(t) \equiv f_i(\mathbf{x}(t)), \quad i \in I(\mathbf{x}^*) \cup \{0\}.$$

Then

$$g_i'(t) = (\mathbf{d} + t\mathbf{z})^T \nabla f_i(\mathbf{x}(t)),$$
$$g_i''(t) = (\mathbf{d} + t\mathbf{z})^T \nabla^2 f_i(\mathbf{x}(t))(\mathbf{d} + t\mathbf{z}) + \mathbf{z}^T \nabla f_i(\mathbf{x}(t)),$$

which in particular implies that

$$g_i'(0) = \nabla f_i(\mathbf{x}^*)^T \mathbf{d},$$
$$g_i''(0) = \mathbf{d}^T \nabla^2 f_i(\mathbf{x}^*) \mathbf{d} + \nabla f_i(\mathbf{x}^*)^T \mathbf{z}.$$

By the quadratic approximation theorem (Theorem 1.25) we have

$$g_i(t) = f_i(\mathbf{x}^*) + (\nabla f_i(\mathbf{x}^*)^T \mathbf{d})t + \left(\mathbf{d}^T \nabla^2 f_i(\mathbf{x}^*)\mathbf{d} + \nabla f_i(\mathbf{x}^*)^T \mathbf{z}\right)\frac{t^2}{2} + o(t^2). \quad (11.22)$$

Therefore, for any $i \in I(\mathbf{x}^*) \cup \{0\}$ there are two cases:

1. $\nabla f_i(\mathbf{x}^*)^T \mathbf{d} < 0$, and in this case $f_i(\mathbf{x}(t)) < f_i(\mathbf{x}^*)$ for small enough $t > 0$.

2. $\nabla f_i(\mathbf{x}^*)^T \mathbf{d} = 0$. In this case, by (11.22), if $\nabla f_i(\mathbf{x}^*)^T \mathbf{z} + \mathbf{d}^T \nabla^2 f_i(\mathbf{x}^*)\mathbf{d} < 0$, then $f_i(\mathbf{x}(t)) < f_i(\mathbf{x}^*)$ for small enough t.

As a conclusion, since $\mathbf{x}^*$ is a local minimum of problem (11.21), the following system of strict inequalities in $\mathbf{z}$ (recall that $\mathbf{d}$ is fixed) does not have a solution:

$$\nabla f_i(\mathbf{x}^*)^T \mathbf{z} + \mathbf{d}^T \nabla^2 f_i(\mathbf{x}^*) \mathbf{d} < 0, \quad i \in J(\mathbf{x}^*) \cup \{0\}, \tag{11.23}$$

where

$$J(\mathbf{x}^*) = \{i \in I(\mathbf{x}^*) : \nabla f_i(\mathbf{x}^*)^T \mathbf{d} = 0\}.$$

Indeed, if there was a solution to system (11.23), then for small enough t, the vector $\mathbf{x}(t)$ would be a feasible solution satisfying $f_0(\mathbf{x}(t)) < f_0(\mathbf{x}^*)$, contradicting the local optimality of $\mathbf{x}^*$. System (11.23) can be written as

$$\mathbf{Az} < \mathbf{b},$$

where $\mathbf{A}$ is the matrix whose components are $\nabla f_i(\mathbf{x}^*)^T, i \in J(\mathbf{x}^*) \cup \{0\}$, and $\mathbf{b}$ is the vector whose components are $-\mathbf{d}^T \nabla^2 f_i(\mathbf{x}^*) \mathbf{d}, i \in J(\mathbf{x}^*) \cup \{0\}$. By the nonhomogenous Gordan's theorem (see Exercise 10.5), we have that there exists $\mathbf{y}$ such that $\mathbf{A}^T \mathbf{y} = \mathbf{0}, \mathbf{b}^T \mathbf{y} \leq 0, \mathbf{y} \geq 0, \mathbf{y} \neq \mathbf{0}$, meaning that there exist $0 \leq y_i, i \in J(\mathbf{x}^*) \cup \{0\}$, not all zeros, such that

$$\sum_{i \in J(\mathbf{x}^*) \cup \{0\}} y_i \nabla f_i(\mathbf{x}^*) = \mathbf{0} \tag{11.24}$$

and

$$\sum_{i \in J(\mathbf{x}^*) \cup \{0\}} y_i (-\mathbf{d}^T \nabla^2 f_i(\mathbf{x}^*) \mathbf{d}) \leq 0,$$

that is,

$$\mathbf{d}^T \left[\sum_{i \in J(\mathbf{x}^*) \cup \{0\}} y_i \nabla^2 f_i(\mathbf{x}^*) \right] \mathbf{d} \geq 0.$$

By the regularity of $\mathbf{x}^*$ and (11.24), we have that $y_0 > 0$, and hence by defining $\lambda_i = \frac{y_i}{y_0}$ for $i \in J(\mathbf{x}^*)$ and $\lambda_i = 0$ for $i \in I(\mathbf{x}^*) \backslash J(\mathbf{x}^*)$, we obtain that

$$\nabla f_0(\mathbf{x}^*) + \sum_{i \in I(\mathbf{x}^*)} \lambda_i \nabla f_i(\mathbf{x}^*) = \mathbf{0}, \tag{11.25}$$

$$\mathbf{d}^T \left[\nabla^2 f_0(\mathbf{x}^*) + \sum_{i \in I(\mathbf{x}^*)} \lambda_i \nabla^2 f_i(\mathbf{x}^*) \right] \mathbf{d} \geq 0.$$

Since $\{\nabla f_i(\mathbf{x}^*)\}_{i \in I(\mathbf{x}^*)}$ are linearly independent, equation (11.25) implies that the multipliers $\lambda_i, i \in I(\mathbf{x}^*)$, do not depend on the initial choice of $\mathbf{d}$. Therefore, by defining $\lambda_i = 0$ for any $i \notin I(\mathbf{x}^*)$, we obtain that

$$\nabla f_0(\mathbf{x}^*) + \sum_{i=1}^m \lambda_i \nabla f_i(\mathbf{x}^*) = \mathbf{0},$$

$$\lambda_i f_i(\mathbf{x}^*) = 0, \quad i = 1, 2, \ldots, m,$$

$$\mathbf{d}^T \left[\nabla^2 f_0(\mathbf{x}^*) + \sum_{i=1}^m \lambda_i \nabla^2 f_i(\mathbf{x}^*) \right] \mathbf{d} \geq 0 \tag{11.26}$$

for any $\mathbf{d} \in D(\mathbf{x}^*)$. All that is left to prove is that (11.26) is satisfied for all $\mathbf{d} \in \Lambda(\mathbf{x}^*)$. Indeed, if $\mathbf{d} \in \Lambda(\mathbf{x}^*)$, then either $\mathbf{d}$ or $-\mathbf{d}$ is in $D(\mathbf{x}^*)$. Thus, $c\mathbf{d} \in D(\mathbf{x}^*)$ for some $c \in \{-1, 1\}$, and as a result

$$(c\mathbf{d})^T \left[\nabla^2 f_0(\mathbf{x}^*) + \sum_{i=1}^m \lambda_i \nabla^2 f_i(\mathbf{x}^*) \right] (c\mathbf{d}) \geq 0,$$

which is the same as

$$\mathbf{d}^T\left[\nabla^2 f_0(\mathbf{x}^*) + \sum_{i=1}^m \lambda_i \nabla^2 f_i(\mathbf{x}^*)\right]\mathbf{d} \geq 0,$$

and the result is established. □

11.5.2 ▪ Necessary Second Order Conditions for Equality and Inequality Constrained Problems

When the problem involves both equality and inequality constraints, a similar result can be proved, and it is stated without a proof below.

Theorem 11.19 (second order necessary conditions for equality and inequality constrained problems). *Consider the problem*

$$\begin{aligned}
\min \quad & f(\mathbf{x}) \\
\text{s.t.} \quad & g_i(\mathbf{x}) \leq 0, \quad i = 1, 2, \ldots, m, \\
& h_j(\mathbf{x}) = 0, \quad j = 1, 2, \ldots, p,
\end{aligned} \quad (11.27)$$

where $f, g_1, \ldots, g_m, h_1, \ldots, h_p$ are twice continuously differentiable over $\mathbb{R}^n$. Let $\mathbf{x}^*$ be a local minimum of problem (11.27), and suppose that $\mathbf{x}^*$ is regular, meaning that $\{\nabla g_i(\mathbf{x}^*), \nabla h_j(\mathbf{x}^*), i \in I(\mathbf{x}^*), j = 1, 2, \ldots, p\}$ are linearly independent, where

$$I(\mathbf{x}^*) = \{i \in \{1, 2, \ldots, m\} : g_i(\mathbf{x}^*) = 0\}.$$

Then there exist $\lambda_1, \lambda_2, \ldots, \lambda_m \geq 0$ and $\mu_1, \mu_2, \ldots, \mu_p \in \mathbb{R}$ such that

$$\nabla f(\mathbf{x}^*) + \sum_{i=1}^m \lambda_i \nabla g_i(\mathbf{x}^*) + \sum_{j=1}^p \mu_j \nabla h_j(\mathbf{x}^*) = 0,$$

$$\lambda_i g_i(\mathbf{x}^*) = 0, \quad i = 1, 2, \ldots, m,$$

and

$$\mathbf{d}^T\left[\nabla^2 f_0(\mathbf{x}^*) + \sum_{i=1}^m \lambda_i \nabla^2 g_i(\mathbf{x}^*) + \sum_{j=1}^p \mu_j \nabla^2 h_j(\mathbf{x}^*)\right]\mathbf{d} \geq 0$$

for all $\mathbf{d} \in \Lambda(\mathbf{x}^*)$ where

$$\Lambda(\mathbf{x}^*) \equiv \{\mathbf{d} \in \mathbb{R}^n : \nabla g_i(\mathbf{x}^*)^T \mathbf{d} = 0, \nabla h_j(\mathbf{x}^*)^T \mathbf{d} = 0, i \in I(\mathbf{x}^*), j = 1, 2, \ldots, p\}.$$

Example 11.20. Consider the problem

$$\begin{aligned}
\min \quad & (2x_1 - 1)^2 + x_2^2 \\
\text{s.t.} \quad & h(x_1, x_2) \equiv -2x_1 + x_2^2 = 0.
\end{aligned}$$

We first note that since the problem consists of minimizing a coercive objective function over a closed set, it follows that an optimal solution does exist. In addition, there are no irregular points to the problem since the gradient of the constraint function h is always different from the zeros vector. Therefore, the optimal solution is one of the KKT points of the problem. The Lagrangian of the problem is

$$L(x_1, x_2, \mu) = (2x_1 - 1)^2 + x_2^2 + \mu(-2x_1 + x_2^2),$$

and the KKT system is

$$\frac{\partial L}{\partial x_1} = 4(2x_1 - 1) - 2\mu = 0,$$

$$\frac{\partial L}{\partial x_2} = 2x_2 + 2\mu x_2 = 0,$$

$$-2x_1 + x_2^2 = 0.$$

By the second equation

$$2(1+\mu)x_2 = 0,$$

and hence there are two cases. In one case, $x_2 = 0$; then by the third equation, $x_1 = 0$, and by the first equation $\mu = -2$. The second case is when $\mu = -1$, and then by the first equation, $4(2x_1 - 1) = -2$, that is, $x_1 = \frac{1}{4}$, and then $x_2 = \pm \frac{1}{\sqrt{2}}$. We thus obtain that there are three KKT points: $(x_1, x_2, \mu) = (0, 0, -2), (\frac{1}{4}, \frac{1}{\sqrt{2}}, -1), (\frac{1}{4}, -\frac{1}{\sqrt{2}}, -1)$.

Note that

$$\nabla^2_{xx} L(x_1, x_2, \mu) = \begin{pmatrix} 8 & 0 \\ 0 & 2(1+\mu) \end{pmatrix}.$$

Note that for the points $(x_1, x_2, \mu) = (\frac{1}{4}, \frac{1}{\sqrt{2}}, -1), (\frac{1}{4}, -\frac{1}{\sqrt{2}}, -1)$ the Hessian of the Lagrangian is positive semidefinite:

$$\nabla^2_{xx} L(x_1, x_2, \mu) = \begin{pmatrix} 8 & 0 \\ 0 & 0 \end{pmatrix} \succeq 0.$$

Therefore, these points satisfy the second order necessary conditions. On the other hand, for the first point $(0,0)$ where $\mu = -2$, the Hessian is given by

$$\nabla^2_{xx} L(x_1, x_2, \mu) = \begin{pmatrix} 8 & 0 \\ 0 & -2 \end{pmatrix},$$

which is not a positive semidefinite matrix. To check the validity of the second order conditions at $(0,0)$, note that

$$\nabla h(x_1, x_2) = \begin{pmatrix} -2 \\ 2x_2 \end{pmatrix},$$

and thus $\nabla h(0,0) = \binom{-2}{0}$. We therefore need to check whether

$$\mathbf{d}^T \nabla^2_{xx} L(x_1, x_2, \mu) \mathbf{d} \geq 0 \text{ for all } \mathbf{d} \text{ satisfying } \nabla h(0,0)^T \mathbf{d} = 0.$$

Since the condition $\nabla h(0,0)^T \mathbf{d} = 0$ translates to $d_1 = 0$, we need to check whether

$$\begin{pmatrix} 0 & d_2 \end{pmatrix} \nabla^2_{xx} L(x_1, x_2, \mu) \begin{pmatrix} 0 \\ d_2 \end{pmatrix} \geq 0$$

for any d_2. However, the latter inequality is equivalent to saying that $-2d_2^2 \geq 0$ for any d_2, which is of course not correct.

The conclusion is that $(0,0)$ does not satisfy the second order necessary conditions and hence cannot be an optimal solution. The optimal solution must be either $(\frac{1}{4}, \frac{1}{\sqrt{2}})$ or $(\frac{1}{4}, -\frac{1}{\sqrt{2}})$ (or both). Since the points have the same objective function value, it follows that they are the optimal solutions of the problem. ∎

11.6 • Optimality Conditions for the Trust Region Subproblem

In Section 8.2.7 we considered the trust region subproblem (TRS) in which one minimizes a (possibly) nonconvex quadratic function subject to a norm constraint:

$$(\text{TRS}): \quad \min\{f(\mathbf{x}) \equiv \mathbf{x}^T \mathbf{A} \mathbf{x} + 2\mathbf{b}^T \mathbf{x} + c : \|\mathbf{x}\|^2 \leq \alpha\},$$

where $\mathbf{A} = \mathbf{A}^T \in \mathbb{R}^{n \times n}, \mathbf{b} \in \mathbb{R}^n, c \in \mathbb{R}$, and $\alpha \in \mathbb{R}_{++}$. Note that here we consider a slight extension of the model given in Section 8.2.7 since the norm constraint has a general upper bound and not 1. We have seen that the problem can be recast as a convex optimization problem. In this section we will look at another aspect of the "easiness" of the problem: the problem possesses necessary and sufficient optimality conditions. We will show that these optimality conditions can be used to develop an algorithm for solving the problem. We begin by stating the necessary and sufficient conditions.

Theorem 11.21 (necessary and sufficient conditions for problem (TRS)). *A vector $\mathbf{x}^*$ is an optimal solution of problem (TRS) if and only if there exists $\lambda^* \geq 0$ such that*

$$(\mathbf{A} + \lambda^* \mathbf{I})\mathbf{x}^* = -\mathbf{b}, \tag{11.28}$$

$$\|\mathbf{x}^*\|^2 \leq \alpha, \tag{11.29}$$

$$\lambda^*(\|\mathbf{x}^*\|^2 - \alpha) = 0, \tag{11.30}$$

$$\mathbf{A} + \lambda^* \mathbf{I} \succeq 0. \tag{11.31}$$

Proof. Sufficiency: To prove the sufficiency, let us assume that $\mathbf{x}^*$ satisfies (11.28)–(11.31) for some $\lambda^* \geq 0$. Define the function

$$h(\mathbf{x}) = f(\mathbf{x}) + \lambda^*(\|\mathbf{x}\|^2 - \alpha) = \mathbf{x}^T(\mathbf{A} + \lambda^* \mathbf{I})\mathbf{x} + 2\mathbf{b}^T \mathbf{x} + c - \alpha \lambda^*. \tag{11.32}$$

Then by (11.31) h is a convex quadratic function. By (11.28) it follows that $\nabla h(\mathbf{x}^*) = 0$, which combined with the convexity of h implies that $\mathbf{x}^*$ is an unconstrained minimizer of h over $\mathbb{R}^n$ (see Proposition 7.8). Let $\mathbf{x}$ be a feasible point, that is, $\|\mathbf{x}\|^2 \leq \alpha$. Then

$$\begin{aligned}
f(\mathbf{x}) &\geq f(\mathbf{x}) + \lambda^*(\|\mathbf{x}\|^2 - \alpha) && (\lambda^* \geq 0, \|\mathbf{x}\|^2 - \alpha \leq 0) \\
&= h(\mathbf{x}) && (\text{by (11.32)}) \\
&\geq h(\mathbf{x}^*) && (\mathbf{x}^* \text{ is a minimizer of } h) \\
&= f(\mathbf{x}^*) + \lambda^*(\|\mathbf{x}^*\|^2 - \alpha) \\
&= f(\mathbf{x}^*) && (\text{by (11.30)})
\end{aligned}$$

and we have established that $\mathbf{x}^*$ is a global optimal solution of the problem.

Necessity: To prove the necessity, note that the second order necessary optimality conditions are satisfied since all the feasible points of (TRS) are regular. Indeed, the regularity condition states that when the constraint is active, that is, when $\|\mathbf{x}^*\|^2 = \alpha$, the gradient of the constraint is not the zeros vector, and indeed, since the gradient in this case is $2\mathbf{x}^*$, it is not equal to the zeros vector (since $\|\mathbf{x}^*\|^2 = \alpha$).

If $\|\mathbf{x}^*\| < \alpha$, then the second order necessary optimality conditions (see Theorem 11.18) are exactly (11.28)–(11.31). If $\|\mathbf{x}^*\|^2 = \alpha$, then by the second order necessary optimality conditions there exists $\lambda^* \geq 0$ such that

$$(\mathbf{A} + \lambda^* \mathbf{I})\mathbf{x}^* = -\mathbf{b} \tag{11.33}$$

$$\|\mathbf{x}^*\|^2 \leq \alpha, \tag{11.34}$$

$$\lambda^*(\|\mathbf{x}^*\|^2 - \alpha) = 0, \tag{11.35}$$

$$\mathbf{d}^T(\mathbf{A} + \lambda^* \mathbf{I})\mathbf{d} \geq 0 \quad \text{for all } \mathbf{d} \text{ satisfying } \mathbf{d}^T \mathbf{x}^* = 0. \tag{11.36}$$

All that is left to show is that the inequality (11.36) is true for any $\mathbf{d}$ and not only for those which are orthogonal to $\mathbf{x}^*$. Suppose on the contrary that there exists a $\mathbf{d}$ such that $\mathbf{d}^T\mathbf{x}^* \neq 0$ and $\mathbf{d}^T(\mathbf{A}+\lambda^*\mathbf{I})\mathbf{d} < 0$. Consider the point $\bar{\mathbf{x}} = \mathbf{x}^* + t\mathbf{d}$, where $t = -2\frac{\mathbf{d}^T\mathbf{x}^*}{\|\mathbf{d}\|^2}$. The vector $\bar{\mathbf{x}}$ is a feasible point since

$$\|\bar{\mathbf{x}}\|^2 = \|\mathbf{x}^* + t\mathbf{d}\|^2 = \|\mathbf{x}^*\|^2 + 2t\mathbf{d}^T\mathbf{x}^* + t^2\|\mathbf{d}\|^2$$
$$= \|\mathbf{x}^*\|^2 - 4\frac{(\mathbf{d}^T\mathbf{x}^*)^2}{\|\mathbf{d}\|^2} + 4\frac{(\mathbf{d}^T\mathbf{x}^*)^2}{\|\mathbf{d}\|^2}$$
$$= \|\mathbf{x}^*\|^2 \leq \alpha.$$

In addition,

$$\begin{aligned}f(\bar{\mathbf{x}}) &= \bar{\mathbf{x}}^T\mathbf{A}\bar{\mathbf{x}} + 2\mathbf{b}^T\bar{\mathbf{x}} + c \\ &= (\mathbf{x}^* + t\mathbf{d})^T\mathbf{A}(\mathbf{x}^* + t\mathbf{d}) + 2\mathbf{b}^T(\mathbf{x}^* + t\mathbf{d}) + c \\ &= \underbrace{(\mathbf{x}^*)^T\mathbf{A}\mathbf{x}^* + 2\mathbf{b}^T\mathbf{x}^* + c}_{f(\mathbf{x}^*)} + t^2\mathbf{d}^T\mathbf{A}\mathbf{d} + 2t\mathbf{d}^T(\mathbf{A}\mathbf{x}^* + \mathbf{b}) \\ &= f(\mathbf{x}^*) + t^2\mathbf{d}^T(\mathbf{A}+\lambda^*\mathbf{I})\mathbf{d} + 2t\mathbf{d}^T\underbrace{((\mathbf{A}+\lambda^*\mathbf{I})\mathbf{x}^* + \mathbf{b})}_{=0 \text{ by } (11.33)} - \lambda^* t \underbrace{\left[t\|\mathbf{d}\|^2 + 2\mathbf{d}^T\mathbf{x}^*\right]}_{=0 \text{ by def. of } t} \\ &= f(\mathbf{x}^*) + t^2\mathbf{d}^T(\mathbf{A}+\lambda^*\mathbf{I})\mathbf{d} \\ &< f(\mathbf{x}^*),\end{aligned}$$

which is a contradiction to the optimality of $\mathbf{x}^*$. $\square$

Theorem 11.21 can be used in order to construct an algorithm for solving the trust region subproblem. We will make the following assumption, which is rather conventional in the literature:

$$-\mathbf{b} \notin \text{Range}(\mathbf{A} - \lambda_{\min}(\mathbf{A})\mathbf{I}). \quad (11.37)$$

Under this condition there cannot be a vector $\mathbf{x}$ for which $(\mathbf{A} - \lambda_{\min}(\mathbf{A})\mathbf{I})\mathbf{x} = -\mathbf{b}$. This means that the multiplier λ^* from the optimality conditions must be different than $-\lambda_{\min}(\mathbf{A})$. The condition (11.37) is considered to be rather mild in the sense that the range space of the matrix $\mathbf{A} - \lambda_{\min}(\mathbf{A})\mathbf{I}$ is of rank which is at most $n-1$. Therefore, at least when $\mathbf{A}$ and $\mathbf{b}$ are generated from a continuous random distribution, the probability that $-\mathbf{b}$ will *not* be in this space is 1.

We will consider two cases.

Case I: $\mathbf{A} \succ 0$. Since in this case the problem is convex, $\mathbf{x}^*$ is an optimal solution of (TRS) if and only if there exists $\lambda^* \geq 0$ such that

$$(\mathbf{A}+\lambda^*\mathbf{I})\mathbf{x}^* = -\mathbf{b}, \quad \lambda^*(\|\mathbf{x}^*\|^2 - \alpha) = 0, \quad \|\mathbf{x}^*\|^2 \leq \alpha.$$

If $\lambda^* = 0$, then $\mathbf{A}\mathbf{x}^* = -\mathbf{b}$, and hence $\mathbf{x}^* = -\mathbf{A}^{-1}\mathbf{b}$. This will be the optimal solution if and only if $\|\mathbf{A}^{-1}\mathbf{b}\|^2 \leq \alpha$. If $\lambda^* > 0$, then $\|\mathbf{x}^*\|^2 = \alpha$, and thus the optimal solution is given by $\mathbf{x}^* = -(\mathbf{A}+\lambda^*\mathbf{I})^{-1}\mathbf{b}$, where λ^* is the unique root of the strictly decreasing function

$$f(\lambda) = \|(\mathbf{A}+\lambda\mathbf{I})^{-1}\mathbf{b}\|^2 - \alpha$$

over $(0, \infty)$.

Case II: $\mathbf{A} \not\succ 0$ In this case, condition (11.31) combined with the nonnegativity of λ^* is the same as $\lambda^* \geq -\lambda_{\min}(\mathbf{A})(\geq 0)$. Under Assumption (11.37), λ^* cannot be equal to

11.6. Optimality Conditions for the Trust Region Subproblem

$-\lambda_{\min}(\mathbf{A})$, and we can thus assume that $\lambda^* > -\lambda_{\min}(\mathbf{A})$. In particular, $\lambda^* > 0$ and hence we have $\|\mathbf{x}^*\|^2 = \alpha$ as well as $\mathbf{A} + \lambda^*\mathbf{I} \succ \mathbf{0}$. Therefore, (11.28) yields

$$\mathbf{x}^* = -(\mathbf{A} + \lambda^*\mathbf{I})^{-1}\mathbf{b}, \quad (11.38)$$

so that $\|\mathbf{x}^*\|^2 = \|(\mathbf{A} + \lambda^*\mathbf{I})^{-1}\mathbf{b}\|^2 = \alpha$. The optimal solution is therefore given by (11.38), where λ^* is chosen as the unique root of the strictly decreasing function $f(\lambda) = \|(\mathbf{A} + \lambda\mathbf{I})^{-1}\mathbf{b}\|^2 - \alpha$ over $(-\lambda_{\min}(\mathbf{A}), \infty)$.

An implementation of the algorithm for solving the trust region subproblem in MATLAB is given below by the function `trs`. The function uses the bisection method in order to find λ^*. Note that the function uses the fact that $f(\lambda) \to \infty$ as $\lambda \to -\lambda_{\min}(\mathbf{A})^+$ by taking the initial lower bound as $-\lambda_{\min}(\mathbf{A}) + \varepsilon$ for some small $\varepsilon > 0$.

```
function x_trs=trs(A,b,alpha)
%INPUT
%=================
%A ................. an nxn matrix
%b ................. an n-length vector
%alpha ............. positive scalar
%OUTPUT
%=================
% x_trs ............ an optimal solution of
%                    min{x'*A*x+2b'*x:||x||^2<=alpha}

n=length(b);
f=@(lam) norm((A+lam*eye(n))\b)^2-alpha;
[L,p]=chol(A,'lower');
% the case when A is positive definite
if (p==0)
    x_naive=-L'\(L\b);
    if(norm(x_naive)^2<=alpha)
        x_trs=x_naive;
    else
        u=1;
        while (f(u)>0)
            u=2*u;
        end
        lam=bisection(f,0,u,1e-7);
        x_trs=-(A+lam*eye(n))\b;
    end
else
    %when A is not positive definite
    u=max(1,-min(eig(A))+1e-7);
    while (f(u)>0)
        u=2*u;
    end
    lam=bisection(f,-min(eig(A))+1e-7,u,1e-7);
    x_trs=-(A+lam*eye(n))\b;
end
```

We can for example use the MATLAB function `trs` in order to solve Example 8.16.

```
>> A=[1,2,3;2,1,4;3,4,3];
>> b=[0.5;1;-0.5];
>> x_trs = trs(A,b,1)
x_trs =

   -0.2300
   -0.7259
    0.6482
```

This is of course the same as the solution obtained in Example 8.16.

11.7 • Total Least Squares

Given an approximate linear system $\mathbf{Ax} \approx \mathbf{b}$ ($\mathbf{A} \in \mathbb{R}^{m \times n}, \mathbf{b} \in \mathbb{R}^m$), the least squares problem discussed in Chapter 3 can be seen as the problem of finding the minimum norm perturbation of the right-hand side of the linear system such that the resulting system is consistent:

$$\begin{array}{ll} \min_{\mathbf{w},\mathbf{x}} & \|\mathbf{w}\|^2 \\ \text{s.t.} & \mathbf{Ax} = \mathbf{b} + \mathbf{w}, \\ & \mathbf{w} \in \mathbb{R}^m. \end{array}$$

This is a different presentation of the least squares problem from the one given in Chapter 3, but it is totally equivalent since plugging the expression for $\mathbf{w}$ ($\mathbf{w} = \mathbf{Ax} - \mathbf{b}$) into the objective function gives the well-known formulation of the problem as one consisting of minimizing the function $\|\mathbf{Ax} - \mathbf{b}\|^2$ over $\mathbb{R}^n$. The least squares problem essentially assumes that the right-hand side is unknown and is subjected to noise and that the matrix $\mathbf{A}$ is known and fixed. However, in many applications the matrix $\mathbf{A}$ is not exactly known and is also subjected to noise. In these cases it is more logical to consider a different problem, which is called *the total least squares problem*, in which one seeks to find a minimal norm perturbation to both the right-hand-side vector and the matrix so that the resulting perturbed system is consistent:

$$\text{(TLS)} \quad \begin{array}{ll} \min_{\mathbf{E},\mathbf{w},\mathbf{x}} & \|\mathbf{E}\|_F^2 + \|\mathbf{w}\|^2 \\ \text{s.t.} & (\mathbf{A} + \mathbf{E})\mathbf{x} = \mathbf{b} + \mathbf{w}, \\ & \mathbf{E} \in \mathbb{R}^{m \times n}, \mathbf{w} \in \mathbb{R}^m. \end{array}$$

Note that we use here the Frobenius norm as a matrix norm. Problem (TLS) is not a convex problem since the constraints are quadratic *equality* constraints. However, despite the nonconvexity of the problem we can use the KKT conditions in order to simplify it considerably and eventually even solve it. The trick is to fix $\mathbf{x}$ and solve the problem with respect to the variables $\mathbf{E}$ and $\mathbf{w}$:

$$(P_\mathbf{x}) \quad \begin{array}{ll} \min_{\mathbf{E},\mathbf{w}} & \|\mathbf{E}\|_F^2 + \|\mathbf{w}\|^2 \\ \text{s.t.} & (\mathbf{A} + \mathbf{E})\mathbf{x} = \mathbf{b} + \mathbf{w}. \end{array}$$

Problem $(P_\mathbf{x})$ is a linearly constrained convex problem and hence the KKT conditions are necessary and sufficient (Theorem 10.7). The Lagrangian of problem $(P_\mathbf{x})$ is given by

$$L(\mathbf{E}, \mathbf{w}, \lambda) = \|\mathbf{E}\|_F^2 + \|\mathbf{w}\|^2 + 2\lambda^T[(\mathbf{A} + \mathbf{E})\mathbf{x} - \mathbf{b} - \mathbf{w}].$$

11.7. Total Least Squares

By the KKT conditions, $(\mathbf{E}, \mathbf{w})$ is an optimal solution of $(P_\mathbf{x})$ if and only if there exists $\lambda \in \mathbb{R}^m$ such that

$$2\mathbf{E} + 2\lambda \mathbf{x}^T = 0 \quad (\nabla_\mathbf{E} L = 0), \tag{11.39}$$

$$2\mathbf{w} - 2\lambda = 0 \quad (\nabla_\mathbf{w} L = 0), \tag{11.40}$$

$$(\mathbf{A} + \mathbf{E})\mathbf{x} = \mathbf{b} + \mathbf{w} \quad \text{(feasibility)}. \tag{11.41}$$

From (11.40) we have $\lambda = \mathbf{w}$. Substituting this in (11.39) we obtain

$$\mathbf{E} = -\mathbf{w}\mathbf{x}^T. \tag{11.42}$$

Combining (11.42) with (11.41) we have $(\mathbf{A} - \mathbf{w}\mathbf{x}^T)\mathbf{x} = \mathbf{b} + \mathbf{w}$, so that

$$\mathbf{w} = \frac{\mathbf{A}\mathbf{x} - \mathbf{b}}{\|\mathbf{x}\|^2 + 1}, \tag{11.43}$$

and consequently, by plugging the above into (11.42) we have

$$\mathbf{E} = -\frac{(\mathbf{A}\mathbf{x} - \mathbf{b})\mathbf{x}^T}{\|\mathbf{x}\|^2 + 1}. \tag{11.44}$$

Finally, by substituting (11.43) and (11.44) into the objective function of problem $(P_\mathbf{x})$ we obtain that the value of problem $(P_\mathbf{x})$ is equal to $\frac{\|\mathbf{A}\mathbf{x}-\mathbf{b}\|^2}{\|\mathbf{x}\|^2+1}$. Consequently, the TLS problem reduces to

$$(\text{TLS}') \quad \min_{\mathbf{x} \in \mathbb{R}^n} \frac{\|\mathbf{A}\mathbf{x} - \mathbf{b}\|^2}{\|\mathbf{x}\|^2 + 1}.$$

We have thus proven the following result.

Theorem 11.22. *$\mathbf{x}$ is an optimal solution of* (TLS') *if and only if* $(\mathbf{x}, \mathbf{E}, \mathbf{w})$ *is an optimal solution of* (TLS) *where* $\mathbf{E} = -\frac{(\mathbf{A}\mathbf{x}-\mathbf{b})\mathbf{x}^T}{\|\mathbf{x}\|^2+1}$ *and* $\mathbf{w} = \frac{\mathbf{A}\mathbf{x}-\mathbf{b}}{\|\mathbf{x}\|^2+1}$.

The new formulation (TLS') is much simpler than the original one. However, the objective function is still nonconvex, and the question that still remains is whether we can efficiently find an optimal solution of this simplified formulation. Using the special structure of the problem, we will show that the problem can actually be solved efficiently by using a *homogenization* argument. Indeed, problem (TLS') is equivalent to

$$\min_{\mathbf{x} \in \mathbb{R}^n, t \in \mathbb{R}} \left\{ \frac{\|\mathbf{A}\mathbf{x} - t\mathbf{b}\|^2}{\|\mathbf{x}\|^2 + t^2} : t = 1 \right\},$$

which is the same as (denoting $\mathbf{y} = \binom{\mathbf{x}}{t}$)

$$f^* = \min_{\mathbf{y} \in \mathbb{R}^{n+1}} \left\{ \frac{\mathbf{y}^T \mathbf{B} \mathbf{y}}{\|\mathbf{y}\|^2} : y_{n+1} = 1 \right\}, \tag{11.45}$$

where

$$\mathbf{B} = \begin{pmatrix} \mathbf{A}^T \mathbf{A} & -\mathbf{A}^T \mathbf{b} \\ -\mathbf{b}^T \mathbf{A} & \|\mathbf{b}\|^2 \end{pmatrix}.$$

Now, let us remove the constraint from problem (11.45) and consider the following problem:

$$g^* = \min_{\mathbf{y} \in \mathbb{R}^{n+1}} \left\{ \frac{\mathbf{y}^T \mathbf{B} \mathbf{y}}{\|\mathbf{y}\|^2} : \mathbf{y} \neq 0 \right\}. \tag{11.46}$$

This is a problem consisting of minimizing the so-called Rayleigh quotient (see Section 1.4) associated with the matrix $\mathbf{B}$, and hence an optimal solution is the vector corresponding to the minimum eigenvalue of $\mathbf{B}$; see Lemma 1.12. Of course, the obtained solution is not guaranteed to satisfy the constraint $y_{n+1} = 1$, but under a rather mild condition, the optimal solution of problem (11.45) can be extracted.

Lemma 11.23. *Let $\mathbf{y}^*$ be an optimal solution of (11.46) and assume that $y_{n+1}^* \neq 0$. Then $\tilde{\mathbf{y}} = \frac{1}{y_{n+1}^*}\mathbf{y}^*$ is an optimal solution of (11.45).*

Proof. Note that since (11.46) is formed from (11.45) by replacing the constraint $y_{n+1} = 1$ with $\mathbf{y} \neq \mathbf{0}$, we have

$$f^* \geq g^*.$$

However, $\tilde{\mathbf{y}}$ is a feasible point of problem (11.45) ($\tilde{y}_{n+1} = 1$), and we have

$$\frac{\tilde{\mathbf{y}}^T \mathbf{B} \tilde{\mathbf{y}}}{\|\tilde{\mathbf{y}}\|^2} = \frac{\frac{1}{(y_{n+1}^*)^2}(\mathbf{y}^*)^T \mathbf{B} \mathbf{y}^*}{\frac{1}{(y_{n+1}^*)^2}\|\mathbf{y}^*\|^2} = \frac{(\mathbf{y}^*)^T \mathbf{B} \mathbf{y}^*}{\|\mathbf{y}^*\|^2} = g^*.$$

Therefore, $\tilde{\mathbf{y}}$ attains the lower bound g^* on the optimal value of problem (11.45), and consequently, it is an optimal solution of (11.45) and the optimal values of the two problems (11.45) and (11.46) are consequently the same. $\square$

All that is left is to find a computable condition under which $y_{n+1}^* \neq 0$. The following theorem presents such a condition and summarizes the solution method of the TLS problem.

Theorem 11.24. *Assume that the following condition holds:*

$$\lambda_{\min}(\mathbf{B}) < \lambda_{\min}(\mathbf{A}^T \mathbf{A}), \tag{11.47}$$

where

$$\mathbf{B} = \begin{pmatrix} \mathbf{A}^T \mathbf{A} & -\mathbf{A}^T \mathbf{b} \\ -\mathbf{b}^T \mathbf{A} & \|\mathbf{b}\|^2 \end{pmatrix}.$$

Then the optimal solution of problem (TLS') *is given by $\frac{1}{y_{n+1}}\mathbf{v}$, where $\mathbf{y} = \begin{pmatrix} \mathbf{v} \\ y_{n+1} \end{pmatrix}$ is an eigenvector corresponding to the minimum eigenvalue of $\mathbf{B}$.*

Proof. By Lemma 11.23 all that we need to prove is that under condition (11.47), an optimal solution $\mathbf{y}^*$ of (11.46) must satisfy $y_{n+1}^* \neq 0$. Assume on the contrary that $y_{n+1}^* = 0$. Then

$$\lambda_{\min}(\mathbf{B}) = \frac{(\mathbf{y}^*)^T \mathbf{B} \mathbf{y}^*}{\|\mathbf{y}^*\|^2} = \frac{\mathbf{v}^T \mathbf{A}^T \mathbf{A} \mathbf{v}}{\|\mathbf{v}\|^2} \geq \lambda_{\min}(\mathbf{A}^T \mathbf{A}),$$

which is a contradiction to (11.47). $\square$

Exercises

11.1. Consider the optimization problem

$$\text{(P)} \quad \begin{array}{ll} \min & x_1 - 4x_2 + x_3 \\ \text{s.t.} & x_1 + 2x_2 + 2x_3 = -2, \\ & x_1^2 + x_2^2 + x_3^2 \leq 1. \end{array}$$

(i) Given a KKT point of problem (P), must it be an optimal solution?

(ii) Find the optimal solution of the problem using the KKT conditions.

11.2. Consider the optimization problem

$$\text{(P)} \quad \min\{\mathbf{a}^T\mathbf{x} : \mathbf{x}^T\mathbf{Q}\mathbf{x} + 2\mathbf{b}^T\mathbf{x} + c \leq 0\},$$

where $\mathbf{Q} \in \mathbb{R}^{n \times n}$ is positive definite, $\mathbf{a}(\neq \mathbf{0}), \mathbf{b} \in \mathbb{R}^n$, and $c \in \mathbb{R}$.

(i) For which values of $\mathbf{Q}, \mathbf{b}, c$ is the problem feasible?

(ii) For which values of $\mathbf{Q}, \mathbf{b}, c$ are the KKT conditions necessary?

(iii) For which values of $\mathbf{Q}, \mathbf{b}, c$ are the KKT conditions sufficient?

(iv) Under the condition of part (ii), find the optimal solution of (P) using the KKT conditions.

11.3. Consider the optimization problem

$$\begin{array}{ll} \min & x_1^4 - 2x_2^2 - x_2 \\ \text{s.t.} & x_1^2 + x_2^2 + x_2 \leq 0. \end{array}$$

(i) Is the problem convex?

(ii) Prove that there exists an optimal solution to the problem.

(iii) Find all the KKT points. For each of the points, determine whether it satisfies the second order necessary conditions.

(iv) Find the optimal solution of the problem.

11.4. Consider the optimization problem

$$\begin{array}{ll} \min & x_1^2 - x_2^2 - x_3^2 \\ \text{s.t.} & x_1^4 + x_2^4 + x_3^4 \leq 1. \end{array}$$

(i) Is the problem convex?

(ii) Find all the KKT points of the problem.

(iii) Find the optimal solution of the problem.

11.5. Consider the optimization problem
$$\begin{align} \min \quad & -2x_1^2 + 2x_2^2 + 4x_1 \\ \text{s.t.} \quad & x_1^2 + x_2^2 - 4 \leq 0, \\ & x_1^2 + x_2^2 - 4x_1 + 3 \leq 0. \end{align}$$

(i) Prove that there exists an optimal solution to the problem.

(ii) Find all the KKT points.

(iii) Find the optimal solution of the problem.

11.6. Use the KKT conditions in order to find an optimal solution of the each of the following problems:

(i)
$$\begin{align} \min \quad & 3x_1^2 + x_2^2 \\ \text{s.t.} \quad & x_1 - x_2 + 8, \leq 0, \\ & x_2 \geq 0. \end{align}$$

(ii)
$$\begin{align} \min \quad & 3x_1^2 + x_2^2 \\ \text{s.t.} \quad & 3x_1^2 + x_2^2 + x_1 + x_2 + 0.1 \leq 0, \\ & x_2 + 10 \geq 0. \end{align}$$

(iii)
$$\begin{align} \min \quad & 2x_1 + x_2 \\ \text{s.t.} \quad & 4x_1^2 + x_2^2 - 2 \leq 0, \\ & 4x_1 + x_2 + 3 \leq 0. \end{align}$$

(iv)
$$\begin{align} \min \quad & x_1^3 + x_2^3 \\ \text{s.t.} \quad & x_1^2 + x_2^2 \leq 1. \end{align}$$

(v)
$$\begin{align} \min \quad & x_1^4 - x_2^2 \\ \text{s.t.} \quad & x_1^2 + x_2^2 \leq 1, \\ & 2x_2 + 1 \leq 0. \end{align}$$

11.7. Let $a > 0$. Find all the optimal solutions of
$$\max\{x_1 x_2 x_3 : a^2 x_1^2 + x_2^2 + x_3^2 \leq 1\}.$$

11.8. (i) Find a formula for the orthogonal projection of a vector $\mathbf{y} \in \mathbb{R}^3$ onto the set
$$C = \{\mathbf{x} \in \mathbb{R}^3 : x_1^2 + 2x_2^2 + 3x_3^2 \leq 1\}.$$

The formula should depend on a single parameter that is a root of a strictly decreasing one-dimensional function.

(ii) Write a MATLAB function whose input is a three-dimensional vector and its output is the orthogonal projection of the input onto C.

11.9. Consider the optimization problem
$$\text{(P)} \quad \begin{aligned} \min \quad & 2x_1 x_2 + \tfrac{1}{2} x_3^2 \\ \text{s.t.} \quad & 2x_1 x_3 + \tfrac{1}{2} x_2^2 \leq 0, \\ & 2x_2 x_3 + \tfrac{1}{2} x_1^2 \leq 0. \end{aligned}$$

(i) Show that the optimal solution of problem (P) is $\mathbf{x}^* = (0,0,0)$.

(ii) Show that $\mathbf{x}^*$ does not satisfy the second order necessary optimality conditions.

11.10. Consider the convex optimization problem

$$(\text{P}) \quad \begin{array}{ll} \min & f_0(\mathbf{x}) \\ \text{s.t.} & f_i(\mathbf{x}) \leq 0, \quad i = 1, 2, \ldots, m, \end{array}$$

where f_0 is a continuously differentiable convex function and $f_1, f_2, \ldots, f_m$ are continuously differentiable *strictly* convex functions. Let $\mathbf{x}^*$ be a feasible solution of (P). Suppose that the following condition is satisfied: there exist $y_i \geq 0, i \in \{0\} \cup I(\mathbf{x}^*)$, which are not all zeros such that

$$y_0 \nabla f_0(\mathbf{x}^*) + \sum_{i \in I(\mathbf{x}^*)} y_i \nabla f_i(\mathbf{x}^*) = 0.$$

Prove that $\mathbf{x}^*$ is an optimal solution of (P).

11.11. Consider the optimization problem

$$\begin{array}{ll} \min & \mathbf{c}^T \mathbf{x} \\ \text{s.t.} & f_i(\mathbf{x}) \leq 0, \quad i = 1, 2, \ldots, m, \end{array}$$

where $\mathbf{c} \neq 0$ and $f_1, f_2, \ldots, f_m$ are continuous over $\mathbb{R}^n$. Prove that if $\mathbf{x}^*$ is a local minimum of the problem, then $I(\mathbf{x}^*) \neq \emptyset$.

11.12. Consider the QCQP problem

$$(\text{QCQP}) \quad \begin{array}{ll} \min & \mathbf{x}^T \mathbf{A}_0 \mathbf{x} + 2 \mathbf{b}_0^T \mathbf{x} \\ \text{s.t.} & \mathbf{x}^T \mathbf{A}_i \mathbf{x} + 2 \mathbf{b}_i^T \mathbf{x} + c_i \leq 0, \quad i = 1, 2, \ldots, m, \end{array}$$

where $\mathbf{A}_0, \mathbf{A}_1, \ldots, \mathbf{A}_m \in \mathbb{R}^{n \times n}$ are symmetric matrices, $\mathbf{b}_0, \mathbf{b}_1, \ldots, \mathbf{b}_m \in \mathbb{R}^n$, and $c_1, c_2, \ldots, c_m \in \mathbb{R}$. Suppose that $\mathbf{x}^*$ satisfies the following condition: there exist $\lambda_1, \lambda_2, \ldots, \lambda_m \geq 0$ such that

$$\left(\mathbf{A}_0 + \sum_{i=1}^m \lambda_i \mathbf{A}_i \right) \mathbf{x}^* + \left(\mathbf{b}_0 + \sum_{i=1}^m \lambda_i \mathbf{b}_i \right) = 0,$$
$$\lambda_i \left[(\mathbf{x}^*)^T \mathbf{A}_i (\mathbf{x}^*) + 2 \mathbf{b}_i^T \mathbf{x}^* + c_i \right] = 0, \quad i = 1, 2, \ldots, m,$$
$$(\mathbf{x}^*)^T \mathbf{A}_i (\mathbf{x}^*) + 2 \mathbf{b}_i^T \mathbf{x}^* + c_i \leq 0, \quad i = 1, 2, \ldots, m,$$
$$\mathbf{A}_0 + \sum_{i=1}^m \lambda_i \mathbf{A}_i \succeq 0.$$

Prove that $\mathbf{x}^*$ is an optimal solution of (QCQP).

Chapter 12

Duality

12.1 • Motivation and Definition

The dual problem, which we will formally define later on, can be motivated as a way to find bounds on a given optimization problem. We will begin with an example.

Example 12.1. Consider the problem

$$(P) \quad \begin{array}{ll} \min & x_1^2 + x_2^2 + 2x_1 \\ \text{s.t.} & x_1 + x_2 = 0. \end{array}$$

Problem (P) is of course not a difficult problem, and it can be solved by reducing it into a one-dimensional problem by eliminating x_2 via the relation $x_2 = -x_1$, thus transforming the objective function to $2x_1^2 + 2x_1$. The (unconstrained) minimizer of the latter function is $x_1 = -\frac{1}{2}$, and the optimal solution is thus $(-\frac{1}{2}, \frac{1}{2})$ with a corresponding optimal value of $f^* = -\frac{1}{2}$.

The theoretical exercise that we wish to make is to find lower bounds on the value of the problem by solving unconstrained problems. For example, the unconstrained problem derived by eliminating the single constraint is

$$(P_0) \quad \min x_1^2 + x_2^2 + 2x_1.$$

The optimal value of (P_0) is a lower bound on the value of the optimal value of (P). We can write this fact by the following notation:

$$\text{val}(P_0) \leq \text{val}(P).$$

The optimal solution of the convex problem (P_0) is attained at the stationary point $x_1 = -1, x_2 = 0$ with a corresponding optimal value of -1 (which is indeed a lower bound on f^*).

In order to find other lower bounds, we use the following trick. Take a real number μ and consider the following problem, which is equivalent to problem (P):

$$\begin{array}{ll} \min & x_1^2 + x_2^2 + 2x_1 + \mu(x_1 + x_2) \\ \text{s.t.} & x_1 + x_2 = 0. \end{array} \quad (12.1)$$

Now we can eliminate the equality constraint and obtain the unconstrained problem

$$(P_\mu) \quad \min x_1^2 + x_2^2 + 2x_1 + \mu(x_1 + x_2).$$

We have that for all $\mu \in \mathbb{R}$
$$\text{val}(P_\mu) \leq \text{val}(P).$$

The optimal solution of (P_μ) is attained at the stationary point $(x_1, x_2) = \left(-1 - \frac{\mu}{2}, -\frac{\mu}{2}\right)$, and the corresponding optimal value, which we denote by $q(\mu)$, is

$$q(\mu) \equiv \text{val}(P_\mu) = \left(-1 - \frac{\mu}{2}\right)^2 + \left(-\frac{\mu}{2}\right)^2 + 2\left(-1 - \frac{\mu}{2}\right) + \mu(-1 - \mu) = -\frac{\mu^2}{2} - \mu - 1.$$

For example, $q(0) = -1$ is the lower bound obtained by (P_0). What interests us the most is the best (i.e., largest), lower bound obtained by this technique. The best lower bound is the solution of the problem

$$(D) \quad \max\{q(\mu) : \mu \in \mathbb{R}\}.$$

This problem will be called *the dual problem*, and by its construction, its optimal value is a lower bound on the optimal value of the original problem (P), which we will call *the primal problem*:

$$\text{val}(D) \leq \text{val}(P).$$

In this case the optimal solution of the dual problem is attained at $\mu = -1$, and the corresponding optimal value of the dual problem is $-\frac{1}{2}$, which is exactly f^*, meaning that the best lower bound obtained by the this technique is actually *equal* to the optimal value f^*. Later on, we will refer to this property as "strong duality" and discuss the conditions under which it holds. ∎

12.1.1 ▪ Definition of the Dual Problem

Consider the general model

$$\begin{aligned} f^* = \min \quad & f(\mathbf{x}) \\ \text{s.t.} \quad & g_i(\mathbf{x}) \leq 0, \quad i = 1, 2, \ldots, m, \\ & h_j(\mathbf{x}) = 0, \quad j = 1, 2, \ldots, p, \\ & \mathbf{x} \in X, \end{aligned} \quad (12.2)$$

where $f, g_i, h_j (i = 1, 2, \ldots, m, j = 1, 2, \ldots, p)$ are functions defined on the set $X \subseteq \mathbb{R}^n$. Problem (12.2) will be referred to as the primal problem. At this point, we do not assume anything on the functions (they are not even assumed to be continuous). The Lagrangian of the problem is

$$L(\mathbf{x}, \lambda, \mu) = f(\mathbf{x}) + \sum_{i=1}^{m} \lambda_i g_i(\mathbf{x}) + \sum_{j=1}^{p} \mu_j h_j(\mathbf{x}) \quad (\mathbf{x} \in X, \lambda \in \mathbb{R}_+^m, \mu \in \mathbb{R}^p),$$

where $\lambda_1, \lambda_2, \ldots, \lambda_m$ are nonnegative Lagrange multipliers associated with the inequality constraints, and $\mu_1, \mu_2, \ldots, \mu_p$ are the Lagrange multipliers associated with the equality constraints. The *dual objective function* $q : \mathbb{R}_+^m \times \mathbb{R}^p \to \mathbb{R} \cup \{-\infty\}$ is defined to be

$$q(\lambda, \mu) = \min_{\mathbf{x} \in X} L(\mathbf{x}, \lambda, \mu). \quad (12.3)$$

Note that we use the "min" notation even though the minimum is not necessarily attained. In addition, the optimal value of the minimization problem in (12.3) is not always finite;

12.1. Motivation and Definition

there may be values of (λ, μ) for which $q(\lambda, \mu) = -\infty$. It is therefore natural to define the domain of the dual objective function as

$$\text{dom}(q) = \{(\lambda, \mu) \in \mathbb{R}_+^m \times \mathbb{R}^p : q(\lambda, \mu) > -\infty\}.$$

The *dual problem* is given by

$$q^* = \max_{} \; q(\lambda, \mu) \quad \text{s.t.} \; (\lambda, \mu) \in \text{dom}(q). \tag{12.4}$$

We begin by showing that the dual problem is always convex; it consists of maximizing a concave function over a convex feasible set.

Theorem 12.2 (convexity of the dual problem). *Consider problem* (12.2) *with* f, g_i, h_j ($i = 1, 2, \ldots, m, j = 1, 2, \ldots, p$) *being functions defined on the set* $X \subseteq \mathbb{R}^n$, *and let q be the function defined in* (12.3). *Then*

(a) $\text{dom}(q)$ *is a convex set*,

(b) q *is a concave function over* $\text{dom}(q)$.

Proof. (a) To establish the convexity of $\text{dom}(q)$, take $(\lambda_1, \mu_1), (\lambda_2, \mu_2) \in \text{dom}(q)$ and $\alpha \in [0, 1]$. Then by the definition of $\text{dom}(q)$ we have that

$$\min_{\mathbf{x} \in X} L(\mathbf{x}, \lambda_1, \mu_1) > -\infty, \tag{12.5}$$

$$\min_{\mathbf{x} \in X} L(\mathbf{x}, \lambda_2, \mu_2) > -\infty. \tag{12.6}$$

Therefore, since the Lagrangian $L(\mathbf{x}, \lambda, \mu)$ is affine with respect to λ, μ, we obtain that

$$\begin{aligned}
q(\alpha \lambda_1 + (1-\alpha)\lambda_2, \alpha \mu_1 + (1-\alpha)\mu_2) &= \min_{\mathbf{x} \in X} L(\mathbf{x}, \alpha \lambda_1 + (1-\alpha)\lambda_2, \alpha \mu_1 + (1-\alpha)\mu_2) \\
&= \min_{\mathbf{x} \in X} [\alpha L(\mathbf{x}, \lambda_1, \mu_1) + (1-\alpha) L(\mathbf{x}, \lambda_2, \mu_2)] \\
&\geq \alpha \min_{\mathbf{x} \in X} L(\mathbf{x}, \lambda_1, \mu_1) + (1-\alpha) \min_{\mathbf{x} \in X} L(\mathbf{x}, \lambda_2, \mu_2) \\
&= \alpha q(\lambda_1, \mu_1) + (1-\alpha) q(\lambda_2, \mu_2) \\
&> -\infty.
\end{aligned}$$

Hence, $\alpha(\lambda_1, \mu_1) + (1-\alpha)(\lambda_2, \mu_2) \in \text{dom}(q)$, and the convexity of $\text{dom}(q)$ is established.

(b) As noted in the proof of part (a), $L(\mathbf{x}, \lambda, \mu)$ is an affine function with respect to (λ, μ). In particular, it is a concave function with respect to (λ, μ). Hence, since $q(\lambda, \mu)$ is the minimum of concave functions, it must be concave. This follows immediately from the fact that the maximum of convex functions is a convex function (Theorem 7.38). □

Note that $-q$ is in fact an extended real-valued convex function over $\mathbb{R}_+^m \times \mathbb{R}^p$ as defined in Section 7.7, and the effective domain of $-q$ is exactly the domain defined in this section. The first important result is closely connected to the motivation of the construction of the dual problem: the optimal dual value is a lower bound on the optimal primal value. This result is called the *weak duality theorem*, and unsurprisingly, its proof is rather simple.

Theorem 12.3 (weak duality theorem). *Consider the primal problem (12.2) and its dual problem (12.4). Then*
$$q^* \leq f^*,$$
where q^*, f^* are the optimal dual and primal values respectively.

Proof. Let us denote the feasible set of the primal problem by
$$S = \{\mathbf{x} \in X : g_i(\mathbf{x}) \leq 0, h_j(\mathbf{x}) = 0, i = 1, 2, \ldots, m, j = 1, 2, \ldots, p\}.$$

Then for any $(\lambda, \mu) \in \mathbb{R}_+^m \times \mathbb{R}^p$ we have
$$\begin{aligned} q(\lambda, \mu) &= \min_{\mathbf{x} \in X} L(\mathbf{x}, \lambda, \mu) \\ &\leq \min_{\mathbf{x} \in S} L(\mathbf{x}, \lambda, \mu) \\ &= \min_{\mathbf{x} \in S} \left[f(\mathbf{x}) + \sum_{i=1}^m \lambda_i g_i(\mathbf{x}) + \sum_{j=1}^p \mu_j h_j(\mathbf{x}) \right] \\ &\leq \min_{\mathbf{x} \in S} f(\mathbf{x}), \end{aligned}$$

where the last inequality follows from the fact that $\lambda_i \geq 0$ and for any $\mathbf{x} \in S$, $g_i(\mathbf{x}) \leq 0, h_j(\mathbf{x}) = 0$ ($i = 1, 2, \ldots, m, j = 1, 2, \ldots, p$). We thus obtain that
$$q(\lambda, \mu) \leq \min_{\mathbf{x} \in S} f(\mathbf{x})$$
for any $(\lambda, \mu) \in \mathbb{R}_+^m \times \mathbb{R}^p$. By taking the maximum over $(\lambda, \mu) \in \mathbb{R}_+^m \times \mathbb{R}^p$, the result follows. □

The weak duality theorem states that the dual optimal value is a lower bound on the primal optimal value. Example 12.1 illustrated that the lower bound can be tight. However, the lower bound does not have to be tight, and the next example shows that it can be totally uninformative.

Example 12.4. Consider the problem
$$\begin{array}{ll} \min & x_1^2 - 3x_2^2 \\ \text{s.t.} & x_1 = x_2^3. \end{array}$$

It is not difficult to solve the problem. Substituting $x_1 = x_2^3$ into the objective function, we obtain that the problem is equivalent to the unconstrained one-dimensional minimization problem
$$\min_{x_2} x_2^6 - 3x_2^2.$$

An optimal solution exists since the objective function is coercive. The stationary points of the latter problem are the solutions to
$$6x_2^5 - 6x_2 = 0,$$
that is,
$$6x_2(x_2^4 - 1) = 0.$$

Hence, the stationary points are $x_2 = 0, \pm 1$, and thence the only candidates for the optimal solutions are $(0, 0), (1, 1), (-1, -1)$. Comparing the corresponding objective function values, it follows that the optimal solutions of the problem are $(1,1),(-1,-1)$ with an optimal value of $f^* = -2$.

Let us construct the dual problem. The Lagrangian function is

$$L(x_1, x_2, \mu) = x_1^2 - 3x_2^2 + \mu(x_1 - x_2^3) = x_1^2 + \mu x_1 - 3x_2^2 - \mu x_2^3.$$

Obviously, for any $\mu \in \mathbb{R}$

$$\min_{x_1, x_2} L(x_1, x_2, \mu) = -\infty,$$

and hence the dual optimal value is $q^* = -\infty$, which is an extremely poor lower bound on the primal optimal value $f^* = -2$. ∎

12.2 • Strong Duality in the Convex Case

In the convex case we can prove under rather mild conditions that strong duality holds; that is, the primal and dual optimal values coincide. Similarly to the derivation of the KKT conditions, we will rely on separation theorems in order to establish the result. The strict separation theorem (Theorem 10.1) from Section 10.1 states that a point can be strictly separated from any closed and convex set. We will require a variation of this result stating that a point can be separated from any convex set, not necessarily closed. Note that the separation is not strict, and in fact it also includes the case in which the point is on the boundary of the convex set and the theorem is hence called *the supporting hyperplane theorem*. Although the theorem holds for any convex set C, we will state and prove it only for convex sets with a nonempty interior.

Theorem 12.5 (supporting hyperplane theorem). *Let $C \subseteq \mathbb{R}^n$ be a convex set with a nonempty interior and let $\mathbf{y} \notin C$. Then there exists $0 \neq \mathbf{p} \in \mathbb{R}^n$ such that*

$$\mathbf{p}^T \mathbf{x} \leq \mathbf{p}^T \mathbf{y} \text{ for any } \mathbf{x} \in C.$$

Proof. Since $\mathbf{y} \notin \text{int}(C)$, it follows that $\mathbf{y} \notin \text{int}(\text{cl}(C))$. (Recall that by Lemma 6.30 $\text{int}(C) = \text{int}(\text{cl}(C))$.) Therefore, there exists a sequence $\{\mathbf{y}_k\}_{k \geq 1}$ satisfying $\mathbf{y}_k \notin \text{cl}(C)$ such that $\mathbf{y}_k \to \mathbf{y}$. Since $\text{cl}(C)$ is convex by Theorem 6.27 and closed by its definition, it follows by the strict separation theorem (Theorem 10.1) that there exists $0 \neq \mathbf{p}_k \in \mathbb{R}^n$ such that

$$\mathbf{p}_k^T \mathbf{x} < \mathbf{p}_k^T \mathbf{y}_k$$

for all $\mathbf{x} \in \text{cl}(C)$. Dividing the latter inequality by $\|\mathbf{p}_k\| \neq 0$, we obtain

$$\frac{\mathbf{p}_k^T}{\|\mathbf{p}_k\|}(\mathbf{x} - \mathbf{y}_k) < 0 \text{ for any } \mathbf{x} \in \text{cl}(C). \tag{12.7}$$

Since the sequence $\{\frac{\mathbf{p}_k}{\|\mathbf{p}_k\|}\}_{k \geq 1}$ is bounded, it follows that there exists a subsequence $\{\frac{\mathbf{p}_k}{\|\mathbf{p}_k\|}\}_{k \in T}$ such that $\frac{\mathbf{p}_k}{\|\mathbf{p}_k\|} \xrightarrow{T} \mathbf{p}$ as $k \xrightarrow{T} \infty$ for some $\mathbf{p} \in \mathbb{R}^n$. Obviously, $\|\mathbf{p}\| = 1$ and hence in particular $\mathbf{p} \neq 0$. Taking the limit as $k \xrightarrow{T} \infty$ in inequality (12.7), we obtain that

$$\mathbf{p}^T(\mathbf{x} - \mathbf{y}) \leq 0 \text{ for any } \mathbf{x} \in \text{cl}(C),$$

which readily implies the result since $C \subseteq \text{cl}(C)$. □

We can now deduce a separation theorem between two disjoint convex sets.

Theorem 12.6 (separation of two convex sets). *Let $C_1, C_2 \subseteq \mathbb{R}^n$ be two convex sets with nonempty interiors such that $C_1 \cap C_2 = \emptyset$. Then there exists $0 \neq \mathbf{p} \in \mathbb{R}^n$ for which*

$$\mathbf{p}^T \mathbf{x} \leq \mathbf{p}^T \mathbf{y} \text{ for any } \mathbf{x} \in C_1, \mathbf{y} \in C_2.$$

Proof. The set $C_1 - C_2$ is a convex set (by part (a) of Theorem 6.8) with a nonempty interior, and since $C_1 \cap C_2 = \emptyset$, it follows that $0 \notin C_1 - C_2$. By the supporting hyperplane theorem (Theorem 12.5), it follows that there exists $0 \neq \mathbf{p} \in \mathbb{R}^n$ such that

$$\mathbf{p}^T (\mathbf{x} - \mathbf{y}) \leq \mathbf{p}^T 0 = 0 \text{ for any } \mathbf{x} \in C_1, \mathbf{y} \in C_2,$$

which is the same as the desired result. □

We will now derive a result which is a nonlinear version of Farkas' lemma. The main difference is that a Slater-type condition must be assumed. Later on, this lemma will be the key in proving the strong duality result.

Theorem 12.7 (nonlinear Farkas' lemma). *Let $X \subseteq \mathbb{R}^n$ be a convex set and let $f, g_1, g_2, \ldots, g_m$ be convex functions over X. Assume that there exists $\hat{\mathbf{x}} \in X$ such that*

$$g_1(\hat{\mathbf{x}}) < 0, \quad g_2(\hat{\mathbf{x}}) < 0, \ldots, g_m(\hat{\mathbf{x}}) < 0.$$

Let $c \in \mathbb{R}$. Then the following two claims are equivalent.

(a) *The following implication holds:*

$$\mathbf{x} \in X, \quad g_i(\mathbf{x}) \leq 0, \quad i = 1, 2, \ldots, m \Rightarrow f(\mathbf{x}) \geq c.$$

(b) *There exist $\lambda_1, \lambda_2, \ldots, \lambda_m \geq 0$ such that*

$$\min_{\mathbf{x} \in X} \left\{ f(\mathbf{x}) + \sum_{i=1}^{m} \lambda_i g_i(\mathbf{x}) \right\} \geq c. \quad (12.8)$$

Proof. The implication (b) $\Rightarrow$ (a) is rather straightforward. Indeed, suppose that there exist $\lambda_1, \lambda_2, \ldots, \lambda_m \geq 0$ such that (12.8) holds, and let $\mathbf{x} \in X$ satisfy $g_i(\mathbf{x}) \leq 0, i = 1, 2, \ldots, m$. Then by (12.8) we have that

$$f(\mathbf{x}) + \sum_{i=1}^{m} \lambda_i g_i(\mathbf{x}) \geq c,$$

and hence, since $g_i(\mathbf{x}) \leq 0, \lambda_i \geq 0$,

$$f(\mathbf{x}) \geq c - \sum_{i=1}^{m} \lambda_i g_i(\mathbf{x}) \geq c.$$

To prove that (a) $\Rightarrow$ (b), let us assume that the implication (a) holds. Consider the following two sets:

$$S = \{\mathbf{u} = (u_0, u_1, \ldots, u_m) : \exists \mathbf{x} \in X \text{ s.t. } f(\mathbf{x}) \leq u_0, g_i(\mathbf{x}) \leq u_i, i = 1, 2, \ldots, m\},$$
$$T = \{(u_0, u_1, \ldots, u_m) : u_0 < c, u_1 \leq 0, u_2 \leq 0, \ldots, u_m \leq 0\}.$$

12.2. Strong Duality in the Convex Case

Note that S, T are convex with nonempty interiors and in addition, by the validity of implication (a), $S \cap T = \emptyset$. Therefore, by Theorem 12.6 (separation of two convex sets), it follows that there exists a vector $\mathbf{a} = (a_0, a_1, \ldots, a_m) \neq \mathbf{0}$, such that

$$\min_{(u_0, u_1, \ldots, u_m) \in S} \sum_{j=0}^{m} a_j u_j \geq \max_{(u_0, u_1, \ldots, u_m) \in T} \sum_{j=0}^{m} a_j u_j. \tag{12.9}$$

First note that $\mathbf{a} \geq \mathbf{0}$. This is due to the fact that if there was a negative component, say $a_i < 0$, then by taking u_i to be a negative number tending to $-\infty$ while fixing all the other components as zeros, we obtain that the right-hand-side maximum in (12.9) is ∞, which is impossible. Since $\mathbf{a} \geq \mathbf{0}$, it follows that the right-hand side is $a_0 c$, and we thus obtain

$$\min_{(u_0, u_1, \ldots, u_m) \in S} \sum_{j=0}^{m} a_j u_j \geq a_0 c. \tag{12.10}$$

Now we will show that $a_0 > 0$. Suppose in contradiction that $a_0 = 0$. Then

$$\min_{(u_0, u_1, \ldots, u_m) \in S} \sum_{j=1}^{m} a_j u_j \geq 0.$$

However, since we can take $u_i = g_i(\hat{\mathbf{x}}), i = 1, 2, \ldots, m$, we can deduce that

$$\sum_{j=1}^{m} a_j g_j(\hat{\mathbf{x}}) \geq 0,$$

which is impossible since $g_j(\hat{\mathbf{x}}) < 0$ for all j, and there exists at least one nonzero component in $(a_1, a_2, \ldots, a_m)$. Since $a_0 > 0$, we can divide (12.10) by a_0 to obtain

$$\min_{(u_0, u_1, \ldots, u_m) \in S} \left\{ u_0 + \sum_{j=1}^{m} \tilde{a}_j u_j \right\} \geq c, \tag{12.11}$$

where $\tilde{a}_j = \frac{a_j}{a_0}$. Define

$$\tilde{S} = \{\mathbf{u} = (u_0, u_1, \ldots, u_m) : \exists \mathbf{x} \in X \text{ s.t. } f(\mathbf{x}) = u_0, g_i(\mathbf{x}) = u_i, i = 1, 2, \ldots, m\}.$$

Then obviously $\tilde{S} \subseteq S$. Therefore,

$$\min_{(u_0, u_1, \ldots, u_m) \in S} \left\{ u_0 + \sum_{j=1}^{m} \tilde{a}_j u_j \right\} \leq \min_{(u_0, u_1, \ldots, u_m) \in \tilde{S}} \left\{ u_0 + \sum_{j=1}^{m} \tilde{a}_j u_j \right\}$$

$$= \min_{\mathbf{x} \in X} \left\{ f(\mathbf{x}) + \sum_{j=1}^{m} \tilde{a}_j g_j(\mathbf{x}) \right\},$$

which combined with (12.11) yields the desired result

$$\min_{\mathbf{x} \in X} \left\{ f(\mathbf{x}) + \sum_{j=1}^{m} \tilde{a}_j g_j(\mathbf{x}) \right\} \geq c. \quad \square$$

We are now able to show a strong duality result in the convex case under a Slater-type condition.

Theorem 12.8 (strong duality of convex problems with inequality constraints). *Consider the optimization problem*

$$f^* = \min \quad f(\mathbf{x}) \\ \text{s.t.} \quad g_i(\mathbf{x}) \leq 0, \quad i = 1, 2, \ldots, m, \\ \mathbf{x} \in X, \tag{12.12}$$

where X is a convex set and $f, g_i, i = 1, 2, \ldots, m$, are convex functions over X and suppose that there exists $\hat{\mathbf{x}} \in X$ for which $g_i(\hat{\mathbf{x}}) < 0, i = 1, 2, \ldots, m$ and suppose that problem (12.12) has a finite optimal value. Then the optimal value of the dual problem

$$q^* = \max\{q(\lambda) : \lambda \in \text{dom}(q)\}, \tag{12.13}$$

where

$$q(\lambda) = \min_{\mathbf{x} \in X} L(\mathbf{x}, \lambda),$$

is attained, and the optimal values of the primal and dual problems are the same:

$$f^* = q^*.$$

Proof. Since $f^* > -\infty$ is the optimal value of (12.12), it follows that the following implication holds:

$$\mathbf{x} \in X, \quad g_i(\mathbf{x}) \leq 0, \quad i = 1, 2, \ldots, m \Rightarrow f(\mathbf{x}) \geq f^*,$$

and hence by the nonlinear Farkas' lemma (Theorem 12.7) we have that there exist $\tilde{\lambda}_1, \tilde{\lambda}_2, \ldots, \tilde{\lambda}_m \geq 0$ such that

$$q(\tilde{\lambda}) = \min_{\mathbf{x} \in X} \left\{ f(\mathbf{x}) + \sum_{j=1}^{m} \tilde{\lambda}_j g_j(\mathbf{x}) \right\} \geq f^*,$$

which combined with the weak duality theorem (Theorem 12.3) yields

$$q^* \geq q(\tilde{\lambda}) \geq f^* \geq q^*.$$

Hence $f^* = q^*$ and $\tilde{\lambda}$ is an optimal solution of the dual problem. □

Example 12.9. Consider the problem

$$\min \quad x_1^2 - x_2 \\ \text{s.t.} \quad x_2^2 \leq 0.$$

The problem is convex but does not satisfy Slater's condition. The optimal solution is obviously $x_1 = x_2 = 0$ and hence $f^* = 0$. The Lagrangian is

$$L(x_1, x_2, \lambda) = x_1^2 - x_2 + \lambda x_2^2 \quad (\lambda \geq 0),$$

and the dual objective function is

$$q(\lambda) = \min_{x_1, x_2} L(x_1, x_2, \lambda) = \begin{cases} -\infty, & \lambda = 0, \\ -\frac{1}{4\lambda}, & \lambda > 0. \end{cases}$$

12.2. Strong Duality in the Convex Case

The dual problem is therefore

$$\max\left\{-\frac{1}{4\lambda} : \lambda > 0\right\}.$$

The dual optimal value is $q^* = 0$, so we do have the equality $f^* = q^*$. However, the strong duality theorem (Theorem 12.8) states that there exists an optimal solution to the dual problem, and this is obviously not the case in this example. The reason why this property does not hold is that in this example Slater's condition is not satisfied—there does not exist $\bar{x}_2$ for which $\bar{x}_2^2 < 0$. ∎

Example 12.10. ([9]) Consider the convex optimization problem

$$\begin{aligned} \min \quad & e^{-x_2} \\ \text{s.t.} \quad & \sqrt{x_1^2 + x_2^2} - x_1 \leq 0. \end{aligned}$$

Note that the feasible set is in fact

$$F = \{(x_1, x_2) : x_1 \geq 0, x_2 = 0\}.$$

The constraint is always satisfied as an equality constraint, and thus Slater's condition is not satisfied. Since x_2 is necessarily 0, it follows that the optimal value is $f^* = 1$. The Lagrangian of the problem is

$$L(x_1, x_2, \lambda) = e^{-x_2} + \lambda(\sqrt{x_1^2 + x_2^2} - x_1) \quad (\lambda \geq 0).$$

The dual objective function is

$$q(\lambda) = \min_{x_1, x_2} L(x_1, x_2, \lambda).$$

We will show that this minimum is 0, no matter what the value of λ is. First of all, $L(x_1, x_2, \lambda) \geq 0$ for all x_1, x_2 and hence $q(\lambda) \geq 0$. On the other hand, for any $\varepsilon > 0$, if we take $x_2 = -\ln \varepsilon$, $x_1 = \frac{x_2^2 - \varepsilon^2}{2\varepsilon}$, we have

$$\begin{aligned} \sqrt{x_1^2 + x_2^2} - x_1 &= \sqrt{\frac{(x_2^2 - \varepsilon^2)^2}{4\varepsilon^2} + x_2^2} - \frac{x_2^2 - \varepsilon^2}{2\varepsilon} \\ &= \sqrt{\frac{(x_2^2 + \varepsilon^2)^2}{4\varepsilon^2}} - \frac{x_2^2 - \varepsilon^2}{2\varepsilon} \\ &= \frac{x_2^2 + \varepsilon^2}{2\varepsilon} - \frac{x_2^2 - \varepsilon^2}{2\varepsilon} \\ &= \varepsilon. \end{aligned}$$

Therefore,

$$L(x_1, x_2, \lambda) = e^{-x_2} + \lambda(\sqrt{x_1^2 + x_2^2} - x_1) = \varepsilon + \lambda\varepsilon = (1 + \lambda)\varepsilon.$$

Consequently, $q(\lambda) = 0$ for all $\lambda \geq 0$. The dual problem is therefore the following "trivial" problem:

$$\max\{0 : \lambda \geq 0\},$$

whose optimal value is obviously $q^* = 0$. We thus obtained that there exists a duality gap $f^* - q^* = 1$, which is a result of the fact that Slater's condition is not satisfied. ∎

We can also derive the complementary slackness conditions under the sole assumption that $q^* = f^*$ (without any convexity assumptions).

Theorem 12.11 (complementary slackness conditions). *Consider problem (12.12) and assume that $f^* = q^*$, where q^* is the optimal value of the dual problem given by (12.13). If $\mathbf{x}^*, \lambda^*$ are optimal solutions of the primal and dual problems respectively, then*

$$\mathbf{x}^* \in \text{argmin} L(\mathbf{x}, \lambda^*),$$
$$\lambda_i^* g_i(\mathbf{x}^*) = 0, i = 1, 2, \ldots, m.$$

Proof. We have

$$q^* = q(\lambda^*) = \min_{\mathbf{x} \in X} L(\mathbf{x}, \lambda^*) \leq L(\mathbf{x}^*, \lambda^*) = f(\mathbf{x}^*) + \sum_{i=1}^{m} \lambda_i^* g_i(\mathbf{x}^*) \leq f(\mathbf{x}^*) = f^*,$$

where the last inequality follows from the fact that $\lambda_i^* \geq 0, g_i(\mathbf{x}^*) \leq 0$. Therefore, since $q^* = f^*$, it follows that all the inequalities in the above chain of inequalities and equalities are satisfied as equalities, meaning that $\mathbf{x}^* \in \text{argmin}_{\mathbf{x} \in X} L(\mathbf{x}, \lambda^*)$ and that $\sum_{i=1}^{m} \lambda_i^* g_i(\mathbf{x}^*) = 0$, which by the fact that $\lambda_i^* \geq 0, g_i(\mathbf{x}^*) \leq 0$, implies that $\lambda_i^* g_i(\mathbf{x}^*) = 0$ for all $i = 1, 2, \ldots, m$. □

Finer analysis can show, for example, the following strong duality theorem that deals with linear equality and inequality constraints as well as nonlinear constraints.

Theorem 12.12. *Consider the optimization problem*

$$\begin{aligned} f^* = \min \quad & f(\mathbf{x}) \\ \text{s.t.} \quad & g_i(\mathbf{x}) \leq 0, \quad i = 1, 2, \ldots, m, \\ & h_j(\mathbf{x}) \leq 0, \quad j = 1, 2, \ldots, p, \\ & s_k(\mathbf{x}) = 0, \quad k = 1, 2, \ldots, q, \\ & \mathbf{x} \in X, \end{aligned} \quad (12.14)$$

where X is a convex set and $f, g_i, i = 1, 2, \ldots, m$, are convex functions over X. The functions $h_j, s_k, j = 1, 2, \ldots, p, k = 1, 2, \ldots, q$, are affine functions. Suppose that there exists $\hat{\mathbf{x}} \in \text{int}(X)$ for which $g_i(\hat{\mathbf{x}}) < 0, i = 1, 2, \ldots, m, h_j(\hat{\mathbf{x}}) \leq 0, j = 1, 2, \ldots, p, s_k(\hat{\mathbf{x}}) = 0, k = 1, 2, \ldots, q$. Then if problem (12.14) has a finite optimal value, then the optimal value of the dual problem

$$q^* = \max\{q(\lambda, \eta, \mu) : (\lambda, \eta, \mu) \in \text{dom}(q)\},$$

where $q : \mathbb{R}_+^m \times \mathbb{R}_+^p \times \mathbb{R}^q \to \mathbb{R} \cup \{-\infty\}$ is given by

$$q(\lambda, \eta, \mu) = \min_{\mathbf{x} \in X} L(\mathbf{x}, \lambda, \eta, \mu) = \min_{\mathbf{x} \in X} \left[f(\mathbf{x}) + \sum_{i=1}^{m} \lambda_i g_i(\mathbf{x}) + \sum_{j=1}^{p} \eta_j h_j(\mathbf{x}) + \sum_{k=1}^{q} \mu_k s_k(\mathbf{x}) \right]$$

is attained, and the optimal values of the primal and dual problems are the same:

$$f^* = q^*.$$

12.3. Examples

Example 12.13. In this example we will demonstrate the fact that there is no "one" dual problem for a given primal problem, and in many cases there are several ways to construct the dual, which result in different dual problems and dual optimal values. Consider the simple two-dimensional problem

$$\begin{aligned} \min \quad & x_1^3 + x_2^3 \\ \text{s.t.} \quad & x_1 + x_2 \geq 1, \\ & x_1, x_2 \geq 0. \end{aligned}$$

It is not difficult to verify (for example, by finding all the KKT points) that the optimal solution of the problem is $(x_1, x_2) = (\frac{1}{2}, \frac{1}{2})$ and the optimal primal value is thus $f^* = (\frac{1}{2})^3 + (\frac{1}{2})^3 = \frac{1}{4}$. We will consider two possible options for constructing a dual problem. If we take the underlying set as

$$X = \{(x_1, x_2) : x_1, x_2 \geq 0\},$$

then the primal problem can be written as

$$\begin{aligned} \min \quad & x_1^3 + x_2^3 \\ \text{s.t.} \quad & x_1 + x_2 \geq 1, \\ & (x_1, x_2) \in X. \end{aligned}$$

Since the objective function is convex over X, it follows that this problem is in fact a convex optimization problem. Therefore, since Slater's condition is satisfied here (e.g., $(x_1, x_2) = (1, 1) \in X$ satisfy $x_1 + x_2 > 1$), we conclude from Theorem 12.8 that strong duality will hold. The dual problem in this case is constructed by associating a single dual variable λ to the linear inequality constraint $x_1 + x_2 \geq 1$. A second option for writing a dual problem is to take the underlying set X as the entire two-dimensional space and associate Lagrange multipliers with each of the three linear constraints. In this case, the assumptions of the strong duality theorem do not hold since $x_1^3 + x_2^3$ is not a convex function over $\mathbb{R}^2$. The Lagrangian is therefore ($\lambda, \eta_1, \eta_2 \in \mathbb{R}_+$)

$$L(x_1, x_2, \lambda, \eta_1, \eta_2) = x_1^3 + x_2^3 - \lambda(x_1 + x_2 - 1) - \eta_1 x_1 - \eta_2 x_2.$$

Since the minimization of a cubic function over the real line is always $-\infty$, it follows that

$$\begin{aligned} \min_{x_1, x_2} L(x_1, x_2, \lambda, \eta_1, \eta_2) &= \min_{x_1}\left[x_1^3 - (\lambda + \eta_1)x_1\right] + \min_{x_2}\left[x_2^3 - (\lambda + \eta_2)x_2\right] + \lambda \\ &= -\infty - \infty + \lambda = -\infty. \end{aligned}$$

Hence, $q^* = -\infty$ and the duality gap is infinite. The conclusion is that the way the dual problem is constructed is extremely important and may result in very different duality gaps. ∎

12.3 • Examples

12.3.1 • Linear Programming

Consider the linear programming problem

$$\begin{aligned} \min \quad & \mathbf{c}^T \mathbf{x} \\ \text{s.t.} \quad & \mathbf{A}\mathbf{x} \leq \mathbf{b}, \end{aligned}$$

where $\mathbf{c} \in \mathbb{R}^n, \mathbf{A} \in \mathbb{R}^{m \times n}$, and $\mathbf{b} \in \mathbb{R}^m$. We will assume that the problem is feasible, and under this condition, Slater's condition given in Theorem 12.12 is satisfied so that strong duality holds. The Lagrangian is ($\lambda \geq 0$)

$$L(\mathbf{x}, \lambda) = \mathbf{c}^T \mathbf{x} + \lambda^T (\mathbf{A}\mathbf{x} - \mathbf{b}) = (\mathbf{c} + \mathbf{A}^T \lambda)^T \mathbf{x} - \mathbf{b}^T \lambda,$$

and the dual objective function is given by

$$q(\lambda) = \min_{\mathbf{x} \in \mathbb{R}^n} L(\mathbf{x}, \lambda) = \min_{\mathbf{x} \in \mathbb{R}^n} (\mathbf{c} + \mathbf{A}^T \lambda)^T \mathbf{x} - \mathbf{b}^T \lambda = \begin{cases} -\mathbf{b}^T \lambda, & \mathbf{c} + \mathbf{A}^T \lambda = 0, \\ -\infty & \text{else.} \end{cases}$$

Therefore, the dual problem is

$$\begin{aligned} \max \quad & -\mathbf{b}^T \lambda \\ \text{s.t.} \quad & \mathbf{A}^T \lambda = -\mathbf{c}, \\ & \lambda \geq 0. \end{aligned}$$

As already mentioned, Slater's condition is satisfied if the primal problem is feasible and under the assumption that the optimal value is finite, strong duality holds.

12.3.2 ▪ Strictly Convex Quadratic Programming

Consider the following general strictly convex quadratic programming problem

$$\begin{aligned} \min \quad & \mathbf{x}^T \mathbf{Q} \mathbf{x} + 2\mathbf{f}^T \mathbf{x} \\ \text{s.t.} \quad & \mathbf{A}\mathbf{x} \leq \mathbf{b}, \end{aligned} \quad (12.15)$$

where $\mathbf{Q} \in \mathbb{R}^{n \times n}$ is positive definite, $\mathbf{f} \in \mathbb{R}^n, \mathbf{A} \in \mathbb{R}^{m \times n}$, and $\mathbf{b} \in \mathbb{R}^m$. The Lagrangian of the problem is

$$(\lambda \in \mathbb{R}_+^m) \quad L(\mathbf{x}, \lambda) = \mathbf{x}^T \mathbf{Q} \mathbf{x} + 2\mathbf{f}^T \mathbf{x} + 2\lambda^T (\mathbf{A}\mathbf{x} - \mathbf{b}) = \mathbf{x}^T \mathbf{Q} \mathbf{x} + 2(\mathbf{A}^T \lambda + \mathbf{f})^T \mathbf{x} - 2\mathbf{b}^T \lambda.$$

To find the dual objective function we need to minimize the Lagrangian with respect to $\mathbf{x}$. The minimizer of the Lagrangian is attained at the stationary point of the Lagrangian which is the solution to

$$\nabla_{\mathbf{x}} L(\mathbf{x}^*, \lambda) = 2\mathbf{Q}\mathbf{x}^* + 2(\mathbf{A}^T \lambda + \mathbf{f}) = 0,$$

and hence

$$\mathbf{x}^* = -\mathbf{Q}^{-1}(\mathbf{f} + \mathbf{A}^T \lambda). \quad (12.16)$$

Substituting this value back into the Lagrangian we obtain that

$$\begin{aligned} q(\lambda) &= L(\mathbf{x}^*, \lambda) \\ &= (\mathbf{f} + \mathbf{A}^T \lambda)^T \mathbf{Q}^{-1} \mathbf{Q} \mathbf{Q}^{-1} (\mathbf{f} + \mathbf{A}^T \lambda) - 2(\mathbf{f} + \mathbf{A}^T \lambda)^T \mathbf{Q}^{-1} (\mathbf{f} + \mathbf{A}^T \lambda) - 2\mathbf{b}^T \lambda \\ &= -(\mathbf{f} + \mathbf{A}^T \lambda)^T \mathbf{Q}^{-1} (\mathbf{f} + \mathbf{A}^T \lambda) - 2\mathbf{b}^T \lambda \\ &= -\lambda^T \mathbf{A} \mathbf{Q}^{-1} \mathbf{A}^T \lambda - 2\mathbf{f}^T \mathbf{Q}^{-1} \mathbf{A}^T \lambda - \mathbf{f}^T \mathbf{Q}^{-1} \mathbf{f} - 2\mathbf{b}^T \lambda \\ &= -\lambda^T \mathbf{A} \mathbf{Q}^{-1} \mathbf{A}^T \lambda - 2(\mathbf{A} \mathbf{Q}^{-1} \mathbf{f} + \mathbf{b})^T \lambda - \mathbf{f}^T \mathbf{Q}^{-1} \mathbf{f}. \end{aligned}$$

The dual problem is

$$\max\{q(\lambda) : \lambda \geq 0\}.$$

12.3. Examples

This problem is also a convex quadratic problem. However, its advantage over the primal problem (12.15) is that its feasible set is "simpler." In fact, we can develop a method for solving problem (12.15) which is based on an orthogonal projection method applied to the dual problem. As an illustration, we develop a method for computing the orthogonal projection onto a polytope defined by a set of inequalities.

Example 12.14. Given a polytope

$$S = \{\mathbf{x} \in \mathbb{R}^n : \mathbf{A}\mathbf{x} \leq \mathbf{b}\},$$

where $\mathbf{A} \in \mathbb{R}^{m \times n}, \mathbf{b} \in \mathbb{R}^m$, we wish to compute the orthogonal projection of a given point $\mathbf{y}$. As opposed to affine spaces, on which the orthogonal projection can be computed by a simple formula like the one derived in Example 10.10 of Section 10.2, there is no simple expression for the orthogonal projection onto polytopes, but using duality we will show that a method finding the projection can be derived. For a given $\mathbf{y} \in \mathbb{R}^n$, the problem of computing $P_S(\mathbf{y})$ can be written as

$$\begin{array}{ll} \min & \|\mathbf{x}-\mathbf{y}\|^2 \\ \text{s.t.} & \mathbf{A}\mathbf{x} \leq \mathbf{b}. \end{array}$$

This problem fits into the general model (12.15) with $\mathbf{Q} = \mathbf{I}$ and $\mathbf{f} = -\mathbf{y}$. Therefore, the dual problem is (omitting constants)

$$\begin{array}{ll} \max & -\boldsymbol{\lambda}^T \mathbf{A}\mathbf{A}^T \boldsymbol{\lambda} - 2(-\mathbf{A}\mathbf{y}+\mathbf{b})^T \boldsymbol{\lambda} \\ \text{s.t.} & \boldsymbol{\lambda} \geq \mathbf{0}. \end{array}$$

We assume that S is nonempty, and under this assumption, by Theorem 12.12, strong duality holds. We can solve the dual problem by the orthogonal projection method (presented in Section 9.4). If we use a constant stepsize, then it can be chosen as $\frac{1}{L}$, where L is the Lipschitz constant of the gradient of the objective function given by $L = 2\lambda_{\max}(\mathbf{A}\mathbf{A}^T)$. The general step of the method would then be

$$\boldsymbol{\lambda}_{k+1} = \left[\boldsymbol{\lambda}_k + \frac{2}{L}(-\mathbf{A}\mathbf{A}^T \boldsymbol{\lambda}_k + \mathbf{A}\mathbf{y} - \mathbf{b})\right]_+.$$

If the method stops at iteration N, then by (12.16) the primal optimal solution (up to a tolerance) is

$$\mathbf{x}^* = \mathbf{y} - \mathbf{A}^T \boldsymbol{\lambda}_N.$$

A MATLAB function implemeting this dual-based method is described below.

```
function x=proj_polytope(y,A,b,N)
%INPUT
%=================
% y ................ n-length vector
% A ................ mxn matrix
% b ................ m-length vector
% N ................ number of iterations
%OUTPUT
%=================
% x ................ x=P_S(y), where S = {x: Ax<=b}

d=size(A);
m=d(1);
n=d(2);
```

```
lam=zeros(m,1);

L=2*max(eig(A*A'));
g=A*y-b;
for k=1:N
    lam=lam+2/L*(-A*(A'*lam)+g);
    lam=max(lam,0);
end
x=y-A'*lam;
```

Consider for example the set

$$S = \{(x_1, x_2) : x_1 + x_2 \leq 1, x_1 \geq 0, x_2 \geq 0\}.$$

This is the triangle in the plane with vertices $(0,0), (0,1), (1,0)$. Suppose that we wish to find the orthogonal projection of $(2,-1)$ onto S. Then taking 100 iterations of the dual-based method gives the solution, which is the vertex $(1,0)$:

```
>> A=[1,1;-1,0;0,-1];
>> b=[1;0;0];
>> y=[2;-1];
>> x=proj_polytope(y,A,b,100)
x =

    1.0000
   -0.0000
```

An interesting property of the method can be seen when applying only a few iterations. For example, after 5 iterations the estimated primal solution is

```
>> x=proj_polytope(y,A,b,5)

x =

    1.5926
   -0.3663
```

This is of course not a feasible solution of the primal problem. In fact, all the iterates are nonfeasible, as illustrated in Figure 12.1, but they converge to the optimal solution, which is of course feasible. This was a demonstration of the fact that dual-based methods generate nonfeasible points with respect to the primal problem. ∎

12.3.3 • Dual of Convex Quadratic Problems

Now we will consider problem (12.15) when $\mathbf{Q}$ is positive semidefinite rather than positive definite. In this case, since $\mathbf{Q}$ is not necessarily invertible, the latter construction of the dual problem is not possible. To write a dual, we will use the fact that since $\mathbf{Q} \succeq \mathbf{0}$, it follows that there exists a matrix $\mathbf{D} \in \mathbb{R}^{n \times n}$ such that $\mathbf{Q} = \mathbf{D}^T \mathbf{D}$. Therefore, the problem can be recast as

$$\begin{aligned} \min \quad & \mathbf{x}^T \mathbf{D}^T \mathbf{D} \mathbf{x} + 2\mathbf{f}^T \mathbf{x} \\ \text{s.t.} \quad & \mathbf{A}\mathbf{x} \leq \mathbf{b}. \end{aligned}$$

12.3. Examples

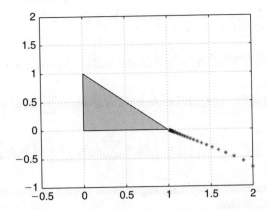

Figure 12.1. *First 30 iterations of the dual-based method (denoted by asterisks) for finding the orthogonal projection.*

We can now rewrite the problem by using an additional variables vector $\mathbf{z}$ as

$$\min_{\mathbf{x},\mathbf{z}} \quad \|\mathbf{z}\|^2 + 2\mathbf{f}^T\mathbf{x}$$
$$\text{s.t.} \quad \mathbf{A}\mathbf{x} \leq \mathbf{b},$$
$$\mathbf{z} = \mathbf{D}\mathbf{x}.$$

The Lagrangian of the new reformulation is

$$(\lambda \in \mathbb{R}_+^m, \mu \in \mathbb{R}^n) \quad L(\mathbf{x},\mathbf{z},\lambda,\mu) = \|\mathbf{z}\|^2 + 2\mathbf{f}^T\mathbf{x} + 2\lambda^T(\mathbf{A}\mathbf{x}-\mathbf{b}) + 2\mu^T(\mathbf{z}-\mathbf{D}\mathbf{x})$$
$$= \|\mathbf{z}\|^2 + 2\mu^T\mathbf{z} + 2(\mathbf{f} + \mathbf{A}^T\lambda - \mathbf{D}^T\mu)^T\mathbf{x} - 2\mathbf{b}^T\lambda.$$

The Lagrangian is separable with respect to $\mathbf{x}$ and $\mathbf{z}$, and we can thus perform the minimizations with respect to $\mathbf{x}$ and $\mathbf{z}$ separately:

$$\min_{\mathbf{x}}(\mathbf{f}+\mathbf{A}^T\lambda-\mathbf{D}^T\mu)^T\mathbf{x} = \begin{cases} 0, & \mathbf{f}+\mathbf{A}^T\lambda-\mathbf{D}^T\mu = 0, \\ -\infty & \text{else}, \end{cases}$$

$$\min_{\mathbf{z}} \|\mathbf{z}\|^2 + 2\mu^T\mathbf{z} = -\|\mu\|^2.$$

Hence,

$$q(\lambda,\mu) = \min_{\mathbf{x},\mathbf{z}} L(\mathbf{x},\mathbf{z},\lambda,\mu) = \begin{cases} -\|\mu\|^2 - 2\mathbf{b}^T\lambda, & \mathbf{f}+\mathbf{A}^T\lambda-\mathbf{D}^T\mu = 0, \\ -\infty & \text{else}. \end{cases}$$

The dual problem is thus

$$\max \quad -\|\mu\|^2 - 2\mathbf{b}^T\lambda$$
$$\text{s.t.} \quad \mathbf{f}+\mathbf{A}^T\lambda-\mathbf{D}^T\mu = 0,$$
$$\lambda \in \mathbb{R}_+^m, \mu \in \mathbb{R}^n.$$

12.3.4 ▪ Convex QCQPs

Consider the QCQP problem

$$\min \quad \mathbf{x}^T\mathbf{A}_0\mathbf{x} + 2\mathbf{b}_0^T\mathbf{x} + c_0$$
$$\text{s.t.} \quad \mathbf{x}^T\mathbf{A}_i\mathbf{x} + 2\mathbf{b}_i^T\mathbf{x} + c_i \leq 0, \quad i=1,2,\ldots,m,$$

where $A_i \succeq 0$ is an $n \times n$ matrix, $b_i \in \mathbb{R}^n, c_i \in \mathbb{R}, i = 0, 1, \ldots, m$. We will consider two cases.

Case I: When $A_0 \succ 0$, then the dual can be constructed as follows. The Lagrangian is

$$(\lambda \in \mathbb{R}_+^m) \quad L(\mathbf{x}, \lambda) = \mathbf{x}^T A_0 \mathbf{x} + 2\mathbf{b}_0^T \mathbf{x} + c_0 + \sum_{i=1}^m \lambda_i (\mathbf{x}^T A_i \mathbf{x} + 2\mathbf{b}_i^T \mathbf{x} + c_i)$$

$$= \mathbf{x}^T \left(A_0 + \sum_{i=1}^m \lambda_i A_i \right) \mathbf{x} + 2 \left(\mathbf{b}_0 + \sum_{i=1}^m \lambda_i \mathbf{b}_i \right)^T \mathbf{x} + c_0 + \sum_{i=1}^m \lambda_i c_i.$$

The minimizer of the Lagrangian with respect to $\mathbf{x}$ is attained at the point in which its gradient is zero:

$$2 \left(A_0 + \sum_{i=1}^m \lambda_i A_i \right) \tilde{\mathbf{x}} = -2 \left(\mathbf{b}_0 + \sum_{i=1}^m \lambda_i \mathbf{b}_i \right).$$

Thus,

$$\tilde{\mathbf{x}} = -\left(A_0 + \sum_{i=1}^m \lambda_i A_i \right)^{-1} \left(\mathbf{b}_0 + \sum_{i=1}^m \lambda_i \mathbf{b}_i \right).$$

Plugging this expression back into the Lagrangian, we obtain the following expression for the dual objective function:

$$q(\lambda) = \min_{\mathbf{x}} L(\mathbf{x}, \lambda) = L(\tilde{\mathbf{x}}, \lambda)$$

$$= \tilde{\mathbf{x}}^T \left(A_0 + \sum_{i=1}^m \lambda_i A_i \right) \tilde{\mathbf{x}} + 2 \left(\mathbf{b}_0 + \sum_{i=1}^m \lambda_i \mathbf{b}_i \right)^T \tilde{\mathbf{x}} + c_0 + \sum_{i=1}^m \lambda_i c_i$$

$$= -\left(\mathbf{b}_0 + \sum_{i=1}^m \lambda_i \mathbf{b}_i \right)^T \left(A_0 + \sum_{i=1}^m \lambda_i A_i \right)^{-1} \left(\mathbf{b}_0 + \sum_{i=1}^m \lambda_i \mathbf{b}_i \right) + c_0 + \sum_{i=1}^m \lambda_i c_i.$$

The dual problem is thus

$$\max \quad -\left(\mathbf{b}_0 + \sum_{i=1}^m \lambda_i \mathbf{b}_i \right)^T \left(A_0 + \sum_{i=1}^m \lambda_i A_i \right)^{-1} \left(\mathbf{b}_0 + \sum_{i=1}^m \lambda_i \mathbf{b}_i \right) + c_0 + \sum_{i=1}^m \lambda_i c_i$$
$$\text{s.t.} \quad \lambda_i \geq 0, \quad i = 1, 2, \ldots, m.$$

Case II: When A_0 is not positive definite but still positive semidefinite, the above dual is not well-defined since the matrix $A_0 + \sum_{i=1}^m \lambda_i A_i$ is not necessarily invertible. However, we can construct a different dual by decomposing the positive semidefinite matrices A_i as $A_i = D_i^T D_i$ ($D_i \in \mathbb{R}^{n \times n}$) and writing the equivalent formulation

$$\min \quad \mathbf{x}^T D_0^T D_0 \mathbf{x} + 2\mathbf{b}_0^T \mathbf{x} + c_0$$
$$\text{s.t.} \quad \mathbf{x}^T D_i^T D_i \mathbf{x} + 2\mathbf{b}_i^T \mathbf{x} + c_i \leq 0, \quad i = 1, 2, \ldots, m,$$

Now, we can add additional variables $\mathbf{z}_i \in \mathbb{R}^n (i = 0, 1, 2, \ldots, m)$ that will be defined to be $\mathbf{z}_i = D_i \mathbf{x}$, giving rise to the formulation

$$\min \quad \|\mathbf{z}_0\|^2 + 2\mathbf{b}_0^T \mathbf{x} + c_0$$
$$\text{s.t.} \quad \|\mathbf{z}_i\|^2 + 2\mathbf{b}_i^T \mathbf{x} + c_i \leq 0, \quad i = 1, 2, \ldots, m,$$
$$\mathbf{z}_i = D_i \mathbf{x}, \quad i = 0, 1, \ldots, m.$$

12.3. Examples

The Lagrangian is ($\lambda \in \mathbb{R}_+^m, \mu_i \in \mathbb{R}^n, i = 0, 1, \ldots, m$)

$$L(\mathbf{x}, \mathbf{z}_0, \ldots, \mathbf{z}_m, \lambda, \mu_0, \ldots, \mu_m)$$
$$= \|\mathbf{z}_0\|^2 + 2\mathbf{b}_0^T\mathbf{x} + c_0 + \sum_{i=1}^m \lambda_i(\|\mathbf{z}_i\|^2 + 2\mathbf{b}_i^T\mathbf{x} + c_i) + 2\sum_{i=0}^m \mu_i^T(\mathbf{z}_i - \mathbf{D}_i\mathbf{x})$$
$$= \|\mathbf{z}_0\|^2 + 2\mu_0^T \mathbf{z}_0 + \sum_{i=1}^m (\lambda_i\|\mathbf{z}_i\|^2 + 2\mu_i^T \mathbf{z}_i) + 2\left(\mathbf{b}_0 + \sum_{i=1}^m \lambda_i \mathbf{b}_i - \sum_{i=0}^m \mathbf{D}_i^T \mu_i\right)^T \mathbf{x}$$
$$+ c_0 + \sum_{i=1}^m c_i \lambda_i.$$

Note that for any $\lambda \in \mathbb{R}_+, \mu \in \mathbb{R}^n$ we have

$$g(\lambda, \mu) \equiv \min_{\mathbf{z}} \lambda\|\mathbf{z}\|^2 + 2\mu^T \mathbf{z} = \begin{cases} -\frac{\|\mu\|^2}{\lambda}, & \lambda > 0, \\ 0, & \lambda = 0, \mu = 0, \\ -\infty, & \lambda = 0, \mu \neq 0. \end{cases}$$

Since the Lagrangian is separable with respect to $\mathbf{z}_i$ and $\mathbf{x}$, we can perform the minimization with respect to each of the variables vectors,

$$\min_{\mathbf{z}_0}\left[\|\mathbf{z}_0\|^2 + 2\mu_0^T \mathbf{z}_0\right] = g(1, \mu_0) = -\|\mu_0\|^2,$$
$$\min_{\mathbf{z}_i}\left[\lambda_i\|\mathbf{z}_i\|^2 + 2\mu_i^T \mathbf{z}_i\right] = g(\lambda_i, \mu_i),$$
$$\min_{\mathbf{x}} \left(\mathbf{b}_0 + \sum_{i=1}^m \lambda_i \mathbf{b}_i - \sum_{i=0}^m \mathbf{D}_i^T \mu_i\right)^T \mathbf{x} = \begin{cases} 0, & \mathbf{b}_0 + \sum_{i=1}^m \lambda_i \mathbf{b}_i - \sum_{i=0}^m \mathbf{D}_i^T \mu_i = 0, \\ -\infty & \text{else,} \end{cases}$$

and deduce that

$$q(\lambda, \mu_0, \ldots, \mu_m) = \min_{\mathbf{x}, \mathbf{z}_0, \ldots, \mathbf{z}_m} L(\mathbf{x}, \mathbf{z}_0, \ldots, \mathbf{z}_m, \lambda, \mu_0, \ldots, \mu_m)$$
$$= \begin{cases} g(1, \mu_0) + \sum_{i=1}^m g(\lambda_i, \mu_i) + c_0 + \sum_{i=1}^m c_i \lambda_i & \text{if } \mathbf{b}_0 + \sum_{i=1}^m \lambda_i \mathbf{b}_i - \sum_{i=0}^m \mathbf{D}_i^T \mu_i = 0, \\ -\infty & \text{else.} \end{cases}$$

The dual problem is therefore

$$\begin{aligned} \max \quad & g(1, \mu_0) + \sum_{i=1}^m g(\lambda_i, \mu_i) + c_0 + \sum_{i=1}^m c_i \lambda_i \\ \text{s.t.} \quad & \mathbf{b}_0 + \sum_{i=1}^m \lambda_i \mathbf{b}_i - \sum_{i=0}^m \mathbf{D}_i^T \mu_i = 0, \\ & \lambda \in \mathbb{R}_+^m, \mu_0, \ldots, \mu_m \in \mathbb{R}^n. \end{aligned}$$

12.3.5 • Nonconvex QCQPs

Consider now the QCQP problem

$$\begin{aligned} \min \quad & \mathbf{x}^T \mathbf{A}_0 \mathbf{x} + 2\mathbf{b}_0^T \mathbf{x} + c_0 \\ \text{s.t.} \quad & \mathbf{x}^T \mathbf{A}_i \mathbf{x} + 2\mathbf{b}_i^T \mathbf{x} + c_i \leq 0, \quad i = 1, 2, \ldots, m, \end{aligned}$$

where $\mathbf{A}_i = \mathbf{A}_i^T \in \mathbb{R}^{n \times n}, \mathbf{b}_i \in \mathbb{R}^n, c_i \in \mathbb{R}, i = 0, 1, \ldots, m$. We do not assume that $\mathbf{A}_i$ are positive semidefinite, and hence the problem is in general nonconvex, and the techniques used in the convex case are not applicable. We begin by forming the Lagrangian ($\lambda \in \mathbb{R}^m_+$):

$$L(\mathbf{x}, \lambda) = \mathbf{x}^T \mathbf{A}_0 \mathbf{x} + 2\mathbf{b}_0^T \mathbf{x} + c_0 + \sum_{i=1}^m \lambda_i \left(\mathbf{x}^T \mathbf{A}_i \mathbf{x} + 2\mathbf{b}_i^T \mathbf{x} + c_i\right)$$

$$= \mathbf{x}^T \left(\mathbf{A}_0 + \sum_{i=1}^m \lambda_i \mathbf{A}_i\right) \mathbf{x} + 2 \left(\mathbf{b}_0 + \sum_{i=1}^m \lambda_i \mathbf{b}_i\right)^T \mathbf{x} + c_0 + \sum_{i=1}^m \lambda_i c_i.$$

The main idea is to use the following presentation of the dual objective function:

$$q(\lambda) = \min_{\mathbf{x}} L(\mathbf{x}, \lambda) = \max_t \{t : L(\mathbf{x}, \lambda) \geq t \text{ for any } \mathbf{x} \in \mathbb{R}^n\}. \quad (12.17)$$

The above equation essentially states that the minimal value of $L(\mathbf{x}, \lambda)$ over $\mathbf{x} \in \mathbb{R}^n$ is in fact the largest lower bound on the function. We will now use Theorem 2.43 on the characterization of the nonnegativity of quadratic functions and deduce that the claim

$$L(\mathbf{x}, \lambda) \geq t \text{ for all } \mathbf{x} \in \mathbb{R}^n$$

is equivalent to

$$\begin{pmatrix} \mathbf{A}_0 + \sum_{i=1}^m \lambda_i \mathbf{A}_i & \mathbf{b}_0 + \sum_{i=1}^m \lambda_i \mathbf{b}_i \\ (\mathbf{b}_0 + \sum_{i=1}^m \lambda_i \mathbf{b}_i)^T & c_0 + \sum_{i=1}^m \lambda_i c_i - t \end{pmatrix} \succeq 0,$$

which combined with (12.17) yields that a dual problem is

$$\begin{array}{ll} \max_{t, \lambda_i} & t \\ \text{s.t.} & \begin{pmatrix} \mathbf{A}_0 + \sum_{i=1}^m \lambda_i \mathbf{A}_i & \mathbf{b}_0 + \sum_{i=1}^m \lambda_i \mathbf{b}_i \\ (\mathbf{b}_0 + \sum_{i=1}^m \lambda_i \mathbf{b}_i)^T & c_0 + \sum_{i=1}^m \lambda_i c_i - t \end{pmatrix} \succeq 0, \\ & \lambda_i \geq 0, \quad i = 1, 2, \ldots, m. \end{array} \quad (12.18)$$

The above problem is convex as a dual problem, but since the primal problem is nonconvex, strong duality is of course not guaranteed. We also note that the form of the dual problem (12.18) is different from all the dual problems derived so far in the sense that not all the constraints are presented as inequality or equality constraints but instead are of the form $\mathbf{B}_0 + \sum_{i=1}^m \lambda_i \mathbf{B}_i \succeq 0$, where $\mathbf{B}_i$ are given matrices. This type of a constraint is called a *linear matrix inequality* (abbreviated LMI). Optimization problems consisting of minimization or maximization of a linear function subject to linear inequalities/equalities and LMIs are called semidefinite programming (SDP) problems, and they are part of a larger class of problems called *conic problems*; see also Section 12.3.9.

12.3.6 ▪ Orthogonal Projection onto the Unit-Simplex

Given a vector $\mathbf{y} \in \mathbb{R}^n$, we would like to compute the orthogonal projection of the vector $\mathbf{y}$ onto Δ_n. The corresponding optimization problem is

$$\begin{array}{ll} \min & \|\mathbf{x} - \mathbf{y}\|^2 \\ \text{s.t.} & \mathbf{e}^T \mathbf{x} = 1, \\ & \mathbf{x} \geq 0. \end{array}$$

We will associate a Lagrange multiplier $\lambda \in \mathbb{R}$ to the linear equality constraint $\mathbf{e}^T \mathbf{x} = 1$ and obtain the Lagrangian function

$$L(\mathbf{x}, \lambda) = \|\mathbf{x} - \mathbf{y}\|^2 + 2\lambda(\mathbf{e}^T \mathbf{x} - 1) = \|\mathbf{x}\|^2 - 2(\mathbf{y} - \lambda \mathbf{e})^T \mathbf{x} + \|\mathbf{y}\|^2 - 2\lambda$$

$$= \sum_{j=1}^n (x_j^2 - 2(y_j - \lambda)x_j) + \|\mathbf{y}\|^2 - 2\lambda.$$

12.3. Examples

The arising problem is therefore saparable with respect to the variables x_j and hence the optimal x_j is the solution to the one-dimensional problem

$$\min_{x_j \geq 0} [x_j^2 - 2(y_j - \lambda)x_j].$$

The optimal solution to the above problem is given by

$$x_j = \begin{cases} y_j - \lambda, & y_j \geq \lambda \\ 0 & \text{else} \end{cases} = [y_j - \lambda]_+,$$

and the optimal value is $-[y_j - \lambda]_+^2$. The dual problem is therefore

$$\max_{\lambda \in \mathbb{R}} \left\{ g(\lambda) \equiv -\sum_{j=1}^{n} [y_j - \lambda]_+^2 - 2\lambda + \|\mathbf{y}\|^2 \right\}.$$

By the basic properties of dual problems, the dual objective function is concave. In order to actually solve the dual problem, we note that

$$\lim_{\lambda \to \infty} g(\lambda) = \lim_{\lambda \to -\infty} g(\lambda) = -\infty.$$

Therefore, since $-g$ is a coercive and differentiable function, it follows that there exists an optimal solution to the dual problem attained at a point λ in which

$$g'(\lambda) = 0,$$

meaning that

$$\sum_{j=1}^{n} [y_j - \lambda]_+ = 1.$$

The function $h(\lambda) = \sum_{j=1}^{n} [y_j - \lambda]_+ - 1$ is a nonincreasing function over $\mathbb{R}$ and is in fact strictly decreasing over $(-\infty, \max_j y_j]$. In addition, by denoting $y_{\max} = \max_{j=1,2,\ldots,n} y_j$, $y_{\min} = \min_{j=1,2,\ldots,n} y_j$, we have

$$h(y_{\max}) = -1,$$

$$h\left(y_{\min} - \frac{2}{n}\right) = \sum_{j=1}^{n} y_j - n y_{\min} + 2 - 1 > 0,$$

and we can therefore invoke a bisection procedure to find the unique root λ of the function h over the interval $[y_{\min} - \frac{2}{n}, y_{\max}]$ and then define $P_{\Delta_n}(\mathbf{y}) = [\mathbf{y} - \lambda \mathbf{e}]_+$. The MATLAB implementation follows.

```
function xp=proj_unit_simplex(y)
% INPUT
% =====================
% y .......a column vector
% OUTPUT
% =====================
% xp .......the orthogonal projection of y onto the unit simplex

f=@(lam) sum(max(y-lam,0))-1;
n=length(y);
lb=min(y)-2/n;
```

```
ub=max(y);
lam=bisection(f,lb,ub,1e-10);
xp=max(y-lam,0);
```

As a sanity check, let us compute the orthogonal projection of the vector $(-1,1,0.3)^T$ onto Δ_3,

```
>> xp=proj_unit_simplex([-1;1;0.3])
xp =
         0
    0.8500
    0.1500
```

and compare it with the result of CVX,

```
cvx_begin
variable x(3)
minimize(norm(x-[-1;1;0.3]))
sum(x)==1
x>=0
cvx_end
```

which unsurprisingly is the same:

```
>> x
x =
    0.0000
    0.8500
    0.1500
```

12.3.7 ▪ Dual of the Chebyshev Center Problem

Recall that in the Chebyshev center problem (see also Section 8.2.4) we are given a set of points $\mathbf{a}_1, \mathbf{a}_2, \ldots, \mathbf{a}_m \in \mathbb{R}^n$, and we seek to find a point $\mathbf{x} \in \mathbb{R}^n$, which is the center of the minimum radius ball containing the points

$$\begin{array}{ll} \min_{\mathbf{x},r} & r \\ \text{s.t.} & \|\mathbf{x} - \mathbf{a}_i\| \leq r, \quad i = 1, 2, \ldots, m. \end{array}$$

Finding a dual problem to this formulation is not an easy task, but we can actually consider a different equivalent formulation to which a dual can be constructed in an easier way. The problem can be recast as

$$\begin{array}{ll} \min_{\mathbf{x},\gamma} & \gamma \\ \text{s.t.} & \|\mathbf{x} - \mathbf{a}_i\|^2 \leq \gamma, \quad i = 1, 2, \ldots, m. \end{array}$$

where γ denotes the squared radius. The problems are equivalent since minimization of the radius is equivalent to minimization of the squared radius. The Lagrangian is ($\boldsymbol{\lambda} \in \mathbb{R}^m_+$)

$$L(\mathbf{x}, \gamma, \boldsymbol{\lambda}) = \gamma + \sum_{i=1}^m \lambda_i (\|\mathbf{x} - \mathbf{a}_i\|^2 - \gamma)$$

$$= \gamma \left(1 - \sum_{i=1}^m \lambda_i\right) + \sum_{i=1}^m \lambda_i \|\mathbf{x} - \mathbf{a}_i\|^2.$$

12.3. Examples

The minimization of the above expression must be $-\infty$ unless $\sum_{i=1}^{m} \lambda_i = 1$, and in this case we have

$$\min_{\gamma} \gamma\left(1 - \sum_{i=1}^{m} \lambda_i\right) = 0.$$

We are therefore left with the task of finding the optimal value of

$$\min_{\mathbf{x}} \sum_{i=1}^{m} \lambda_i \|\mathbf{x} - \mathbf{a}_i\|^2$$

when $\lambda \in \Delta_m$. Since the objective function of the above minimization can be written as

$$\sum_{i=1}^{m} \lambda_i \|\mathbf{x} - \mathbf{a}_i\|^2 = \|\mathbf{x}\|^2 - 2\left(\sum_{i=1}^{m} \lambda_i \mathbf{a}_i\right)^T \mathbf{x} + \sum_{i=1}^{m} \lambda_i \|\mathbf{a}_i\|^2, \quad (12.19)$$

it follows that the minimum is attained at the point in which the gradient vanishes, meaning at

$$\mathbf{x}^* = \sum_{i=1}^{m} \lambda_i \mathbf{a}_i = \mathbf{A}\lambda, \quad (12.20)$$

where $\mathbf{A}$ is the $n \times m$ matrix whose columns are the vectors $\mathbf{a}_1, \mathbf{a}_2, \ldots, \mathbf{a}_m$. Substituting this expression back into (12.19), we have that the dual objective function is

$$q(\lambda) = \|\mathbf{A}\lambda\|^2 - 2(\mathbf{A}\lambda)^T(\mathbf{A}\lambda) + \sum_{i=1}^{m} \lambda_i \|\mathbf{a}_i\|^2 = -\|\mathbf{A}\lambda\|^2 + \sum_{i=1}^{m} \lambda_i \|\mathbf{a}_i\|^2.$$

The dual problem is therefore

$$\begin{array}{ll} \max & -\|\mathbf{A}\lambda\|^2 + \sum_{i=1}^{m} \lambda_i \|\mathbf{a}_i\|^2 \\ \text{s.t.} & \lambda \in \Delta_m. \end{array}$$

We can actually write a MATLAB function that solves this problem. For that, we will use the gradient projection method with a constant stepsize $\frac{1}{L}$, where $L = 2\lambda_{\max}(\mathbf{A}^T\mathbf{A})$ is the Lipschitz constant of the gradient of the objective function. At each iteration we will also use the MATLAB function proj_unit_simplex to find the orthogonal projection onto the unit-simplex. Note that the derived method is also a dual-based method and that it incorporates another dual-based method for computing the projection.

```
function [xp,r]=chebyshev_center(A)
% INPUT
% =======================
% A ........ a matrix whose columns are points in R^m
% OUTPUT
% =======================
% xp ........ the Chebyshev center of the points
% r ........ radius of the Chebyshev circle

d=size(A);
m=d(2);
Q=A'*A;
L=2*max(eig(Q));
b=sum(A.^2)';
%initialization with the uniform vector
lam=1/m*ones(m,1);
```

```
old_lam=zeros(m,1);
while (norm(lam-old_lam)>1e-5)
    old_lam=lam;
    lam=proj_unit_simplex(lam+1/L*(-2*Q*lam+b));
end
xp=A*lam;
r=0;
for i=1:m
    r=max(r,norm(xp-A(:,i)));
end
```

Example 12.15. Returning to Example 8.14, suppose that we wish to find the Chebyshev center of the 5 points

$$(-1,3), \quad (-3,10), \quad (-1,0), \quad (5,0), \quad (-1,-5).$$

For that, we can invoke the MATLAB function `chebyshev_center` that was just described:

```
A=[-1,-3,-1,5,-1;3,10,0,0,-5];
[xp,r]=chebyshev_center(A);
```

The Chebyshev center and radius are

```
>> xp

xp =

   -2.0000
    2.5000

>> r

r =

    7.5664
```

These are the exact same results obtained by CVX in Example 8.14. ∎

12.3.8 • Minimization of Sum of Norms

Consider the problem

$$\min_{\mathbf{x} \in \mathbb{R}^n} \sum_{i=1}^{m} \|\mathbf{A}_i \mathbf{x} + \mathbf{b}_i\|, \tag{12.21}$$

where $\mathbf{A}_i \in \mathbb{R}^{k_i \times n}, \mathbf{b}_i \in \mathbb{R}^{k_i}, i = 1, 2, \ldots, m$. At a first glance, it seems that it is not possible to find a dual of an unconstrained problem. However, we can use a technique of variable decoupling to reformulate the problem as a problem with affine constraints. Specifically, problem (12.21) is the same as

$$\begin{aligned}
\min_{\mathbf{x}, \mathbf{y}_i} & \quad \sum_{i=1}^{m} \|\mathbf{y}_i\| \\
\text{s.t.} & \quad \mathbf{y}_i = \mathbf{A}_i \mathbf{x} + \mathbf{b}_i, \quad i = 1, 2, \ldots, m.
\end{aligned}$$

12.3. Examples

Associating a Lagrange multiplier vector $\lambda_i \in \mathbb{R}^{k_i}$ with the ith set of constraints $\mathbf{y}_i = \mathbf{A}_i \mathbf{x} + \mathbf{b}_i$, we obtain the following Lagrangian:

$$L(\mathbf{x}, \mathbf{y}_1, \mathbf{y}_2, \ldots, \mathbf{y}_m, \lambda_1, \lambda_2, \ldots, \lambda_m) = \sum_{i=1}^{m} \|\mathbf{y}_i\| + \sum_{i=1}^{m} \lambda_i^T (\mathbf{y}_i - \mathbf{A}_i \mathbf{x} - \mathbf{b}_i)$$

$$= \sum_{i=1}^{m} \left[\|\mathbf{y}_i\| + \lambda_i^T \mathbf{y}_i \right] - \left(\sum_{i=1}^{m} \mathbf{A}_i^T \lambda_i \right)^T \mathbf{x} - \sum_{i=1}^{m} \mathbf{b}_i^T \lambda_i.$$

By the separability of the Lagrangian with respect to $\mathbf{x}, \mathbf{y}_1, \mathbf{y}_2, \ldots, \mathbf{y}_m$, it follows that the dual objective function is given by

$$q(\lambda_1, \lambda_2, \ldots, \lambda_m) = \sum_{i=1}^{m} \min_{\mathbf{y}_i \in \mathbb{R}^{k_i}} \left[\|\mathbf{y}_i\| + \lambda_i^T \mathbf{y}_i \right] + \min_{\mathbf{x} \in \mathbb{R}^n} \left[-\left(\sum_{i=1}^{m} \mathbf{A}_i^T \lambda_i \right)^T \mathbf{x} \right] - \sum_{i=1}^{m} \mathbf{b}_i^T \lambda_i.$$

Obviously,

$$\min_{\mathbf{x} \in \mathbb{R}^n} \left[-\left(\sum_{i=1}^{m} \mathbf{A}_i^T \lambda_i \right)^T \mathbf{x} \right] = \begin{cases} 0, & \sum_{i=1}^{m} \mathbf{A}_i^T \lambda_i = 0, \\ -\infty & \text{else.} \end{cases} \quad (12.22)$$

In addition, we have for any $\mathbf{a} \in \mathbb{R}^k$

$$\min_{\mathbf{y} \in \mathbb{R}^k} \|\mathbf{y}\| + \mathbf{a}^T \mathbf{y} = \begin{cases} 0, & \|\mathbf{a}\| \leq 1, \\ -\infty, & \|\mathbf{a}\| > 1. \end{cases} \quad (12.23)$$

To prove this result, note that when $\|\mathbf{a}\| \leq 1$, we have by the Cauchy–Schwarz inequality that for any $\mathbf{y} \in \mathbb{R}^k$

$$\mathbf{a}^T \mathbf{y} \geq -\|\mathbf{a}\| \cdot \|\mathbf{y}\| \geq -\|\mathbf{y}\|$$

and hence $\|\mathbf{y}\| + \mathbf{a}^T \mathbf{y} \geq 0$ for any $\mathbf{y} \in \mathbb{R}^k$, and in addition $\|\mathbf{0}\| + \mathbf{a}^T \mathbf{0} = 0$, implying that $\min_{\mathbf{y} \in \mathbb{R}^k} \|\mathbf{y}\| + \mathbf{a}^T \mathbf{y} = 0$. If $\|\mathbf{a}\| > 1$, then taking $\mathbf{y}_\alpha = -\alpha \mathbf{a}$ we obtain that $\|\mathbf{y}_\alpha\| + \mathbf{a}^T \mathbf{y}_\alpha = \alpha \|\mathbf{a}\|(1 - \|\mathbf{a}\|) \to -\infty$ as $\alpha \to \infty$, establishing the result (12.23). We thus conclude that for any $i = 1, 2, \ldots, m$

$$\min_{\mathbf{y}_i \in \mathbb{R}^{k_i}} \left[\|\mathbf{y}_i\| + \lambda_i^T \mathbf{y}_i \right] = \begin{cases} 0, & \|\lambda_i\| \leq 1, \\ -\infty & \text{else,} \end{cases}$$

which combined with (12.22) implies that the dual objective function is

$$q(\lambda_1, \lambda_2, \ldots, \lambda_m) = \begin{cases} -\sum_{i=1}^{m} \lambda_i^T \mathbf{b}_i, & \|\lambda_i\| \leq 1, i = 1, 2, \ldots, m, \\ & \sum_{i=1}^{m} \mathbf{A}_i^T \lambda_i = 0, \\ -\infty & \text{else.} \end{cases}$$

The dual problem is therefore

$$\begin{aligned} \max \quad & -\sum_{i=1}^{m} \mathbf{b}_i^T \lambda_i \\ \text{s.t.} \quad & \sum_{i=1}^{m} \mathbf{A}_i^T \lambda_i = 0, \\ & \|\lambda_i\| \leq 1, \quad i = 1, 2, \ldots, m. \end{aligned}$$

12.3.9 • Conic Duality

A conic optimization problem is a convex optimization problem of the form

$$\text{(C)} \quad \begin{aligned} \min \quad & \mathbf{a}^T\mathbf{x} \\ \text{s.t.} \quad & \mathbf{A}\mathbf{x} = \mathbf{b}, \\ & \mathbf{x} \in K, \end{aligned}$$

where $K \subseteq \mathbb{R}^n$ is a closed and convex cone and $\mathbf{a} \in \mathbb{R}^n, \mathbf{A} \in \mathbb{R}^{m \times n}, \mathbf{b} \in \mathbb{R}^m$. Our aim is to find an expression for the dual problem. For that, let us associate a Lagrange multipliers vector $\mathbf{y} \in \mathbb{R}^m$ with the equality constraints $\mathbf{A}\mathbf{x} = \mathbf{b}$, leading to the Lagrangian

$$L(\mathbf{x},\mathbf{y}) = \mathbf{a}^T\mathbf{x} - \mathbf{y}^T(\mathbf{A}\mathbf{x} - \mathbf{b}) = (\mathbf{a} - \mathbf{A}^T\mathbf{y})^T\mathbf{x} + \mathbf{b}^T\mathbf{y}.$$

To construct the dual problem, we need to minimize the Lagrangian over $\mathbf{x} \in K$:

$$q(\mathbf{y}) = \min_{\mathbf{x} \in K} L(\mathbf{x},\mathbf{y}) = \mathbf{b}^T\mathbf{y} + \min_{\mathbf{x} \in K}(\mathbf{a} - \mathbf{A}^T\mathbf{y})^T\mathbf{x}.$$

We will now use the following easily verifiable fact: for any $\mathbf{d} \in \mathbb{R}^n$ one has

$$\min_{\mathbf{x} \in K} \mathbf{d}^T\mathbf{x} = \begin{cases} 0, & \mathbf{d} \in K^*, \\ -\infty, & \mathbf{d} \notin K^*. \end{cases}$$

where K^* is the dual cone defined in Exercise 6.11 as

$$K^* = \{\mathbf{z} \in \mathbb{R}^n : \mathbf{z}^T\mathbf{x} \geq 0 \text{ for any } \mathbf{x} \in K\}.$$

The latter fact is almost a tautology. If $\mathbf{d} \in K^*$, then by the definition of the dual cone, $\mathbf{d}^T\mathbf{x} \geq 0$ for any $\mathbf{x} \in K$ and also $\mathbf{d}^T\mathbf{0} = 0$ (recall that $\mathbf{0} \in K$ since K is a closed cone), and hence $\min_{\mathbf{x} \in K} \mathbf{d}^T\mathbf{x} = 0$. On the other hand, if $\mathbf{d} \notin K^*$, then it means that there exists $\mathbf{x}_0 \in K$ such that $\mathbf{d}^T\mathbf{x}_0 < 0$. Therefore, taking any $\alpha > 0$ we have that $\alpha \mathbf{x}_0 \in K$, and we obtain that $\mathbf{d}^T(\alpha \mathbf{x}_0) = \alpha(\mathbf{d}^T\mathbf{x}_0) \to -\infty$ as α tends to ∞, and hence $\min_{\mathbf{x} \in K} \mathbf{d}^T\mathbf{x} = -\infty$. To conclude, the dual objective function is

$$q(\mathbf{y}) = \begin{cases} \mathbf{b}^T\mathbf{y}, & \mathbf{a} - \mathbf{A}^T\mathbf{y} \in K^*, \\ -\infty, & \mathbf{a} - \mathbf{A}^T\mathbf{y} \notin K^*. \end{cases}$$

The dual problem is thus

$$\text{(DC)} \quad \begin{aligned} \max \quad & \mathbf{b}^T\mathbf{y} \\ \text{s.t.} \quad & \mathbf{a} - \mathbf{A}^T\mathbf{y} \in K^*. \end{aligned}$$

We can now invoke the strong duality theorem for convex problems (Theorem 12.12) and state one of the versions of the so-called *conic duality theorem*.

Theorem 12.16 (conic duality theorem). *Consider the primal and dual problems (C) and (DC). Suppose that there exists $\mathbf{x} \in \text{int}(K)$ such that $\mathbf{A}\mathbf{x} = \mathbf{b}$ and that problem (C) is bounded below. Then the dual problem (DC) has an optimal solution, and we have*

$$val(C) = val(DC).$$

12.3.10 • Denoising

Suppose that we are given a signal contaminated with noise. In mathematical terms the model is

$$\mathbf{y} = \mathbf{x} + \mathbf{w},$$

where $\mathbf{x} \in \mathbb{R}^n$ is the noise-free signal, $\mathbf{w} \in \mathbb{R}^n$ is the unknown noise vector, and $\mathbf{y} \in \mathbb{R}^n$ is the observed and known vector. An example of "clean" and "noisy" signals can be found in Figure 12.2.

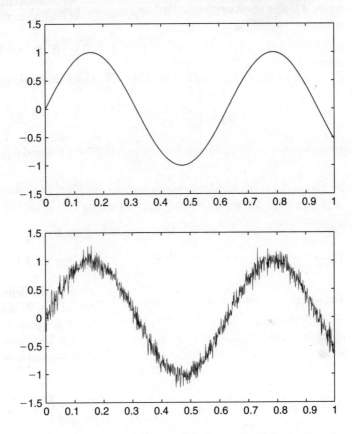

Figure 12.2. *True signal (top image) and noisy signal (bottom image).*

The plots were created by the MATLAB commands

```
randn('seed',314);
t=linspace(0,1,1000)';
n=length(t);
x=sin(10*t);
figure(1)
plot(t,x)
axis([0,1,-1.5,1.5]);
y=x+0.1*randn(size(t));
figure(2)
plot(t,y)
```

The objective is to reconstruct the true signal from the observed vector **y**. A common approach for denoising is to use some prior information on the true image. A natural information is the smoothness of the signal. This information can be incorporated by adding a quadratic penalty function that measures in some sense the smoothness of the signal. For example, a standard approach is to solve the optimization problem

$$\min \|\mathbf{x} - \mathbf{y}\|^2 + \lambda \sum_{i=1}^{n-1} (x_i - x_{i+1})^2,$$

where $\lambda > 0$ is some predetermined regularization parameter. The problem can also be written as

$$\min \|\mathbf{x} - \mathbf{y}\|^2 + \lambda \|\mathbf{L}\mathbf{x}\|^2, \tag{12.24}$$

where

$$\mathbf{L} = \begin{pmatrix} 1 & -1 & 0 & 0 & 0 & \cdots & 0 \\ 0 & 1 & -1 & 0 & 0 & \cdots & 0 \\ 0 & 0 & 1 & -1 & 0 & \cdots & 0 \\ 0 & 0 & 0 & 1 & -1 & \cdots & 0 \\ \vdots & \vdots & \vdots & \vdots & \vdots & \vdots & \vdots \\ 0 & 0 & 0 & 0 & 0 & 1 & -1 \end{pmatrix}.$$

Problem (12.24) is a regularized least squares problem (see Section 3.3), and its optimal solution can be derived by writing the stationarity condition

$$2(\mathbf{x} - \mathbf{y}) + 2\lambda \mathbf{L}^T \mathbf{L} \mathbf{x} = \mathbf{0}.$$

Thus,

$$\mathbf{x} = (\mathbf{I} + \lambda \mathbf{L}^T \mathbf{L})^{-1} \mathbf{y}. \tag{12.25}$$

The solution of the problem in the case of $\lambda = 1$ can thus be obtained by the MATLAB commands

```
L=sparse(n-1,n);
for i=1:n-1
    L(i,i)=1;
    L(i,i+1)=-1;
end

lambda=100;
xde=(speye(n)+lambda*L'*L)\y;
figure(3)
plot(t,xde);
```

resulting in the relatively good reconstruction given in Figure 12.3. The quadratic regularization method does not work so well for all types of signals. Suppose, for example, that we are given a noisy step signal generated by the MATLAB commands

```
randn('seed',314);
x=zeros(1000,1);
x(1:250)=1;
x(251:500)=3;
x(501:750)=0;
x(751:1000)=2;
```

```
figure(1)
plot(1:1000,x,'.')
axis([0,1000,-1,4]);
y=x+0.05*randn(size(x));
figure(2)
plot(1:1000,y,'.')
```

The "true" and "noisy" step signals are given in Figure 12.4.

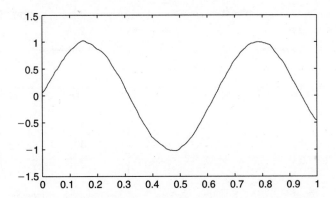

Figure 12.3. *Denoising of the sine signal via quadratic regularization.*

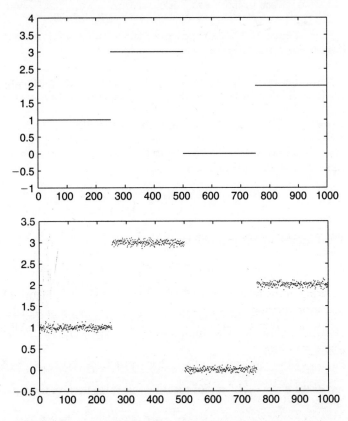

Figure 12.4. *True signal (top image) and noisy signal (bottom image).*

Unfortunately, the quadratic regularization approach does not yield good results, no matter what the value of the chosen regularization parameter λ is. Indeed, the regularized least squares solution (12.25) is not a good reconstruction since it is unable to deal correctly with the three breakpoints. The reason is that the jumps contribute large values to the penalty function $\|\mathbf{Lx}\|^2$ since their values are squared. Therefore, in a sense the regularized least squares solution tries to "smooth" the jumps, resulting in the reconstructions in Figure 12.5.

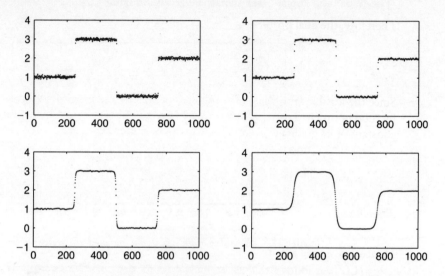

Figure 12.5. *Quadratic regularization with* $\lambda = 0.1$ *(top left),* $\lambda = 1$ *(top right),* $\lambda = 10$ *(bottom left),* $\lambda = 100$ *(bottom right).*

Another approach for denoising that is able to overcome this disadvantage is to solve the following problem in which the regularization term is in the l_1 norm:

$$\min \|\mathbf{x}-\mathbf{y}\|^2 + \lambda\|\mathbf{Lx}\|_1. \tag{12.26}$$

We would like to construct a dual to problem (12.26). For that, note that the problem is equivalent to the optimization problem

$$\begin{aligned} \min_{\mathbf{x},\mathbf{z}} \quad & \|\mathbf{x}-\mathbf{y}\|^2 + \lambda\|\mathbf{z}\|_1 \\ \text{s.t.} \quad & \mathbf{z} = \mathbf{Lx}. \end{aligned}$$

The Lagrangian of the problem is

$$\begin{aligned} L(\mathbf{x},\mathbf{z},\mu) &= \|\mathbf{x}-\mathbf{y}\|^2 + \lambda\|\mathbf{z}\|_1 + \mu^T(\mathbf{Lx}-\mathbf{z}) \\ &= \|\mathbf{x}-\mathbf{y}\|^2 + (\mathbf{L}^T\mu)^T\mathbf{x} + \lambda\|\mathbf{z}\|_1 - \mu^T\mathbf{z}. \end{aligned}$$

The Lagrangian is separable with respect to $\mathbf{x}$ and $\mathbf{z}$ and thus we can perform the minimization separately. The minimum of $\|\mathbf{x}-\mathbf{y}\|^2 + (\mathbf{L}^T\mu)^T\mathbf{x}$ over $\mathbf{x}$ is attained when the gradient vanishes,

$$2(\mathbf{x}-\mathbf{y}) + \mathbf{L}^T\mu = \mathbf{0},$$

and hence $\mathbf{x} = \mathbf{y} - \frac{1}{2}\mathbf{L}^T\mu$. Substituting this value back to the $\mathbf{x}$-part of the Lagrangian, we obtain

$$\min_{\mathbf{x}} \|\mathbf{x}-\mathbf{y}\|^2 + (\mathbf{L}^T\mu)^T\mathbf{x} = -\frac{1}{4}\mu^T\mathbf{LL}^T\mu + \mu^T\mathbf{Ly}.$$

12.3. Examples

In addition,

$$\min_{\mathbf{z}} \lambda \|\mathbf{z}\|_1 - \mu^T \mathbf{z} = \begin{cases} 0, & \|\mu\|_\infty \leq \lambda, \\ -\infty & \text{else.} \end{cases}$$

To conclude, the dual objective function is given by

$$q(\mu) = \min_{\mathbf{x},\mathbf{z}} L(\mathbf{x},\mathbf{z},\mu) = \begin{cases} -\frac{1}{4}\mu^T \mathbf{LL}^T \mu + \mu^T \mathbf{Ly}, & \|\mu\|_\infty \leq \lambda, \\ -\infty & \text{else.} \end{cases}$$

Therefore, the dual problem is

$$\begin{aligned} \max \quad & -\tfrac{1}{4}\mu^T \mathbf{LL}^T \mu + \mu^T \mathbf{Ly} \\ \text{s.t.} \quad & \|\mu\|_\infty \leq \lambda. \end{aligned} \qquad (12.27)$$

Since the feasible set of the dual problem is a box, we can employ the gradient projection method in order to solve it. For that, we need to know an upper bound on the Lipschitz constant of its gradient. To find such an upper bound, note that

$$\|\mathbf{Lx}\|^2 = \sum_{i=1}^{n-1}(x_i - x_{i+1})^2 \leq 2\left(\sum_{i=1}^{n-1} x_i^2 + \sum_{i=1}^{n-1} x_{i+1}^2\right) \leq 4\|\mathbf{x}\|^2.$$

Therefore

$$\lambda_{\max}(\mathbf{LL}^T) = \lambda_{\max}(\mathbf{L}^T \mathbf{L}) \leq 4.$$

Hence, since the Lipschitz constant of the gradient of the objective function of (12.27) is $\frac{1}{2}\lambda_{\max}(\mathbf{LL}^T)$, it follows that an upper bound on the Lipschitz constant is 2. The consequence is that we can employ the gradient projection method on problem (12.27) with constant stepsize $\frac{1}{2}$, and the convergence is guaranteed by Theorems 9.16 and 9.18. Explicitly, the method will read as

$$\mu_{k+1} = P_C\left(\mu_k - \frac{1}{4}\mathbf{LL}^T\mu_k + \frac{1}{2}\mathbf{Ly}\right),$$

where

$$C = \{\mathbf{z} \in \mathbb{R}^{n-1} : -\lambda \leq z_i \leq \lambda, i = 1, 2, \ldots, n-1\}.$$

If the result of the gradient projection method is μ^*, the primal optimal solution (up to some tolerance) will be $\mathbf{x}^* = \mathbf{y} - \frac{1}{2}\mathbf{L}^T \mu^*$. Following is a short MATLAB code that employs 1000 iterations of the gradient projection method:

```
lambda=1;
mu=zeros(n-1,1);
for i=1:1000
    mu=mu-0.25*L*(L'*mu)+0.5*(L*y);
    mu=lambda*mu./max(abs(mu),lambda);
    xde=y-0.5*L'*mu;
    end
figure(5)
plot(t,xde,'.');
axis([0,1,-1,4])
```

and the result is given in Figure 12.6. This result is much better than any of the quadratic regularization reconstructions, and it captures the breakpoints very well.

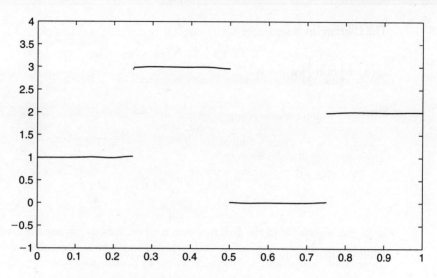

Figure 12.6. *Result of denoising via an l_1 norm regularization.*

12.3.11 ▪ Dual of the Linear Separation Problem

In Section 8.2.3 we considered the problem of finding a maximal margin hyperplane that separates two sets of points. We will assume that the given classified points are $\mathbf{x}_1, \mathbf{x}_2, \ldots, \mathbf{x}_m \in \mathbb{R}^n$. For each i, we are given a scalar y_i which is equal to 1 if $\mathbf{x}_i$ is in class A or -1 if it is in class B. The linear separation problem is given by

$$\begin{array}{ll} \min & \frac{1}{2}\|\mathbf{w}\|^2 \\ \text{s.t.} & y_i(\mathbf{w}^T\mathbf{x}_i + \beta) \geq 1, \quad i = 1, 2, \ldots, m. \end{array} \qquad (12.28)$$

The disadvantage of the formulation (12.28) is that it is only relevant when the two classes of points are linearly separable. However, in many practical situations the two classes are not linearly separable, and in this case we need to find a formulation in which violation of the constraints is allowed and at the same time a penalty term is added to the objective function that is equal to the sum of the violations of the constraints. The new formulation is as follows:

$$\begin{array}{ll} \min & \frac{1}{2}\|\mathbf{w}\|^2 + C\sum_{i=1}^m \xi_i \\ \text{s.t.} & y_i(\mathbf{w}^T\mathbf{x}_i + \beta) \geq 1 - \xi_i, \quad i = 1, 2, \ldots, m, \\ & \xi_i \geq 0, \quad i = 1, 2, \ldots, m, \end{array}$$

where $C > 0$ is a given parameter. We will rewrite the problem in a slightly different form:

$$\begin{array}{ll} \min & \frac{1}{2}\|\mathbf{w}\|^2 + C(\mathbf{e}^T\xi) \\ \text{s.t.} & \mathbf{Y}(\mathbf{X}\mathbf{w} + \beta\mathbf{e}) \geq \mathbf{e} - \xi, \\ & \xi \geq 0, \end{array}$$

where $\mathbf{Y} = \text{diag}(y_1, y_2, \ldots, y_m)$ and $\mathbf{X}$ is the $m \times n$ matrix whose rows are $\mathbf{x}_1^T, \mathbf{x}_2^T, \ldots, \mathbf{x}_m^T$. We begin by constructing the Lagrangian ($\alpha \in \mathbb{R}_+^m$)

$$\begin{aligned} L(\mathbf{w}, \beta, \xi, \alpha) &= \frac{1}{2}\|\mathbf{w}\|^2 + C(\mathbf{e}^T\xi) - \alpha^T[\mathbf{Y}\mathbf{X}\mathbf{w} + \beta\mathbf{Y}\mathbf{e} - \mathbf{e} + \xi] \\ &= \frac{1}{2}\|\mathbf{w}\|^2 - \mathbf{w}^T[\mathbf{X}^T\mathbf{Y}\alpha] - \beta(\alpha^T\mathbf{Y}\mathbf{e}) + \xi^T(C\mathbf{e} - \alpha) + \alpha^T\mathbf{e}. \end{aligned}$$

The Lagrangian is separable with respect to $\mathbf{w}, \beta$ and ξ and therefore

$$q(\alpha) = \left[\min_{\mathbf{w}} \frac{1}{2}\|\mathbf{w}\|^2 - \mathbf{w}^T[\mathbf{X}^T\mathbf{Y}\alpha]\right] + \left[\min_{\beta}(-\beta(\alpha^T\mathbf{Y}\mathbf{e}))\right] + \left[\min_{\xi \geq 0} \xi^T(C\mathbf{e} - \alpha)\right] + \alpha^T\mathbf{e}.$$

Since

$$\min_{\mathbf{w}} \frac{1}{2}\|\mathbf{w}\|^2 - \mathbf{w}^T[\mathbf{X}^T\mathbf{Y}\alpha] = -\frac{1}{2}\alpha^T\mathbf{Y}\mathbf{X}\mathbf{X}^T\mathbf{Y}\alpha,$$

$$\min_{\beta}(-\beta(\alpha^T\mathbf{Y}\mathbf{e})) = \begin{cases} 0, & \alpha^T\mathbf{Y}\mathbf{e} = 0, \\ -\infty & \text{else,} \end{cases}$$

$$\min_{\xi \geq 0} \xi^T(C\mathbf{e} - \alpha) = \begin{cases} 0, & \alpha \leq C\mathbf{e}, \\ -\infty & \text{else,} \end{cases}$$

it follows that the dual objective function is given by

$$q(\alpha) = \begin{cases} \alpha^T\mathbf{e} - \frac{1}{2}\alpha^T\mathbf{Y}\mathbf{X}\mathbf{X}^T\mathbf{Y}\alpha, & \alpha^T\mathbf{Y}\mathbf{e} = 0, 0 \leq \alpha \leq C\mathbf{e}, \\ -\infty & \text{else.} \end{cases}$$

The dual problem is therefore

$$\begin{array}{ll} \max & \alpha^T\mathbf{e} - \frac{1}{2}\alpha^T\mathbf{Y}\mathbf{X}\mathbf{X}^T\mathbf{Y}\alpha \\ \text{s.t.} & \alpha^T\mathbf{Y}\mathbf{e} = 0, \\ & 0 \leq \alpha \leq C\mathbf{e}. \end{array}$$

We can also write the dual problem in the following way:

$$\begin{array}{ll} \max & \sum_{i=1}^{m}\alpha_i - \frac{1}{2}\sum_{i=1}^{m}\sum_{j=1}^{m}\alpha_i\alpha_j y_i y_j (\mathbf{x}_i^T\mathbf{x}_j) \\ \text{s.t.} & \sum_{i=1}^{m} y_i\alpha_i = 0, \\ & 0 \leq \alpha_i \leq C, \quad i = 1, 2, \ldots, m. \end{array}$$

12.3.12 • A Geometric Programming Example

A *geometric programming* problem is an optimization problem of the form

$$\begin{array}{ll} \min & f(\mathbf{x}) \\ \text{s.t.} & g_i(\mathbf{x}) \leq 1, \quad i = 1, 2, \ldots, m, \\ & h_j(\mathbf{x}) = 1, \quad j = 1, 2, \ldots, p, \\ & \mathbf{x} \in \mathbb{R}_{++}^n, \end{array}$$

where $f, g_1, g_2, \ldots, g_m$ are *posynomials* and $h_1, h_2, \ldots, h_p$ are *monomials*. In the context of geometric programming, a monomial is a function $\phi : \mathbb{R}_{++}^n \to \mathbb{R}$ of the form $\phi(\mathbf{x}) = c\Pi_{j=1}^n x_j^{\alpha_j}$, where $c > 0$ and $\alpha_1, \alpha_2, \ldots, \alpha_n \in \mathbb{R}$. A posynomial is a sum of monomials. Geometric programming problems are not convex, but they can be easily transformed into convex optimization problems. In addition, their dual can be explicitly derived by an elegant argument. Instead of showing the derivations in the most general setting, we will illustrate them using a simple example. Consider the geometric programming problem

$$\begin{array}{ll} \min & \frac{1}{t_1 t_2 t_3} + t_2 t_3 \\ \text{s.t.} & 2t_1 t_3 + t_1 t_2 \leq 4, \\ & t_1, t_2, t_3 > 0. \end{array}$$

We will make the transformation $t_i = e^{x_i}$, which transforms the problem into

$$\begin{aligned} \min_{x_1, x_2, x_3} \quad & e^{-x_1-x_2-x_3} + e^{x_2+x_3} \\ \text{s.t.} \quad & e^{\ln 2 + x_1 + x_3} + e^{x_1+x_2} \leq 4. \end{aligned}$$

The transformed problem is convex, and we have thus shown how to transform the nonconvex geometric programming problem into a convex problem. To find a dual of this problem, we will consider the following equivalent problem:

$$\begin{aligned} \min \quad & e^{y_1} + e^{y_2} \\ \text{s.t.} \quad & e^{y_3} + e^{y_4} \leq 4, \\ & y_1 = -x_1 - x_2 - x_3, \\ & y_2 = x_2 + x_3, \\ & y_3 = x_1 + x_3 + \ln 2, \\ & y_4 = x_1 + x_2. \end{aligned} \qquad (12.29)$$

We will now construct the Lagrangian. The first constraint will be associated with the nonnegative multiplier w, and with the four linear constraints, we associate the Lagrange multipliers u_1, u_2, u_3, u_4:

$$\begin{aligned} L(\mathbf{y}, \mathbf{x}, w, \mathbf{u}) &= e^{y_1} + e^{y_2} + w\left(e^{y_3} + e^{y_4} - 4\right) - u_1(y_1 + x_1 + x_2 + x_3) \\ &\quad - u_2(y_2 - x_2 - x_3) - u_3(y_3 - x_1 - x_3 - \ln 2) - u_4(y_4 - x_1 - x_2) \\ &= [e^{y_1} - u_1 y_1] + [e^{y_2} - u_2 y_2] \\ &\quad + [we^{y_3} - u_3 y_3] + [we^{y_4} - u_4 y_4] \\ &\quad - x_1(u_1 - u_3 - u_4) - x_2(u_1 - u_2 - u_4) - x_3(u_1 - u_2 - u_3) \\ &\quad + (\ln 2)u_3 - 4w. \end{aligned}$$

We will use the following simple and technical lemma. Note that we use the convention that $0 \ln 0 = 0$.

Lemma 12.17. *Let $\lambda \geq 0$ and $a \in \mathbb{R}$. Then*

$$\min_{y \in \mathbb{R}} [\lambda e^y - ay] = \begin{cases} a - a \ln\left(\frac{a}{\lambda}\right), & \lambda > 0, \quad a \geq 0, \\ 0, & \lambda = a = 0, \\ -\infty, & \lambda \geq 0, \quad a < 0, \\ -\infty, & \lambda = 0, \quad a > 0. \end{cases}$$

If $\lambda > 0$ and $a > 0$, then the optimal y is $y = \ln\left(\frac{a}{\lambda}\right)$.

Proof. If $\lambda = 0$, then obviously, the minimum is 0 if and only if $a = 0$, and otherwise it is $-\infty$. If $\lambda > 0$, then

$$\min_{y \in \mathbb{R}} [\lambda e^y - ay] = \lambda \min_{y \in \mathbb{R}} \left[e^y - \frac{a}{\lambda} y \right],$$

and the optimal solutions of both minimization problems are the same. If $a < 0$, then the minimum is $-\infty$ since taking $y \to -\infty$ we obtain that the objective function goes to $-\infty$. If $a = 0$ the (unattained) minimal value is 0. If $a > 0$, then the optimal solution is attained at the stationary point which is the solution to $e^y = \frac{a}{\lambda}$, that is, at $y = \ln\left(\frac{a}{\lambda}\right)$. Substituting this expression back to the objective function we obtain that the optimal value is

$$\lambda \left[\frac{a}{\lambda} - \frac{a}{\lambda} \ln\left(\frac{a}{\lambda}\right) \right] = a - a \ln\left(\frac{a}{\lambda}\right). \quad \square$$

12.3. Examples

Based on Lemma 12.17 and the relations

$$\min_{x_1}[-x_1(u_1-u_3-u_4)] = \begin{cases} 0, & u_1-u_3-u_4=0, \\ -\infty & \text{else,} \end{cases}$$

$$\min_{x_2}[-x_2(u_1-u_2-u_4)] = \begin{cases} 0, & u_1-u_2-u_4=0, \\ -\infty & \text{else,} \end{cases}$$

$$\min_{x_3}[-x_3(u_1-u_2-u_3)] = \begin{cases} 0, & u_1-u_2-u_3=0, \\ -\infty & \text{else,} \end{cases}$$

we obtain that the dual problem is given by

$$\begin{aligned}
\max \quad & u_1 - u_1 \ln u_1 + u_2 - u_2 \ln u_2 + u_3 - u_3 \ln\left(\tfrac{u_3}{w}\right) + u_4 - u_4 \ln\left(\tfrac{u_4}{w}\right) + (\ln 2)u_3 - 4w \\
\text{s.t.} \quad & u_1 - u_3 - u_4 = 0, \\
& u_1 - u_2 - u_4 = 0, \\
& u_1 - u_2 - u_3 = 0, \\
& u_1, u_2, u_3, u_4, w \geq 0.
\end{aligned}$$

To make the problem well-defined and in order to be consistent with Lemma 12.17, the function $-u \ln\left(\tfrac{u}{w}\right)$ has the value 0 when $u = w = 0$ and the value $-\infty$ when $u > 0, w = 0$.

It is interesting to note that we can actually solve the dual problem. Indeed, noting that the expression in the objective function that depends on w is

$$(u_3 + u_4)\ln w - 4w,$$

we deduce that at an optimal solution $w = \tfrac{u_3+u_4}{4}$. The constraints of the dual problem imply that $u_2 = u_3 = u_4$ and that $u_1 = 2u_2$. Therefore, denoting the joint value of u_2, u_3, u_4 by $\alpha(\geq 0)$, we conclude that $u_1 = 2\alpha$ and $w = \tfrac{\alpha}{2}$. Plugging this into the dual, we obtain that the dual problem is reduced to the one-dimensional problem

$$\max\{3(1-\ln 2)\alpha - 3\alpha \ln(\alpha) : \alpha \geq 0\}.$$

The optimal solution of this problem is attained at the point at which the derivative vanishes,

$$3(1-\ln 2) - 3 - 3\ln \alpha = 0,$$

that is, at $\alpha = \tfrac{1}{2}$, and hence

$$u_1 = 1, \quad u_2 = u_3 = u_4 = \frac{1}{2}, \quad w = \frac{1}{4}.$$

We can also find the optimal solution of the primal problem. For that, we will first compute the optimal y_1, y_2, y_3, y_4:

$$y_1 = \ln u_1 = 0, \quad y_2 = \ln u_2 = -\ln 2, \quad y_3 = \ln\left(\tfrac{u_3}{w}\right) = \ln 2, \quad y_4 = \ln\left(\tfrac{u_4}{w}\right) = \ln 2.$$

Hence, by the constraints of (12.29) we have

$$\begin{aligned}
x_1 + x_2 + x_3 &= 0, \\
x_2 + x_3 &= -\ln 2, \\
x_1 + x_3 &= 0, \\
x_1 + x_2 &= \ln 2,
\end{aligned}$$

whose solution is $x_1 = \ln 2, x_2 = 0, x_3 = -\ln 2$. Therefore, the optimal solution of the primal problem is

$$t_1 = e^{x_1} = 2, \quad t_2 = e^{x_2} = 1, \quad t_3 = e^{x_3} = \frac{1}{2}.$$

Exercises

12.1. Find a dual problem to the convex problem

$$\begin{aligned}\min \quad & x_1^2 + 0.5x_2^2 + x_1 x_2 - 2x_1 - 3x_2 \\ \text{s.t.} \quad & x_1 + x_2 \leq 1.\end{aligned}$$

Find the optimal solutions of both the dual and primal problems.

12.2. Write a dual problem to the problem

$$\begin{aligned}\min \quad & x_1 - 4x_2 + x_3^4 \\ \text{s.t.} \quad & x_1 + x_2 + x_3^2 \leq 2 \\ & x_1 \geq 0 \\ & x_2 \geq 0.\end{aligned}$$

Solve the dual problem.

12.3. Consider the problem

$$\begin{aligned}\min \quad & x_1^2 + 2x_2^2 + 2x_1 x_2 + x_1 - x_2 - x_3 \\ \text{s.t.} \quad & x_1 + x_2 + x_3 \leq 1 \\ & x_3 \leq 1.\end{aligned}$$

(i) Is the problem convex?

(ii) Find an optimal solution of the problem.

(iii) Find a dual problem and solve it.

12.4. Consider the primal optimization problem

$$\begin{aligned}\min \quad & x_1^4 - 2x_2^2 - x_2 \\ \text{s.t.} \quad & x_1^2 + x_2^2 + x_2 \leq 0.\end{aligned}$$

(i) Is the problem convex?

(ii) Does there exist an optimal solution to the problem?

(iii) Write a dual problem. Solve it.

(iv) Is the optimal value of the dual problem equal to the optimal value of the primal problem? Find the optimal solution of the primal problem.

12.5. Consider the problem

$$\begin{aligned}\min \quad & 3x_1^2 + x_1 x_2 + 2x_2^2 \\ \text{s.t.} \quad & 3x_1^2 + x_1 x_2 + 2x_2^2 + x_1 - x_2 \geq 1 \\ & x_1 \geq 2x_2.\end{aligned}$$

(i) Is the problem convex?

(ii) Find a dual problem. Is the dual problem convex?

Exercises

12.6. Find a dual to the convex optimization problem

$$\begin{aligned}\min \quad & \sum_{i=1}^{n}(x_i \ln x_i - x_i)\\ \text{s.t.} \quad & \mathbf{Ax} \le \mathbf{b},\\ & \mathbf{x} > 0,\end{aligned}$$

where $\mathbf{A} \in \mathbb{R}^{m \times n}, \mathbf{b} \in \mathbb{R}^m$.

12.7. Find a dual problem to the following convex minimization problem:

$$\begin{aligned}\min \quad & \sum_{i=1}^{n}(a_i x_i^2 + 2b_i x_i + e^{\alpha_i x_i})\\ \text{s.t.} \quad & \sum_{i=1}^{n} x_i = 1,\end{aligned}$$

where $\mathbf{a}, \boldsymbol{\alpha} \in \mathbb{R}_{++}^n, \mathbf{b} \in \mathbb{R}^n$.

12.8. Consider the convex optimization problem

$$\begin{aligned}\min \quad & \sum_{j=1}^{n} x_j \ln \frac{x_j}{c_j}\\ \text{s.t.} \quad & \mathbf{Ax} \ge \mathbf{b},\\ & \sum_{j=1}^{n} x_j = 1\\ & \mathbf{x} > 0,\end{aligned}$$

where $\mathbf{c} > 0, \mathbf{A} \in \mathbb{R}^{m \times n}, \mathbf{b} \in \mathbb{R}^m$. Find a dual problem.

12.9. Consider the following problem (also called *second order cone programming*):

$$\begin{aligned}\min \quad & \mathbf{g}^T \mathbf{x}\\ \text{s.t.} \quad & \|\mathbf{A}_i \mathbf{x} + \mathbf{b}_i\| \le \mathbf{c}_i^T \mathbf{x} + d_i, \quad i = 1, 2, \ldots, k,\end{aligned}$$

where $\mathbf{g} \in \mathbb{R}^n, \mathbf{A}_i \in \mathbb{R}^{m_i \times n}, \mathbf{b}_i \in \mathbb{R}^{m_i}, \mathbf{c}_i \in \mathbb{R}^n, d_i \in \mathbb{R}, i = 1,2,\ldots,k$. Find a dual problem.

12.10. Consider the primal optimization problem

$$\begin{aligned}\min \quad & \sum_{j=1}^{n} \frac{c_j}{x_j}\\ \text{s.t.} \quad & \mathbf{a}^T \mathbf{x} \le b,\\ & \mathbf{x} > 0,\end{aligned}$$

where $\mathbf{a} \in \mathbb{R}_{++}^n, \mathbf{c} \in \mathbb{R}_{++}^n, b \in \mathbb{R}_{++}$.

(i) Find a dual problem with a single dual decision variable.

(ii) Solve the dual and primal problems.

12.11. Consider the following optimization problem in the variables $\alpha \in \mathbb{R}$ and $\mathbf{q} \in \mathbb{R}^n$:

$$\text{(P)} \quad \begin{aligned}\min \quad & \alpha\\ \text{s.t.} \quad & \mathbf{Aq} = \alpha \mathbf{f}\\ & \|\mathbf{q}\|^2 \le \varepsilon,\end{aligned}$$

where $\mathbf{A} \in \mathbb{R}^{m \times n}, \mathbf{f} \in \mathbb{R}^m, \varepsilon \in \mathbb{R}_{++}$. Assume in addition that the rows of $\mathbf{A}$ are linearly independent.

(i) Explain why strong duality holds for problem (P).

(ii) Find a dual problem to problem (P). (Do not assign a Lagrange multiplier to the quadratic constraint.)

(iii) Solve the dual problem obtained in part (ii) and find the optimal solution of problem (P).

12.12. Let $\mathbf{a} \in \mathbb{R}^n_{++}$ and consider the problem

$$\begin{array}{ll} \min & \sum_{i=1}^n -\log(x_i + a_i) \\ \text{s.t.} & \sum_{i=1}^n x_i = 1 \\ & \mathbf{x} \geq \mathbf{0}. \end{array}$$

Find a dual problem with one dual decision variable. Is strong duality satisfied?

12.13. Let $\mathbf{a}_1, \mathbf{a}_2, \mathbf{a}_3 \in \mathbb{R}^n$, and consider the Fermat-Weber problem

$$\min \{\|\mathbf{x} - \mathbf{a}_1\| + \|\mathbf{x} - \mathbf{a}_2\| + \|\mathbf{x} - \mathbf{a}_3\|\}.$$

Find a dual problem.

12.14. Find a dual problem to the following primal one:

$$\begin{array}{ll} \min & \sum_{i=1}^n x_i \ln\left(\frac{x_i}{\alpha_i}\right) + \|\mathbf{x}\|^2 + 2\mathbf{a}^T \mathbf{x} \\ \text{s.t.} & \mathbf{x}^T \mathbf{A} \mathbf{x} + 2\mathbf{b}^T \mathbf{x} + c \leq 0, \end{array}$$

where $\boldsymbol{\alpha} \in \mathbb{R}^n_{++}, \mathbf{a} \in \mathbb{R}^n, \mathbf{A} \succ \mathbf{0}, \mathbf{b} \in \mathbb{R}^n, c \in \mathbb{R}$. Under what condition is strong duality guaranteed to hold? Find a condition that is written explicitly in terms of the data.

12.15. Let $\mathbf{a}_1, \mathbf{a}_2, \ldots, \mathbf{a}_m \in \mathbb{R}^n$ and $b_1, b_2, \ldots, b_m \in \mathbb{R}$ and consider the problem of finding the so-called analytic center of the polytope $S = \{\mathbf{x} \in \mathbb{R}^n : \mathbf{a}_i^T \mathbf{x} < b_i, i = 1, 2, \ldots, m\}$ given by

$$\text{(A)} \quad \min\left\{-\sum_{i=1}^m \ln(b_i - \mathbf{a}_i^T \mathbf{x}) : \mathbf{x} \in S\right\}.$$

Find a dual problem to (A).

12.16. Let $f : \mathbb{R}^n \to \mathbb{R}$ be defined as (k is a positive integer smaller than n)

$$f(\mathbf{x}) = \sum_{i=1}^k x_{[i]},$$

where $x_{[i]}$ is the ith largest values in the vector $\mathbf{x}$. We have seen in Example 7.27 that f is convex.

(i) Show that for any $\mathbf{x} \in \mathbb{R}^n$, we have that $f(\mathbf{x})$ is the optimal value of the problem

$$\text{(P)} \quad \begin{array}{ll} \max & \mathbf{x}^T \mathbf{y} \\ \text{s.t.} & \mathbf{e}^T \mathbf{y} = k \\ & \mathbf{0} \leq \mathbf{y} \leq \mathbf{e}. \end{array}$$

(ii) For any $\alpha \in \mathbb{R}$ show that $f(\mathbf{x}) \leq \alpha$ if and only if there exist $\boldsymbol{\lambda} \in \mathbb{R}^n_+$ and $u \in \mathbb{R}$ such that

$$ku + \mathbf{e}^T \boldsymbol{\lambda} \leq \alpha, u\mathbf{e} + \boldsymbol{\lambda} \geq \mathbf{x}.$$

(iii) Let $\mathbf{Q} \in \mathbb{R}^{n \times n}$ be a positive definite matrix. Find a dual to the problem

$$\begin{array}{ll} \min & \mathbf{x}^T \mathbf{Q} \mathbf{x} \\ \text{s.t.} & f(\mathbf{x}) \leq \alpha. \end{array}$$

12.17. Consider the inequality constrained problem

$$f^* = \min_{\text{s.t.}} \ f(\mathbf{x})$$
$$\text{s.t.} \quad g_i(\mathbf{x}) \leq 0, \quad i = 1, 2, \ldots, m,$$

where $f, g_1, g_2, \ldots, g_m$ are convex functions over $\mathbb{R}^n$. Suppose that there exists $\hat{\mathbf{x}} \in \mathbb{R}^n$ such that

$$g_i(\hat{\mathbf{x}}) < 0, \quad i = 1, 2, \ldots, m,$$

and that the problem is bounded below, i.e., $f^* > -\infty$. Consider also the dual problem given by

$$\max\{q(\lambda) : \lambda \in \text{dom}(q)\},$$

where $q(\lambda) = \min_{\mathbf{x}} L(\mathbf{x}, \lambda), \text{dom}(q) = \{\lambda \in \mathbb{R}_+^m : \min L(\mathbf{x}, \lambda) > -\infty\}$. Let λ^* be an optimal solution of the dual problem. Prove that

$$\sum_{i=1}^m \lambda_i^* \leq \frac{f(\hat{\mathbf{x}}) - f^*}{\min_{i=1,2,\ldots,m}(-g_i(\hat{\mathbf{x}}))}.$$

12.18. Consider the optimization problem (with the convention that $0 \ln 0 = 0$)

$$\min \ x_1 + 2x_2 + 3x_3 + 4x_4 + \sum_{i=1}^4 x_i \ln x_i$$
$$\text{s.t.} \ \sum_{i=1}^4 x_i = 1$$
$$x_1, x_2, x_3, x_4 \geq 0.$$

(i) Show that the problem cannot have more than one optimal solution.

(ii) Find a dual problem in one dual decision variable.

(iii) Solve the dual problem.

(iv) Find the optimal solution of the primal problem.

12.19. Find a dual problem for the optimization problem

$$\min \ x_1 + 2x_2^2 + x_3^2 + \sqrt{4x_1^2 + x_1 x_3 + x_3^2 + 2}$$
$$\text{s.t.} \ x_1 + x_2^2 + 4x_3^2 \leq 5$$
$$x_1 + x_2 + x_3 \leq 15.$$

12.20. Consider the problem

$$\min \ x_1^2 - x_2^4 + x_3^2 + \sqrt{x_1^2 + x_3^2}$$
$$\text{s.t.} \ x_1^2 + x_2^4 + x_3^2 + x_3 \leq 4.$$

(i) Find a dual problem.

(ii) Is the dual problem convex?

12.21. Consider the optimization problem

$$\text{(P)} \quad \begin{aligned} \min \ & -6x_1 + 2x_2 + 4x_3^2 \\ \text{s.t.} \ & 2x_1 + 2x_2 + x_3 \leq 0 \\ & -2x_1 + 4x_2 + x_3^2 = 0 \\ & x_2 \geq 0 \end{aligned}$$

(i) Is the problem convex?

(ii) Find a dual problem for (P). Do not assign a Lagrange multiplier to the constraint $x_2 \geq 0$.

(iii) Find the optimal solution of the dual problem.

12.22. Let $\alpha \in \mathbb{R}_{++}$ and define the set

$$T_\alpha = \left\{ \mathbf{x} \in \mathbb{R}^n : \sum_{j=1}^n x_j = 1, 0 \leq x_j \leq \alpha \right\}.$$

(i) For which values of α is the set T_α nonempty?

(ii) Find a dual problem with one dual decision variable to the problem of finding the orthogonal projection of a given vector $\mathbf{y} \in \mathbb{R}^n$ onto T_α.

(iii) Write a MATLAB function for computing the orthogonal projection onto the set T_α based on the dual problem found in part (ii). The call to the function will be in the form

```
function xp=proj_bound(y,alpha)
```

where y is the point which should be projected onto T_α and xp is the resulting projection. The function should check whether the set T_α is nonempty.

(iv) Write a function for finding the orthogonal projection onto T_α which is based on CVX. The function call will be in the form

```
function xp=proj_bound_cvx(y,alpha)
```

(v) Compute the orthogonal projection of the vector $(0.5, 0.7, 0.1, 0.3, 0.1)^T$ onto $T_{0.3}$ using the MATLAB functions constructed in parts (iii) and (iv).

(vi) Compare the CPU times of the two functions on a problem with 10000 variables using the following commands:

```
rand('seed',314);
x=rand(10000,1);
tic,y=proj_bound(x,0.01);toc
tic,y=proj_bound_cvx(x,0.01);toc
```

12.23. Consider the following optimization problem:

(P) $\min \left\{ h(\mathbf{x}) \equiv \|\mathbf{x} - \mathbf{d}\|^2 + \sqrt{x_1^2 + x_2^2} + \sqrt{x_2^2 + x_3^2} + \sqrt{x_3^2 + x_4^2} + \sqrt{x_4^2 + x_5^2} : \mathbf{x} \in \mathbb{R}^5 \right\}$,

where $\mathbf{d} = (1, 2, 3, 2, 1)^T$.

(a) Find an explicit dual problem of (P) with a "simple" constraint set (meaning a set on which it is easy to compute the orthogonal projection operator).

(b) Run 10 iterations of the gradient projection method on the derived dual problem. Use a constant stepsize $\frac{1}{L}$, where L is the corresponding Lipschitz constant. You need to show at each iteration both the dual objective function value as well as the objective function of the primal problem (use the relations between the optimal primal and dual solutions to derive at each iteration a primal solution). Finally, write explicitly the optimal solution of the problem.

Bibliographic Notes

Chapter 1 For a comprehensive treatment of multidimensional calculus and linear algebra, the reader can refer to [24, 29, 33, 34, 35] and also Appendix A of [9].

Chapter 2 The topic of optimality conditions in Sections 2.1–2.3 is classical and can also be found in many other books such as [1, 30]. The principal minors criterion is also known as "Sylvester's criterion," and its proof can be found for example in [24].

Chapter 3 A comprehensive study of least squares methods and generalizations can be found in [11]. The discussion on circle fitting follows [4]. An excellent guide for MATLAB is the book [20].

Chapter 4 The gradient method is discussed in many books; see, for example, [9, 27, 31]. More background and convergence results on the Gauss–Newton method can be found in the book [28]. The original paper of Weiszfeld appeared in [39]. The analysis in Section 4.6 follows [6].

Chapter 5 More details and further extensions of Newton's method can be found, for example, in [9, 15, 17, 27, 28]. The hybrid Newton's method can be found in [38]. An excellent reference for an abundant of optimization methods is the book [28].

Chapters 6 and 7 A classical reference for convex analysis is [32]. The book [10] also contains a comprehensive treatment of the subject.

Chapter 8 A large variety of examples of convex optimization problems can be found in [13] and also in [8]. The original paper of Markowitz describing the portfolio optimization model is [23]. The CVX MATLAB software as well as a user guide can be found in [19]. The CVX software uses two conic optimization solvers: SeDuMi [36] and SDPT3 [37]. The reformulation of the trust region subproblem as a convex problem described in Section 8.2.7 follows the paper [11].

Chapter 9 Stationarity is a basic concept in optimization that can be found in many sources such as [9]. The gradient mapping is extensively studied in [27]. The analysis of the convergence gradient projection method follows [5, 6]. The discussion on sparsity constrained problems is based on [3]. The iterative hard thresholding method was initially presented and studied in [12].

Chapter 10 Generalization and extensions of separation and alternative theorems can be found in [32, 10, 21]. The discussion on the orthogonal regression problem follows [4].

Chapter 11 A comprehensive treatment of the KKT conditions, including variants that were not presented in this book, can be found in [1, 9]. The derivation of the second order necessary conditions follows the paper [7]. Optimality conditions and algorithms for solving the trust region subproblem and its generalizations can be found in [26, 25, 16]. The total least squares problem was initially presented in [18] and was extensively studied and generalized by many authors; see the review paper [22] and references therein. The derivation of the reduced form of the total least squares problem via the KKT conditions follows [2].

Chapter 12 Duality is a classic topic and is covered in many other optimization books; see, for example, [1, 9, 10, 13, 21, 32]. The dual algorithm for solving the denoising problem is based on Chambolle's algorithm for solving two-dimensional denoising problems with total variation regularization [14]. More on duality in geometric programming can be found in [30].

Bibliography

[1] M. S. Bazaraa, H. D. Sherali, and C. M. Shetty. *Nonlinear programming. Theory and algorithms*. Wiley-Interscience [John Wiley & Sons], Hoboken, NJ, third edition, 2006.

[2] A. Beck and A. Ben-Tal. On the solution of the Tikhonov regularization of the total least squares problem. *SIAM J. Optim.*, 17(1):98–118, 2006.

[3] A. Beck and Y. C. Eldar. Sparsity constrained nonlinear optimization: Optimality conditions and algorithms. *SIAM J. Optim.*, 23(3):1480–1509, 2013.

[4] A. Beck and D. Pan. On the solution of the GPS localization and circle fitting problems. *SIAM J. Optim.*, 22(1):108–134, 2012.

[5] A. Beck and M. Teboulle. A fast iterative shrinkage-thresholding algorithm for linear inverse problems. *SIAM J. Imaging Sci.*, 2(1):183–202, 2009.

[6] A. Beck and M. Teboulle. Gradient-based algorithms with applications to signal-recovery problems. In *Convex optimization in signal processing and communications*, pages 42–88. Cambridge University Press, Cambridge, 2010.

[7] A. Ben-Tal. Second-order and related extremality conditions in nonlinear programming. *J. Optim. Theory Appl.*, 31(2):143–165, 1980.

[8] A. Ben-Tal and A. Nemirovski. *Lectures on modern convex optimization. Analysis, algorithms, and engineering applications*. MPS/SIAM Series on Optimization. Society for Industrial and Applied Mathematics (SIAM), Philadelphia, PA, 2001.

[9] D. P. Bertsekas. *Nonlinear programming*. Athena Scientific, Belmont, MA, second edition, 1999.

[10] D. P. Bertsekas. *Convex analysis and optimization*. Athena Scientific, Belmont, MA, 2003. With Angelia Nedić and Asuman E. Ozdaglar.

[11] A. Björck. *Numerical methods for least squares problems*. Society for Industrial and Applied Mathematics (SIAM), Philadelphia, PA, 1996.

[12] T. Blumensath and M. E. Davies. Iterative thresholding for sparse approximations. *J. Fourier Anal. Appl.*, 14(5-6):629–654, 2008.

[13] S. Boyd and L. Vandenberghe. *Convex optimization*. Cambridge University Press, Cambridge, 2004.

[14] A. Chambolle. An algorithm for total variation minimization and applications. *J. Math. Imaging Vision*, 20(1-2):89–97, 2004. Special issue on mathematics and image analysis.

[15] R. Fletcher. *Practical methods of optimization. Vol. 1. Unconstrained optimization*. John Wiley & Sons, Ltd., Chichester, 1980. A Wiley-Interscience Publication.

[16] C. Fortin and H. Wolkowicz. The trust region subproblem and semidefinite programming. *Optim. Methods Softw.*, 19(1):41–67, 2004.

[17] P. E. Gill, W. Murray, and M. H. Wright. *Practical optimization*. Academic Press, Inc. [Harcourt Brace Jovanovich, Publishers], London-New York, 1981.

[18] G. H. Golub and C. F. Van Loan. An analysis of the total least squares problem. *SIAM J. Numer. Anal.*, 17(6):883–893, 1980.

[19] M. Grant and S. Boyd. CVX: MATLAB software for disciplined convex programming, version 2.0 beta. http://cvxr.com/cvx, September 2013.

[20] D. J. Higham and N. J. Higham. *MATLAB guide*. Society for Industrial and Applied Mathematics (SIAM), Philadelphia, PA, second edition, 2005.

[21] O. L. Mangasarian. *Nonlinear programming*. McGraw-Hill Book Co., New York, 1969.

[22] I. Markovsky and S. Van Huffel. Overview of total least-squares methods. *Signal Process.*, 87(10):2283–2302, October 2007.

[23] H. Markowitz. Portfolio selection. *J. Finance*, 7:77–91, 1952.

[24] C. Meyer. *Matrix analysis and applied linear algebra*. Society for Industrial and Applied Mathematics (SIAM), Philadelphia, PA, 2000. With 1 CD-ROM (Windows, Macintosh and UNIX) and a solutions manual (iv+171 pp.).

[25] J. J. Moré. Generalizations of the trust region subproblem. *Optim. Methods Software*, 2:189–209, 1993.

[26] J. J. Moré and D. C. Sorensen. Computing a trust region step. *SIAM J. Sci. Statist. Comput.*, 4(3):553–572, 1983.

[27] Y. Nesterov. *Introductory lectures on convex optimization*, volume 87 of Applied Optimization. Kluwer Academic Publishers, Boston, MA, 2004. A basic course.

[28] J. Nocedal and S. J. Wright. *Numerical optimization*. Springer Series in Operations Research and Financial Engineering. Springer, New York, second edition, 2006.

[29] J. M. Ortega and W. C. Rheinboldt. *Iterative solution of nonlinear equations in several variables*, volume 30 of Classics in Applied Mathematics. Society for Industrial and Applied Mathematics (SIAM), Philadelphia, PA, 2000. Reprint of the 1970 original.

[30] A. L. Peressini, F. E. Sullivan, and J. J. Uhl, Jr. *The mathematics of nonlinear programming*. Undergraduate Texts in Mathematics. Springer-Verlag, New York, 1988.

[31] B. T. Polyak. *Introduction to optimization*. Translations Series in Mathematics and Engineering. Optimization Software Inc. Publications Division, New York, 1987. Translated from the Russian, With a foreword by Dimitri P. Bertsekas.

[32] R. T. Rockafellar. *Convex analysis*. Princeton Mathematical Series, No. 28. Princeton University Press, Princeton, N.J., 1970.

[33] W. Rudin. *Real analysis*. McGraw-Hill, N. Y., 1976.

[34] S. Strang. *Calculus*. Wellesley-Cambridge Press, 1991.

[35] S. Strang. *Introduction to linear algebra*. Wellesley-Cambridge Press, fourth edition, 2009.

[36] J. F. Sturm. Using SeDuMi 1.02, a MATLAB toolbox for optimization over symmetric cones. *Optim. Methods Softw.*, 11-12:625–653, 1999.

[37] K. C. Toh, M. J. Todd, and R. H. Tütüncü. SDPT3—a MATLAB software package for semi-definite programming, version 1.3. *Optim. Methods Softw.*, 11/12(1–4):545–581, 1999. Interior point methods.

[38] L. Vandenberghe. *Applied numerical computing*. 2011. Lecture notes, in collaboration with S. Boyd.

[39] E. V. Weiszfeld. Sur le point pour lequel la somme des distances de n points donnes est minimum. *The Tohoku Mathematical Journal*, 43:355–386, 1937.

Index

active constraints, 195, 208
affine function, 118
argmax, 14
argmin, 14
arithmetic geometric mean
    inequality, 139

backtracking, 51, 178
basic feasible solution, 107
bisection, 219
boundary point, 7
boundary set, 7
bounded below, 36
bounded set, 8

Carathéodory theorem, 102
Cauchy–Schwarz, 3
Chebyshev ball, 153
Chebyshev center, 153, 162, 256
Cholesky factorization, 89
circle fitting, 45
classification, 150, 266
closed ball, 6
closed line segment, 2
closed set, 7
closure, 8
coercive, 25
compact set, 8
complementary slackness, 196, 246
concave function, 118
condition number, 58, 60
cone, 104
conic combination, 106
conic duality theorem, 260
conic hull, 106
conic problem, 254
conic representation theorem, 106
constant stepsize, 51
constrained least squares, 218
constraint qualification, 210

continuously differentiable, 9
contour set, 7
convex combination, 101
convex function, 117
convex hull, 102
convex optimization, 147
convex polytope, 100
convex problem, 147
convex set, 97
CVX, 158

damped Gauss–Newton method, 67, 68
damped Newton's method, 88
data fitting, 39
denoising, 42
descent direction, 49
descent directions method, 50
descent lemma, 74
descent method, 50
diagonally dominant matrix, 22
directional derivative, 8
distance function, 130, 157
dot product, 2
dual cone, 114, 205
dual objective function, 238
dual problem, 237

effective domain, 135
eigenvalue, 5
eigenvector, 5
ellipsoid, 99
epigraph, 136
Euclidean norm, 3
exact line search, 51
extended real-valued function, 135
extreme point, 111

facility location, 69
Farkas lemma, 192
feasible descent direction, 207
feasible set, 13

feasible solution, 13
Fejér monotonicity, 183
Fermat's theorem, 16
Fermat–Weber problem, 68, 80
first projection theorem, 157
Fritz–John conditions, 208
Frobenius norm, 4
full column rank, 37
function
    distance, 130, 157
    dual objective, 238
    Huber, 189
    quadratic, 32
    Rosenbrock, 60

generalized Slater's condition, 216
geometric programming, 267
global maximum and minimum, 13
global optimum, 13
Gordan's theorem, 194, 205, 208
gradient, 9
gradient inequality, 119
gradient mapping, 177
gradient method, 52
gradient projection method, 175

Hessian, 10
hidden convexity, 155, 165
Hölder's inequality, 140
Huber function, 189
hyperplane, 98

identity matrix, 2
ill-conditioned matrices, 60
indefinite matrix, 18
indicator function, 135
induced matrix norm, 4
inequality
    Jensen's, 118
    Kantorovich, 59
    linear matrix, 254

Minkowski's, 140
strongly convex, 117
inner product, 2
interior, 7
interior point, 7
iterative hard-thresholding, 187

Jacobian, 67
Jensen's inequality, 118

Kantorovich inequality, 59
KKT conditions, 195, 207
KKT point, 198, 211
Krein–Milman theorem, 113

Lagrange multipliers, 196
Lagrangian function, 198
least squares, 37
    linear, 45
    regularized, 41, 219, 262
    total, 230
level set, 7, 130
line, 97
line search, 50
line segment principle, 108
linear approximation theorem, 10
linear fitting, 39
linear function, 118
linear least squares, 45
linear matrix inequality, 254
linear programming, 107, 149, 247
linear rate, 60
Lipschitz continuity of the gradient, 73
local maximum point, 15
local minimum point, 15
log-sum-exp, 124
Lorentz cone, 105

matrix norm, 3
maximal value, 13
maximizer, 13
minimial value, 13
minimizer, 13
Minkowski functional, 113
Minkowski's inequality, 140
monomial, 267
Motzkin's theorem, 205

negative definite, 18
negative semidefinite, 18
neighborhood, 7
Newton direction, 83
Newton's method, 64, 83
    damped, 88

pure, 83
nonexpansive, 174
nonlinear Farkas' lemma, 242
nonlinear fitting, 40
nonlinear least squares, 45, 67
nonnegative orthant, 1
nonnegative part, 157
norm
    diagonally dominant, 22
    identity, 2
    indefinite, 18
    induced matrix, 4
    square root of, 21
    strictly diagonally dominant, 22
normal cone, 114
normal system, 37

open ball, 6
open line segment, 2
open set, 7
optimal set, 148
orthogonal projection, 156
orthogonal regression, 203
overdetermined system, 37

partial derivative, 8
pointed cone, 114
positive definite, 17
positive orthant, 2
positive semidefinite, 17
posynomial, 267
primal problem, 238
principal minors criterion, 21
proper, 135
pure Newton's method, 83

Q-norm, 35
QCQP, 155, 235
quadratic approximation theorem, 10
quadratic function, 32
quadratic problem, 150
quadratic-over-linear, 125, 127
quasi-convex function, 131

Rayleigh quotient, 5, 204, 232
real-valued, 135
regularity, 211
regularization parameter, 41
regularized least squares, 41, 219, 262
robust regression, 163
Rosenbrock function, 60

saddle point, 24
scaled gradient method, 63
second projection theorem, 173
semidefinite programming, 254
separation of two convex sets, 242
separation theorem, 191
set
    bounded, 8
    open, 7
sgn, 156
singleton, 187
Slater's condition, 214
source localization problem, 80
sparsity constrained problems, 183
spectral decomposition, 5
spectral norm, 4
square root of a matrix, 21
standard basis, 1
stationary point, 17, 169
strict global maximum, 13
strict global minimum, 13
strict local maximum, 15
strict local minimum, 15
strict separation theorem, 191
strictly concave function, 118
strictly convex function, 117
strictly diagonally dominant matrix, 22
strong convexity parameter, 144
strong duality, 241
strongly convex function, 144, 190
sublinear rate, 182
sufficient decrease lemma, 75, 176
support, 184
support function, 136
supporting hyperplane theorem, 241

total least squares, 230
trust region subproblem, 155, 227

unit-simplex, 2, 254
unit-sum set, 171

Vandermonde, 41

weak duality theorem, 239
Weierstrass theorem, 25
well-conditioned matrices, 60

Young's inequality, 140